FLIES AND DISEASE, Volume II

Biology and Disease Transmission

FLIES AND DISEASE

Volume II   BIOLOGY AND DISEASE TRANSMISSION

BY BERNARD GREENBERG

Princeton University Press, Princeton, New Jersey · 1973

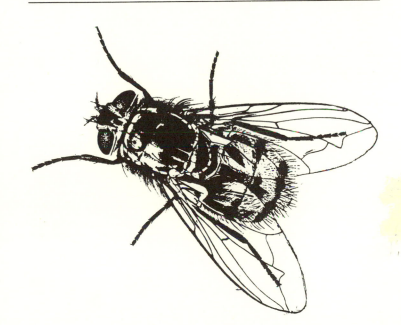

RA641
F6
G7
v.2

Publication of this book has been aided by grants DA-MD-59-193-67-G-9225, DADA-17-69-G-9225, and DADA-17-70-G-9313 from the U.S. Army
Medical Research and Development Command.

To my children, Gary, Linda, Debbie, and Danny

Preface

THE FIRST VOLUME on *Flies and Disease* was actually an outgrowth of the present one. Before getting down to the diseases in which flies are involved, I felt there ought to be a foundation of ecology, classification, and biotic associations of flies. While the first volume serves as a detailed reference work, this volume attempts to integrate the myriad facts and observations into a full portrait of the fly as an instrument of disease. Since we are dealing for the most part with contaminative diseases, sober reality has dictated caution in assigning major or minor roles to the players—fingers, feces, food, and flies. Generally, fly-borne diseases are a blend of these and other factors, the total orchestration varying with season, locale, culture, sanitation, technology, living standard, and the nutritional state of the individual. While human malaria depends on *Anopheles* for transmission, and urban yellow fever depends on *Aedes aegypti*, the eradication of synanthropic flies would not, of itself, eliminate most of the diseases we shall be discussing in this book. For this reason, we have tried to give a balanced view of fly involvement in human and animal diseases. Undoubtedly, some entomologists and epidemiologists at opposite poles will fault this middle course, but I feel that a partisan approach—in this case, an overemphasis on flies and underplay of other factors—would be all too easy and untrue. Those readers who can live with doubt and uncertainty will not mistake complexity for irresolution. At the same time, I do not intend to convey the impression that the evidence for fly involvement is weak. After two decades of experimentation and sifting the data of others, I am convinced that flies are unquestionably guilty of disease transmission, but not in every case or place. How much more satisfying it is to present the fly as a vector of the nematode diseases, habronemiasis and thelaziosis, and to stand on the firmer ground of biological transmission.

The opening chapter is a survey of attitudes—annoyance, awe, fear, repulsion—toward flies through recorded history. These creatures have had too impressive an impact—through art, religion, poem, proscription, and myth—to warrant further neglect of this subject. It may surprise the reader as much as it did me to discover that the fly danger was clearly apprehended in the Talmud, a millennium before Mercurialis penned his frequently quoted observation in 1577. Chapter 2 presents as much of the natural history and distribution of a number of important species as

I feel the scope of the book allows. Obviously, this subject is worth its own volume. Included are the overwintering habits, and a rather complete treatment of fly dispersal, for reason of its epidemiologic importance. Chapter 3 regards the fly as a microecosystem whose morphology and physiology control the fate of pathogens. The intake, survival, and output of disease agents are considered and also the influence of metamorphosis, fly immunity factors, and microbial interactions. Sections on gnotobiotic techniques and tissue culture are included here. The final two chapters deal with specific human and animal diseases. Some of these diseases have no real association with flies but they were once thought to. They are discussed with the sole purpose of putting the fly association to final rest. Salmonellosis is primarily a disease of animals involving man tangentially. We include it among the more strictly human enteric infections in Chapter 4 because it illuminates the similarities and differences in the way these diseases are spread. The reader will find other instances where the rigidity of the chapter title has been somewhat loosened.

If this book makes any contribution, hopefully it will be as a stimulus and guide to further research, as Luther West's "The Housefly" was to me. I have tried to point out specific research needs where these were apparent to me. I am certain that the alert reader will perceive many others.

I acknowledge with appreciation the efforts of Mrs. Mary Grant, manuscript typist, Mrs. Sharon Pendola and James Janicke, bibliographic researchers, and thank Ron Jendryaszek and Janice Rajecki for some of the illustrations. I am especially indebted to Major Moufied A. Moussa and Lt. Colonels Bruce F. Eldridge and Wallace P. Murdoch (Ret.) of the U.S. Army Medical Research and Development Command.

Bernard Greenberg
University of Illinois at Chicago Circle
December 1, 1971

Contents

Illustrations

FLIES AND DISEASE, Volume II
Biology and Disease Transmission

1
Flies Through History

"Busy, curious, thirsty fly!
Drink with me, and drink as I!
Freely welcome to my cup,
Coulds't thou sip and sip it up;
Make the most of life you may;
Life is short and wears away!

"Both alike, both mine and thine,
Hasten quick to their decline!
Thine's a summer; mine no more,
Though repeated to threescore!
Threescore summers, when they're gone,
Will appear as short as one!"

WILLIAM OLDYS, *The Fly—*
Occasioned by a Fly Drinking
Out of the Author's Cup.

WHAT PREHISTORIC MAN thought about flies will never be known. Yet, the enduring affinity of synanthropic flies for man and domestic animals presupposes an ancient and intense association. The impact of these insects as annoyers and as despoilers of food is still felt in many regions where primitive systems of sanitation and waste disposal prevail. The urbane reader whose exposure to flies is happily limited to "kamikazi" attacks on his morning sleep cannot appreciate the ubiquity of these insects, nor their potential for explosive multiplication during war, famine, or other disaster. Hopefully, in time, flies will join cockroaches, bedbugs, and lice as vanquished symbols of poverty, apathy, and disease. At present, however, a pathetically small part of humanity is reasonably free of these pests.

ANNOYERS

Buzzing swarms of flies—the greater their numbers the more demoniac they seem to be—must have attracted attention wherever they occurred. We have selected a few historical passages to underscore the timelessness of the fly problem and its global aspect. Probably best known is the passage from Exodus 8:21, 24: "Else if thou wilt not let my people go, behold, I will send swarms of flies upon thee, and upon thy servants, and upon thy people, and into thy houses: and the houses of the Egyptians shall be full of swarms of flies, and also the ground whereon they are. And the Lord did so; and there came a grievous swarm of flies into the house of Pharaoh, and into his servants' houses, and into all the land of Egypt: the land was corrupted by reason of the swarm of flies."

Ecologically, it is reasonable that a plague of frogs preceded the flies, "and they gathered them [the dead frogs] together upon heaps: and the land stank." For flies would have bred in enormous numbers in these heaps, more so than from single frogs, which would have quickly dried in the heat. In a normal succession, the fly plague might also be expected to follow the murrain that killed the livestock of the Egyptians. Instead, the Bible places the outbreak of flies before the destruction of livestock. Was this an intuition of possible cause and effect? We shall see that later writers often associated eruptions of disease with prior outbreaks of flies.

Flies wax numerous on the wastes of war and other calamities. In this personal account of the aftermath of "The Battel of S. Quintin, 1557" by the great military surgeon Amboise Paré, which appeared in a con-

FIG. 1. Flies biting man. *Hortus Sanitatis*, Mainz, 1491 (National Library of Medicine, Bethesda, Maryland).

temporary translation, we can recognize at least two genera of blow flies: "We saw more than halfe a league about us the earth covered with dead bodyes; neither could we abide long there, for the cadaverous scents which did arise from the dead bodyes, as well of men, as of horses. And I thinke we were the cause, that so great a number of flyes, arose from the dead bodees, which were procreated by their humidity and the heate of the Sunne, having their tayles greene and blew; that being up in the ayre made a shaddow in the Sunne. We heard them buzze, or humme, which was much mervaile to us. And I think it was enough to cause the Plague, where they alighted."

In explaining the hasty adoption of the American Declaration of Independence by the Continental Congress on July 4, 1776, Jefferson writes: "The weather was oppressively warm, and the room occupied by the deputies was hard by a stable, whence the hungry flies swarmed thick and fierce, alighting on their legs and biting hard through their thin silk stockings. Treason was preferable to discomfort." No doubt the fly in this account was *Stomoxys calcitrans* (Fuller, 1913; see Fig. 1).

Lieutenant George Grey describes his experiences in western Australia in December 1837: "Sleep after sunrise was impossible, on account of the number of flies which kept buzzing about the face. To open our mouths was dangerous—in they flew and mysteriously disappeared, to be rapidly ejected again in a violent fit of coughing; and into the eyes, when unclosed, they soon found their way, and by inserting their proboscis, and sucking, speedily made them sore; neither were the nostrils safe from their attacks, which were made simultaneously on all points, and in multitudes. This was a very troublesome annoyance, but I afterwards found it to be a very general one throughout all the unoccupied portions of Australia; although in general, the further north you go in this continent, the more intolerable does the fly nuisance become." Cleland (1913), in quoting this passage, adds the comment, "Those who have had experience of the bush today know that the plague has in no way decreased. In many situations in the drier parts of the interior it is almost impossible to eat one's food on account of myriads of these insects hovering around the mouth." These accounts probably refer to *Musca sorbens*, the eye fly, which is also troublesome throughout North Africa and southern Asia.

Jean Paul Sartre in his play "The Flies" uses these insects as a symbol of the enduring and terrible torment that Orestes and Electra must forever endure in the face of the brutish indifference of their fellow citizens of Argos:

"Electra: Listen! The sound of their wings is like a roaring furnace. They're all around us, Orestes, watching, biding their time. Presently they'll swoop down on us and I shall feel thousands of tiny clammy feet crawling over me. Oh, Look! They're growing bigger, bigger; now they're as big as bees. We'll never escape them; they'll follow us everywhere in a dense cloud. Oh, God, now I can see their eyes, millions of beady eyes all staring at us!

"Orestes: What do the flies matter to us?

"Electra: They're the Furies, Orestes, the goddesses of remorse."

FOLKLORE AND SUPERSTITION

The Middle East, which cradled civilization, was and still is a fly haven. The mild climate has provided warmth and humidity, and man has unintentionally contributed food and shelter for the flies. In Bible lands the fly nuisance had to be endured in all seasons. Animal domestication, the rise of villages, and fly commensalization went hand in hand. Henceforth, certain species, e.g. *Musca domestica*, would never again be found outside their indigenous areas except in close association with man. Other species, which were less inclined than the house fly to enter dwellings, displayed varying degrees of affinity for man (synanthropy). Nevertheless, they readily exploited human and animal feces, carrion, and kitchen and other wastes. Production of potential fly breeding materials mounted as human and domestic animal populations increased.

It is therefore not surprising that the ubiquitous fly is already present in art and in writing at the beginning of written history, about 3000 B.C.

The 14th Tablet of the series Har-ra-Hubulla is a systematic list of names of wild terrestrial animals that dates from the time of Hammurabi, some 3600 years ago and that is based on even more ancient Sumerian lists. This compilation is written in Akkadian cuneiform on clay tablets and is the oldest known book in zoology. There are 396 animals listed, of which 111 are insects, and about 10 are flies. Here we find mentioned for the first time the "green fly" (*Phaenicia*), the "blue fly" (*Calliphora*), the "toothed fly" and the "biting fly" (*Stomoxys* and *Haematobia*), and the "trembling fly" (probably a syrphid).

The British Museum contains a cylinder seal of the Uruk III period (3000 B.C.) from Mesopotamia (Fig. 2). It shows a reclining gazelle between two ibexes with a perfectly drawn fly above it. The habitus of the wings and the proportions of head, thorax, and abdomen are an accurate rendering of what can easily pass for a house fly.

The close observation and accurate depiction of such small creatures as flies suggest some level of entomological sophistication. It is interesting to contrast that sophistication with the level of certain Arab groups living in the same region today who do not differentiate between earthworms, small wasps, and flies. In fact the larval stages of all insects are called "dud." This seems contrary to the idea that those living close to nature tend to be familiar with their flora and fauna. But it is reasonable that a group's life style determines what is relevant and, in the case of such marginal existence, only environmental factors of singular importance are individualized. Before we conclude that the ancients were more sophisticated naturalists than their sympatric moderns we should consider that ancient records probably reflect the thinking of a learned few. The average man's zoology in those times may have been primitive indeed.

Nevertheless, the great variety of flies recognized in the Har-ra-Hubulla is evidence of the relevance of flies—biting and nonbiting—to the ancients. Under circumstances that produced huge populations of flies, writes the Assyriologist, Elizabeth van Buren, ". . . flies . . . would have been regarded with disfavor, if not with horror: nevertheless, in the texts comparisons with flies carry no stigma, and the gods themselves are compared to these pestilential insects. In the Epic of Gilgamesh (Tab. XI, 162) we are told that the gods gathered like flies around Utnapistim when he offered sacrifice." Another passage relates: "The gods of strong-walled Uruk are changed into flies and buzz around the streets."

Fly amulets skillfully carved from lapis lazuli, shell, frit, and stone have been unearthed from various levels in ancient Babylonian sites. Pendants in the shape of a fly formed the central beads of necklaces found at Babylon and at Ur. Similar adornments were found in Egypt, and it is generally believed that their widespread use in the ancient world served an apotropaic function, viz. to ward off evil spirits. Flies often occur on cylinder seals associated with the gods (Fig. 3), and this led to the 19th-century notion, perpetuated by contemporary medical historians, that the fly was the symbol of Nergal, the Mesopotamian god of disease and death. Some modern scholars feel there is no evidence for a fly god in Mesopotamia, nor of a link between Nergal, flies, and disease. Instead, they suggest that fly symbols in this early period served a purely decorative function, as did lizards, mammals, and flowers. We are inclined to favor van Buren's view that fly symbols served to ward off evil and ensure prosperity. Such common and intrusive insects, which swarmed on wounds and feces, got into one's eyes, food, and mouth, attacked carcasses and corpses, and avariciously attended at the sacrificial altar, undoubtedly made an impact far beyond the decorative. That ten kinds of flies were already distinguished in man's first writings would seem to underscore their significance. However, the scanty evidence limits useful speculation on prevailing attitudes toward flies at the dawn of history.

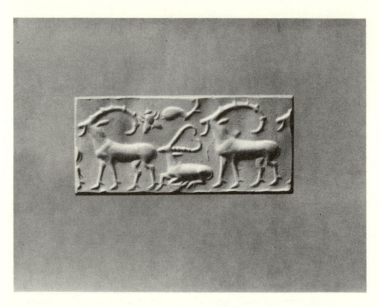

FIG. 2. Mesopotamian cylinder seal, ca. 3,000 B.C., showing fly. Courtesy British Museum.

FIG. 3. Babylonian seal showing the god Mardok with fly. Courtesy the Pierpont Morgan Library.

8 FLIES THROUGH HISTORY

The hieroglyphic ←→ stands for fly, wasp, or bee. To the Egyptians, the fly symbolized both courage and impudence. Amenemhab, one of Thutmose's victorious commanders, is rewarded in the presence of the army with two decorations for distinguished service: "a lion of the finest gold" and "two flies" (ca. 1450 B.C.). A slip of paper found in the mouth of a mummy contains the inscription: "The maggots will not turn into flies within you" (Papyrus Gizeh No. 18026:4:14). This shows clearly that the metamorphosis of the fly was known to the Egyptians, yet it is interesting that several thousand years were to pass before a simple experiment by Redi (1667) proved the reverse—that only flies can produce maggots (Figs. 4 and 5).

In Zoroastran Persia, about 1000 B.C., we find definite evidence of a dual attitude toward flies. The Chaldean god, Baalzebub, originated during this period and was to survive until modern times as Beelzebub, Satan's First Lieutenant. The name derives from the Assyrian bel (lord) and bel dababi (an adversary in a lawsuit, hence an enemy), and combines zumbu (fly) to become Baalzebub, Lord of Flies. He was an important god to the Phoenicians and Philistines, and later to the Greeks as well. A special temple at Ekron was erected to him, and the Hebrew King Ahazia, when he fell ill, sent to this god to learn if he might recover, for Baalzebub was endowed with oracular powers. He often appeared as a fly and could either ward off plagues of the insect or send his winged legions as punishment.

The Greeks transferred a component of Baalzebub to Apollo, defender against all evils. As pasture god (Apollon Nomios) and protector of herds (Apollon Karneios), the function of Myiagros, or fly-chaser, belonged to him. "When Myiagros is appeased all the flies die," says Pliny, and for this reason a steer was sacrificed to Apollo at the Olympic games, so that the contests might be watched in comfort. There were times, however, when Zeus, the almighty lord of the world, insisted on performing the task of chasing flies away himself in the form of Zeus Apomyios. It was also Zeus who saved Hercules when the latter was almost vanquished by flies. Scholars believe that the classical Greek worship of Hera had probably absorbed the cult of a primitive cow-goddess, hence the Homeric reference to Hera as "the ox-eyed" and her connection with flies.

The Greek fly, if we may impart to it an ethnic character, was omnipresent and mischievous. Plautus remarks: "This man is a fly, my father, nothing can be concealed from him, whether secret or public, he is presently there and knows all the matter." In Greek mythology, Musca was a beautiful and talented songstress, although a bit loquacious. Diana could not endure her as a rival for the love of Endymion and changed

Fig. 4. Flies attacking a dead or dying animal. *Hortus Sanitatis*, Antwerp, 1521.

Fig. 5. Flies swarming on maggot-infested beef. Knowledge of the life cycle is clearly implied. *Hortus Sanitatis*, Antwerp, 1521.

her into a fly. Driven by her love for Endymion, Musca continues to disturb with restless flight the dreams of those who slumber, particularly adolescents. True to form, Roman soldiers converted the lovely Greek myth into something more practical. They called the stones they used in their catapults "Musca" because they buzzed as they moved through the air.

Flies· were also associated with death and the devil, but only when the dead were buried. In cultures in which the dead were burned, this concept of the fly did not develop. Thus, the Greek demon of decay, Eurynomos, appears both as a vulture and as a fly. In the Persian book Vendidad it is written that, as soon as a person dies, the unclean demon of death throws himself upon the corpse in the form of a fly, bringing about decay and destruction.

Widely divergent cultures share this antipathy for flies. In Teutonic myth, Loki, god of the underworld, crawls through the keyhole in the form of a fly to gain entry into people's homes. Witches appeared as flies in Hungary. The Mic-Mac Indians of the northeastern United States ascribed the origin of flies to the dust of the bones of a wicked giant that once lived. And among Arabs today, the expression "father of the fly" signifies a person with a foul breath; an unlucky man is a "fly man." From these qualities, the Arab word for fly, dhubab, has come to signify evil or mischief.

FLIES AND MYIASIS

The Furies: We are the flies, the suckers of pus,
We shall have open house with you,
We shall gather our food from your mouths,
And our light from the depths of your eyes.
All your life we shall be with you,
Until we make you over to the worms.
("The Flies," JEAN PAUL SARTRE)

It is futile to associate flies with specific diseases in ancient writings because syndromes were not defined and such notable symptoms as fever, skin lesions, or diarrhea are common to many illnesses. Medical historians are convinced that many persons who were ostracized as lepers were probably suffering from some other disease.

Myiasis, however, is an exception. The presence of maggots in a wound or their escape from someone's body is unmistakable and would tend to be accurately reported. The role of flies in myiasis is not treated later in the book because excellent monographs are available. Since a comprehensive historical treatment is generally not available, I have included it in this history of flies and disease.

Herodotus ends his description of the Persian expedition against Barca in Libya (ca. 520 B.C.) with this passage: "A fitting conclusion to this story is the manner of Pheretima's death—for the web of her life was not woven happily to the end. No sooner had she returned to Egypt after her revenge upon the people of Barca, than she died a horrible death, her body seething with maggots while she was still alive. Thus this daughter of Battus, by the nature and severity of her punishment of the Barcaeans, showed how true it is that all excess in such things draws down upon men the anger of the gods."

The Bible alludes several times to maggots infesting human flesh, for this could not have been an uncommon affliction. Intense competition among flies for breeding sites would have encouraged carrion feeders to turn opportunistically to the wounds of living hosts. Strictly speaking, those maggots that confine their appetites to tissue debris and secretions, abjuring living tissue, are not parasites. But the victim can hardly be expected to appreciate this fine distinction. Job, the prototype of the endless sufferer, is not spared this affliction:

"My flesh is clothed with maggots and clods of dust,
My skin rotted and fouled afresh." (*Job* VII, 5)

and again:

"If I have said to corruption: 'Thou are my father.'
To the maggots: 'Thou are my mother and my sister.' "

God smote Antiochus for his cruelty to the Jews, and he fell from his chariot: ". . . so that the maggots rose up out of the body of this wicked man and while he lived in sorrow and pain his flesh fell away and the filthiness of his smell was noisome to all his army" (Maccabees II, 9:9, Apocrypha, Oxford Ed., p. 276). Josephus recounts that King Herod died five days after being smitten with gangrene during which the maggots bred in the gangrenous mass (Book of Acts XII, 23: New Testament). In the spring of the year 70, Titus led his Roman legions against Judea. Jerusalem fell after a protracted siege, and the Temple of Solomon was sacked. Legend relates that his punishment for the destruction of the temple was that a small fly would gnaw in his brain for seven years. The reader appreciates that maggots, and not adults, are involved in myiasis and that the period of seven years is apocryphal rather than scientific. We recently heard of a derelict in Chicago's Skid Row who appeared at a clinic with a complaint of rumblings in his head, perhaps like Titus' complaint. Myiasis was amply confirmed when 35 maggots were washed from the patient's ear.

According to Plutarch, the Persian kings employed myiasis in criminal justice. The most serious offenders were exposed with honey-smeared faces to sun and flies; the outcome was certain and the torture was exceedingly painful.

FLIES AND DISEASE

Long and intimate contact with Babylon helped shape Jewish myths and attitudes. Bocharti (1712) states that Israelites considered flies as contaminated flying reptiles, born from filth and flying among filth, then to the food of man and to the sacrificial altars, which they dirty. We have not been able to verify this wording in any original source. The Talmud, written between 200 and 500 A.D., does contain exhaustive commentaries on already ancient Jewish customs and practices, in which the attitude toward flies is unequivocally negative, if one may judge by the following statements:

A fly in food is offensive, and its presence there is ground for divorce (Git. 6b). If a fly falls into a cup of wine and is removed, the wine is still fit to drink; fastidious people, however, do not drink it, though the vulgar even eat a dish into which a fly has fallen (Tosef., Sotah, V.9, Yer. 17a). The fly is extremely annoying when one is eating, and since it persistently returns after being driven away it is the symbol of evil desires (Ber. 10b, 61a; Targ. Eccl. X,1). The Egyptian fly (Isa. VII, 18) [*Musca sorbens*?] is so dangerous that it may be killed even on the Sabbath (Shab. 121b). To what can Amalek be compared? To a fly which is enthused for the sore . . . (Pes. d'Rabh Kah., 3ak., 26b). And perhaps the most significant of all: Beware of the flies of the man afflicted with ra-athan [a skin disease] (Ket. 77b). The concept of the fly as a contaminator of food and transmitter of disease is clearly expressed possibly for the first time in history, thus anticipating by more than a millennium the rise of similar ideas in Europe.

These attitudes were carried into Christian, Moslem, and other lore, although lacking the forcefulness of the Talmudic proscriptions. Flies, according to St. Augustine, were created by God to punish our arrogance. Mohammed said: "If a fly falls into a container of one of you have him remove all of it then throw it away because certainly in one of its wings is remedy and in the other disease." This dual concept persisted in Arab thinking and later appeared in "Animal Life," the most comprehensive Arab zoology of the time, written by Kamal ad Damiri (1349-1391). In it he says, "I have carefully observed flies and found that they always defend themselves with their left wing which is also supposed to carry the cause of disease, just as the right wing is supposed to have the cure for this disease." This statement sums up a general ambivalence toward flies as creatures of filth, disease, and death, on one hand and as source for a variety of cures on the other.

Before the modern era, any extensive outbreak of disease might erroneously have been called plague. Muffet (1658) refers to an outbreak in Rome: "Hither may be referred that which Strabo reports, *lib. Georg.* 3. That amongst the Romans a Plague did often happen by reason of them [flies], insomuch that they were fain to hire men of purpose to catch them, who were payed according to the quantity more or less that they caught." The awesome sweep and terror of the actual disease inspired great speculation as to its cause and spread. Some of the earliest associations between flies and a specific disease refer to plague, although we now know it is spread by flea bite or by the respiratory route and not by flies. Again we quote from Muffet: "In the year 1348, great numbers of flies dropping out of the air, did cause in the Eastern Countreys incredible noisomness and putrefaction; upon which followed such a Plague among the people, that scarce the tenth man among them was left alive." In 1498, Bishop Knud of Aarhus, Denmark, noted in *De regimine pestilentico* that the increase of flies is one of the first signs of the approach of a plague. Varwich, in Tractatlin von der Pestilenz, 1577, observed that the summer of 1576, a plague year, was exceptionally hot and that large numbers of flies were seen.

Around 1577, the Italian physician Mercurialis penned his famous observation: "There can be no doubt that flies, saturated with the juice of the dead or of the diseased, then visit neighboring houses and infect the food, and persons who eat of it are infected." This clearly expresses the role of flies as mechanical contaminators.

Athanasius Kircher, a German Jesuit priest and distinguished natural philosopher, accepted the current belief that dogs, cats, and other domestic animals spread plague, but he also included flies. The following discourse was written in 1668, shortly after a severe visitation of the disease:

> During the recent Neapolitan plagues a certain noble, although I do not know what he observed at the window, beheld a certain hornet flying which sat on his nose and sitting still, biting with its sting, it made a swelling, and gradually by the rising within the skin, as poison from a snake, within two days after the plague was contracted, he died—for there is no doubt from the contagious fluid that the fly brought it from the carcass.

Besides including wasps under *musca*, Kircher adds nothing to Mercurialis' idea. Later, Paullinus borrowed the same story of the Neapolitan nobleman to fit an outbreak of dysentery. We may never know the true effect of that fatal bite on the unfortunate nobleman's nose, but it advanced the cause of medical entomology, just the same.

Thomas Sydenham, the great English physician, had this to say about dysentery and other diseases: "I have remarked that, if swarms of insects, especially house flies, were abundant in the summer, the succeeding autumn was unhealthy. This I observed to be the case during the whole summer of the aforesaid year (1661); whilst in the summers of the two following years, which were very healthy, the insects were very few." He goes on to express judicious doubts concerning the validity of such a correlation, "Still I must remark, that at the approach of even so severe a disease as the plague itself, they were not observed to be very abundant," and concludes by saying, "With these two exceptions, I have observed that all prognostics are fallacious. . . ."

In his *Observationes Medico-Physicae*, Paullinus (1706) attributes the death of a young lady in a monastery to the bite of a fly, and he then attempts to link an outbreak of similar deaths in a nearby but isolated village to flies in the following remarkable passage: "When you say that dysentery was going about then in nearby places, it is as if the fly [Paullinus called it *Musca dysenteriae genitrix*] had settled on the excrements or the corpses of infected men, and had carried the contagious fluid from these sources which it was able to transfer easily to another. For truly at the same time you may hear that it is either dysentery or the gripping plague." Paullinus describes dysentery as a terminal symptom of a poisoning, which the fly disseminates, since there was no other contact between the villagers and the sick in the monastery.

Sir John Pringle writes on the causes of a dysentery epidemic he had observed in London in 1762: "And there is an old observation, that such seasons as produce most flies, caterpillars and other insects (whose increase depends so much on heat and moisture and consequently on corruption) have likewise been most productive of the dysentery." This is simply Sydenham re-stated.

The following quotation from a letter by a Mr. Balle appeared in a book by R. Bradley, early in the 18th century. The correspondent links flies and other insects to poor urban sanitation when he says: "Experience shows us how much insects delight in Stinking places, and that they increase much faster in uncleanly Cities, such as London was formerly, than in cleaner Places; but the city of London having been for the most Part, burnt the year after the Pestilence, its streets were enlarg'd, many drains were made and good Laws were put in execution for keeping the City clean, and it has not had any Plague ever since."

In the United States in the early years of this century, flies suffered from a particularly bad public image. This was the result of massive campaigns to expose them as a health menace (Fig. 6). Though these attacks have diminished along with the fly nuisance in this country, they have been revived elsewhere, notably in China (Fig. 7).

THERAPEUTIC USES OF FLIES

Thomas Muffet, in "The Theater of Insects" (1658), details the uses of flies, since antiquity, for wasp stings, spider bites, epilepsy, constipation, and a number of other ailments. As late as 1751, an Italian Pharmacopeia still listed flies as a cure for baldness and suggested an aqueous distillation of flies for the treatment of eye ailments. In the treatment of stye, it follows Galen's homeopathic remedy: "Rub with flies whose heads have been cut off and make compresses with white wax." The Roman Consul Mucianus carried a live fly in a linen sack to protect him from eye disease, probably in deference to *Musca sorbens*, the eye fly that is so common throughout the Mediterranean region.

Some have cast Amboise Paré as the forerunner of maggot therapy, but his writings suggest only a traditional revulsion for maggots as agents of corruption. He conscientiously made every effort to keep them out of the wounds of his patients. Following the Battle of San Quentin in 1557, Paré relates that many wounded were neglected for a time. As a consequence, their wounds "were greatly stinking and full of wormes with Gangreene and putrefaction. . . . To correct and stay the putrefaction, and to kill the wormes which were entered into their wounds; I washed them with *Egyptiacum* dissolved in wine and *Aqua vitae* . . ." (Paré, Keynes Edition, p. 69, 1952). In another place Paré describes an infected wound beneath a skull bone from which, after several months, a great number of maggots issued. A bone the size of the palm was lost, yet the patient recovered "beyond all men's expectation" (Paré, Johnson Edition, 1678). He did not attribute the patient's recovery to the maggots.

The first positive statement concerning the role of maggots in the healing of wounds is D. V. Larrey's, Napoleon's military surgeon, inspired by his experiences during the campaigns of 1799:

> There remains an unusual matter which we do not believe ought to be passed without mention; it concerns that which we have had occasion to observe in Syria, during the Egyptian expedition, among the majority of our wounded. While their sores were suppurating, these wounds were bothered by worms or larvae of the blue fly, common in this climate. These insects formed in a few hours, developing with such rapidity that, from one day to the next, they were the size of a small quill which frightened our soldiers very much, despite all

THE HOUSE-FLY AT THE BAR

FIG. 6. An illustration that appeared in 1909 in a publication of the Merchants' Association of New York, which was active in anti-fly campaigns.

FIG. 7. A Chinese view of the fly menace is shown in this poster depicting the insects as bombers raining cholera, typhoid and other deadly germs on the populace. The poster declares: "Destroy the fly. Our weapons are the swatter, the net and sticky paper. Keep flies out with screens." (China Journal, V. 27, 1937.)

FLIES THROUGH HISTORY 17

we could do to reassure them in this regard: there had been the experience to convince them that, far from being harmful to their wounds, these insects by accelerating cicatrization shortened the work of nature, and also by lessening the cellular eschars which they devoured. These larvae, in effect, have an avidity only for putrefying matter, always sparing the living parts; also I have never seen, in these circumstances, evidence of hemorrhage, the insects are carried only to that depth which is the extent of the wound.

The first therapeutic use of maggots is credited to J. F. Zacharias, a surgeon in the Confederate army, who wrote: "During my service in the hospital at Danville, Virginia, I first used maggots to remove the decayed tissue in hospital gangrene and with eminent satisfaction. In a single day, they would clean a wound much better than any agents we had at our command. I used them afterwards at various places. I am sure I saved many lives by their use, escaped septicemia, and had rapid recoveries."

Paramonov (1933) mentions that the beneficial influence of dipterous larvae upon wounds was reported by Pirogov in 1886, but no details are given and we have not seen the original publication. It remained for Baer (1931) to apply his own experiences in World War I to the scientific establishment of maggot therapy in our time. It continued in international vogue throughout the 1930s and passed out of the picture with the advent of the sulfa drugs and antibiotics.

Medical corpsmen stationed in northern Burma during World War II observed the use of maggot therapy by the isolated hill people of the area. Maggots were placed on the wound, which was covered with mud and dressed with wet grass. This practice extended into Yunan Province, China, and has also been reported among the aborigines of New South Wales, Australia (Lee, 1968).

2
Biology of Flies

The fly "is the most perfectly free and republican of creatures . . . he does not care whether it is a king or clown whom he teases, and in every step of his swift mechanical march, and in every pause of its resolute observation there is one and the same perfect expression of perfect egotism, perfect independence and self-confidence and conviction of the world having been made for flies. Your fly free in the air, free in the chamber, a black incarnation of caprice, wandering, investigating, fleeting, flitting, feasting at his will with rich variety of feast from the heaped sweets in the grocer's window to those of the butcher's back yard, and from the galled place on your horse's neck to the brown spot on the road from which, as the hoof disturbs him, he rises with angry republican buzz; what freedom is like this?"

John Ruskin

IN CHAPTER THREE of the first volume of this work, we gave summaries of the bionomics of important synanthropic flies. For the first part of this chapter, we have selected a number of these species for treatment in greater detail with the intention of illustrating the typical life cycles of manure, carrion, and other flies. For the most part, these reviews bring together the old and more recent investigations on the biology of a species for the first time. We have included developmental periods, the behavior of the active stages, the breeding habits, materials, and preferences, the adult feeding habits and attractants, seasonal population features, and overwintering habits.

We have omitted discussions of the biology of certain important flies because they are well covered elsewhere in the literature or because little is known. For example, the house fly, *Musca domestica*, was admirably treated by West (1951). The bionomics of the Australian blow flies is thoroughly discussed by various authors. We have not gone into the details of myiasis because this was done by James (1947) and Zumpt (1965). There is little biological information about sarcophagids, *Muscina stabulans* and others, in the literature at present.

The second part of this chapter is a discussion of fly dispersal. The various types of fly movements, marking and trapping techniques, population density estimation and the influence of meteorological factors are considered as well as the significance of fly dispersal patterns in disease dissemination.

HIPPELATES EYE GNATS

The eye gnats of the family Chloropidae are important synanthropes that frequent the face, eyes, mucous and sebaceous secretions, and the wounds, pus, and blood of their victims. The species in the southern and southwestern United States have long been considered possible vectors of conjunctivitis and pinkeye or sore eye of man and animals. The significant species in these areas are *H. collusor* in the southwest and Mexico and *H. pusio* in the south and southeast, and they have been extensively studied in the last several years. Sabrosky (1941) reviewed the classification of the United States *Hippelates*; his *pusio* group includes the medically important *pusio*, *flavipes*, *pallipes*, *bishoppi*, and *collusor* and other species. Sabrosky (1951) further considered the nomenclature, and Graham-Smith (1930) reviewed the disease implications of the eye gnat group to that date. In Jamaica, *H. flavipes* has been implicated in yaws transmission and feeds avidly on animals. The seasonal prevalence of *Siphunculina funicola*, which is distributed in India, Ceylon, Java, and other parts of the Orient, was found to coincide with the incidence of epidemic conjunctivitis in Assam, India (Roy, 1938). We will treat the biology of the *pusio* group and separately discuss the biology of *S. funicola*.

ADULT HABITS

Burgess (1951) found *pusio* (= *collusor*) present all year around in the desert and foothills of the Coachella Valley of southern California and annoying man from April through November. During the seasonal peaks in the spring and fall, the gnats were noticeable in the early morning and the late afternoon in shaded places. When the weather was cool, he sometimes found them seeking shelter and warmth in clods of earth and in warm manure heaps. In the San Joaquin Valley area, Womeldorf and Mortenson (1962, 1963) used egg-bait traps to determine the seasonal prevalence of eye gnats. They reported: a general rise in the population in March; a peak in late June and early July; a reduction through July and early August and a second peak in late August; the numbers declined until mid-October and then fell off rapidly. No gnats were trapped after December 5, and they reappeared in early February. *H. collusor* was the predominant and nuisance gnat in this area and the most frequently attracted to man. *H. pusio*, considered by Womeldorf to be distinct from *collusor* on the basis of behavior and color, was generally distributed in the valley and more numerous in the fall but was attracted to man in smaller numbers compared to *collusor*; *robertsoni* occurred mostly in the mountain collections and was not responsible for any great number of complaints; *dorsalis* and *microcentrus* occurred in low numbers.

Studies in southern California show that *Hippelates* gnats are strong fliers, resisting wind speeds of 2-5 mph. They usually fly throughout the day and infest residential areas, sprinkled lawns, car washes, golf courses, school yards, wineries, irrigated areas, orchards, grapefruit plantings, date gardens, vineyards, crop fields, and densely packed shrubbery. Hall (1932) observed that they are attracted to the eyes and natural orifices of man and animals, feeding about these openings and on the exudates of sores and cuts. Mulla (1959) observed that the gnats approach their host quietly, alighting at some distance from the feeding site; they crawl or fly intermittently to the site. These gnats are nonbiting but possibly can produce minute incisions with their labellar spines.

Hall (1932) found the greatest annoyance on warm and humid days in recently irrigated and moist situations. On cool days in early spring and fall, gnats were more noticeable near midday at about 70°F; they were not noticeable at below 70°F. During summer days with temperatures around 100°F, adults exhibited bimodal activity, being more noticeable in the early morning and late afternoon in shaded areas. They were not attracted to lights at night but were positively heliotropic. In insectary cages, the adults fed freely on sugar water, meat juice, fruit juice, and decaying vegetable matter.

Dow et al. (1951) found that *pusio* and *bishoppi* have similar periods of development and a wide distribution in plowed areas. Their data indicated that, in the Orlando, Florida, region, *pusio* was more or less abundant throughout the year, whereas *bishoppi*, although taken as an adult in each month of the year, bred mostly from March to May and was more abundant in citrus-growing areas than where most of the land was in truck. Fifteen collections of gnats associated with various animals contained *pusio* but not *bishoppi*.

Acute conjunctivitis is endemic in various sections of the southern United States and reaches a high level of incidence each summer. Dow and Hines (1957) stated that the incidence of this disease seemed closely related to the seasonal and geographic abundance of eye gnats and considered them possible vectors. Earlier, Bigham (1941) had taken a survey trip through the southeastern United States to locate areas of gnat prevalence and sore eye incidence. He encountered mostly *pusio*. Gnats were abundant in the rolling, sandy country of southeastern Alabama and southwestern Georgia, in the ridge section (a citrus-growing region) and several truck farming areas of Florida. Generally, gnats were found in great numbers wherever there was extensive farming or truck raising in sandy or muck soils. There were also yearly outbreaks of sore eye infections in the same areas. The gnats were low in numbers during times of low agricultural activity and increased in areas where the soil had been freshly turned.

Kumm (1935) reported on the adult habits of *H. flavipes* (= *pallipes*) in Jamaica in connection with his studies on yaws. The gnats were attracted to and fed on ulcers on the legs and bodies of donkeys, horses, dogs, mules, cows, goats, and pigs and were observed to feed around the penis of a dog. The adults were seen often to feed intermittently on several persons and on different lesions and sometimes to take in an exceptional amount of pus or serum. They apparently did not feed on manure or decaying fruit or vegetable matter. They were abundantly trapped using decaying liver in urea as bait. They fed well in direct sunlight and chose resting sites in the open under the surface of leaves on trees and shrubs.

Dispersal studies in southwestern Georgia by Dow (1959) using ^{32}P tagged flies showed that the release of gnats one-half and one mile from a rural population center resulted in the almost complete penetration of the small town on the day of release. In another test, traps more than one mile from the release point caught 15 tagged gnats in less than 3½ hours. In the Coachella Valley, using the same technique, Mulla and March (1959) reported gnats traveling over 4 miles from the release point and crossing ½ mile of desert to reach such suitable habitats as agricultural lands and residential sites (Table 3). They also found that they rest at night on dry or damp ground, in soil clods, on dried rootlets protruding above the ground and on low growing foliage.

BREEDING HABITS

Burgess (1951) observed the mating of *pusio* (= *collusor*) in the field in cool shady locations such as the lower leaves of shrubs. In laboratory studies, Schwartz (1965) found that 39% of the females laid eggs from 5-8 days after emergence; by 12 days 83% had oviposited. Both adults required 36 hours for sexual maturation before insemination occurred; males could inseminate as many as four females in a 24-hour period. The females laid an average of 26 eggs in 18 days. Mulla (1962) found a preoviposition period of 5-7 days in the lab and determined that proteins and carbohydrates were essential for the production of a maximum number of hatchable eggs. Burgess (1951) captured females that then laid 1-50 eggs; a second, smaller batch was laid about a week later.

Although more recent studies indicate soil situations to be the primary breeding sites, Hall (1932) reported oviposition in the field in the Coachella Valley on decaying meats, various excrements, and decaying fruit and vegetable matter. The eggs were laid singly and at random. He found human excrement to be the most favorable larval medium under insectary conditions and thought fecal contamination of moist soil was important in the field. The eggs had an average incubation period of 3.7 days, a 2-day optimum and a maximum of 30 days at low temperatures. A lack of moisture was fatal to eggs, and a deficiency was detrimental to development. The larvae were negatively heliotropic, tunneling into the food or burrowing into sand. They apparently ate the products of decay rather than larger portions of decaying material; the presence of moisture was critical in development. The larval period averaged 11.4 days on human feces; 8.7 days on dog manure; and 17 days on decaying oranges. Pupation took place in tunnels or in the surrounding sand; the pupal period averaged 9.8 days (4-20), depending on the media and temperature. Under lab conditions, Burgess (1951) found these periods:

egg, 3 days; larval, 11 days; prepupal, 1 day; pupal, 6 days; preovi-positional adult, 7 days; and a total egg-to-egg period of 28 days. He reported that larvae would not develop in closely packed soil or putrid material, nor in excrement unless mixed with loose earth.

During the winter months, the larval and pupal stages may last for many weeks, and gnats may overwinter in the immature stages. Burgess (1951) stated that some adults survive the winter, but his experiments indicated that 75.7% of all gnats emerging during the entire year came from eggs deposited during October-November.

Mulla's (1962, 1963, 1966) investigations in southern California have clearly demonstrated the breeding niches of these flies in this range. They breed primarily in loose, moist, sandy, well-aerated, tilled soil mixed with quantities of green organic matter, in situations where intensive farming is done such as citrus groves, date gardens, vineyards, peach and apricot orchards and vegetable and field-crop fields. He studied the problem by placing emergence cages over flood-irrigated soil that was recently disked and by washing soil samples for pupae and puparia. He observed increased oviposition activity after disturbance or disking of the soil; the gnats swarmed and laid eggs on the soil as soon as it was turned. Most of the oviposition activity was manifested within a few hours after disking. Within 24 hours, all the eggs to be deposited in a disturbed area were laid. Most Coachella Valley eye gnat emergence took place 2-4 weeks after disking; there was usually a longer period during cool weather. Tilled farmlands produced the greatest numbers, but it was found that gnats lay eggs and breed to a lesser extent on non-disked, moist, or irrigated alfalfa fields, golf courses, lawns, and flower beds. Recent studies have led to the observation that some eye gnat species also may breed in river basins, ditch banks, reservoir edges, lake shores, and seepage grounds.

Earlier, in 1941, Bigham had reported that it appeared that little or no oviposition occurred after the first week in the plowed-land breeding sites of these gnats in the southeastern United States and that most eggs were laid within the first few hours after plowing. During the summer, gnats began emerging within the third week after plowing and completed emergence in another two weeks; some flies emerged in 12 days.

In laboratory experiments, Mulla (1963) added known amounts of food to Coachella fine sand and found that 2% food (organic matter) in the soil was optimum for maximum emergence. In field experiments, he killed off the weed cover of both tilled and nontilled plots of ground, and this resulted in almost complete absence of gnat breeding. Bay and Legner (1963) successfully reared eye gnat larvae on the living roots of barley and wheat in sand culture and demonstrated that larvae may feed on living grass roots in nature. Probably, decaying fruits and vegetables, plant rootlets, and fecal matter all may contribute to the organic matter in the soil on which gnat larvae feed.

ADULT HABITS

Ayyar (1917) reported this fly widely distributed through India, along the seacoasts and at fairly high elevations. It was found all year around except during the cold November-February months, with maximum abundance in summer and in short periods of warm weather after the southwest monsoon set in. Roy (1928) stated that it was a serious pest in Assam, India, during May-July, and it was generally held responsible for Naga sore and epidemic conjunctivitis, the incidences of which closely coincided with its peak prevalence.

Patton (1921) noted the hematophagous habit of the adults in Madras; they were common on horses and cattle, following biting flies. They also fed on the serous discharges from the eyes and sores and were an intolerable nuisance during hot weather in the plains of India. Roy (1928) reported adults feeding on warm human excretions and on the margins of animal ulcers and rectums. Large numbers of flies occurred in cow sheds in Assam, and adults were also observed to feed on fresh and decomposing cow and horse dung. Ayyar (1917) observed these gnats hovering about the face and settling at the corners of eyes of people and frequently getting into children's eyes. They especially were attracted to lachrymal secretions and perspiration and fed greedily on blood from razor wounds. Patton (1921) saw large numbers on human sores and cuts.

The adults are very active on hot, sunny days; none can be found when the air is cool and the sky cloudy (Roy, 1928). Syddiq (1938) found them likely to collect in very large numbers on hanging strings, cords and in cobwebs; they commonly rested on thatched roofs, strips of cane and in grass on sunny mornings (Roy, 1928). An electric light cord that was frequented by the flies was painted with a fluorescein solution by Syddiq (1938), and gnats that picked up the dye were detected 100 yards away.

BREEDING HABITS

Ayyar (1917) could not find the natural breeding sites of *S. funicola*, but observed a few gnats emerging from earth brought in from his garden for another purpose; gnats also appeared in jars in which grasshopper eggs were kept in damp sand in the laboratory. However, he field-collected adults and reared larvae successfully in the lab on fresh cow dung; the eggs hatched in 3 days, the larvae were full grown in a week and left the liquid medium to pupate in dry dung or soil. The pupal period was about a week. Roy (1928) bred a few flies on a few occasions from earth sodden with dung and urine in cattle sheds, but did not collect them in traps placed over human feces, cow or horse dung, decaying fruit or leaves. Syddiq (1938) reported that the natural breeding place is in moist mud contaminated by decomposing organic matter. Damp, soiled earth around improperly kept pail-latrines is a favored site; the gnats also breed in badly kept cattle stables and more frequently in connection with badly kept and contaminated surface drains. Antonipulle (1957) found six types of breeding places of this fly in Ceylon. They were all characterized by organic pollution: cattle droppings, human excrement, pumpkin plant beds, saturated soil from latrine washings, household fish dressing areas (where the soil becomes impregnated with decaying fish parts), and palmyra and other fruits (negligible).

Dissections of females showed them capable of laying about 40-50 eggs. Syddiq (1938) also found that the larvae hatched from eggs in 3 days; the three larval instars took 4-5 days to complete development, and the pupal stage lasted about 2 days, which is exceptionally brief.

FANNIA

The biology of the *Fannia* group has been described primarily by observers in Great Britain, North America, and Soviet Central Asia (Samarkand). Most of our information comes from extensive studies on *F. canicularis*, the lesser house fly, and *F. scalaris*, the latrine fly.

ADULT HABITS

In England and in the United States, *canicularis* occurs more abundantly in the cool spring and early summer (Hewitt, 1912a; Hansens, 1963). In Massachusetts, heavy populations sometimes occur on poultry farms, and the flies invade homes in nearby residential sections (Steve, 1960). In northern California, a similar situation exists with *canicularis* but with midsummer seasonal peaks and with *F. femoralis* also present and peaking between August and October (Anderson and Poorbaugh, 1964). Legner and Brydon (1966) found *femoralis* to be the predominant species breeding in chicken droppings in southern California, comprising over 95% of all flies sampled at any given time of the year. In Utah dairy barns, *canicularis* was the predominant fly species observed (Ogden and Kilpatrick, 1958).

The adults normally comprise 2 to 25% of the usual fly populations in houses (Hewitt, 1912a) and in late fall may make up to 100% of the flies in upper-story rooms (Chillcott, 1960). House counts of 7824 flies in Samarkand showed 83% of these to be *Fannia* and 95.3% of these *canicularis*; this was related to the high relative humidity and low temperatures outdoors (Sychevskaĩa, 1954). This species sometimes vies with *Musca domestica* as the most important pest fly on the basis of complaints by householders in certain areas. As a general rule, the higher temperatures and the dryness of the breeding media probably suppress *Fannia* populations during the midsummer months (Fig. 8).

The very active *F. canicularis* males are collected with greater frequency in buildings. They show a distinctive aimless, hovering, and zig-zagging flight of long duration. The females apparently are less active and have shorter flight periods, which are somewhat similarly patterned but more or less directed. Females are mostly found outside or inside near breeding places. Preferred outdoor resting sites are shaded areas, weeds, tree trunks and branches, and very often the sides of buildings (Lewallen, 1954). Adults alight on walls and ceilings inside buildings, where they show an irregular dancing flight with reference to ceiling fixtures, or under a tree outdoors. Observations by Steve (1960) showed a preference for cool environments. With the increasing temperatures of summer, *Fannia* spp. tend to rest more often (Hansens, 1963).

Sychevskaĩa (1954) observed and described the synanthropic *Fannia* of Soviet Central Asia. These were *F. canicularis, scalaris, leucosticta,* and *subscalaris,* which was rare. *F. canicularis* bred the year around in frostless winters. Adult *leucosticta* were usually observed from the end of April until the end of October, and larvae and pupae were found throughout the year (Fig. 9). Most of the *Fannia* occurring in houses, basements, and cattle sheds were *canicularis* (Fig. 10). On fruits in market places, the males of *canicularis* and *scalaris* predominated. Preferred day resting sites were under the leaves of woody and grassy growths, hollows and cracks in tree trunks, holes in walls of buildings, and shaded toilets and pits. In summer all the *Fannia* species spend the night outdoors in the same places as the day resting sites.

As is the usual case in North America and Europe, total numbers of outdoor *Fannia* decreased daily with the onset of hot summer temperatures in Samarkand; this was reflected by decreased migration into buildings in July. Seasonal population changes have been described by Sychevskaĩa (1954). From March to mid-May the maximum numbers of *canicularis* and *scalaris* are collected at the hottest hour. In later spring beyond 20-25°C, a two-peaked curve is described with the hottest hours avoided. These summer curve peaks begin to grow together at the end of October until a single peak is evident by mid-November. With *leucosticta* there is almost always one peak at the hottest hour.

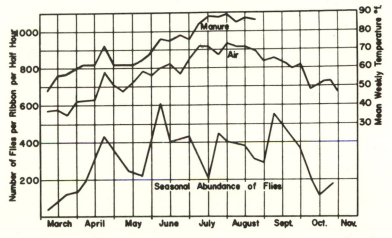

Fɪɢ. 8. Seasonal distribution of *F. canicularis* adults on a poultry farm during 1958, as determined by the numbers caught on fly ribbons exposed for half-hour periods. (From Steve, 1960.)

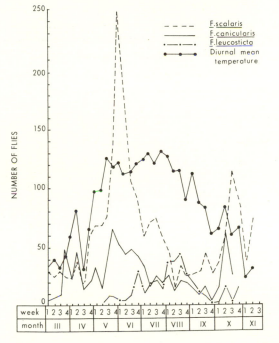

Fɪɢ. 9. Seasonal curve of *Fannia* spp. at an outdoor point in Samarkand in 1950 (adapted from Sychevskaĩa, 1954). No scale was given for the daily mean temperature.

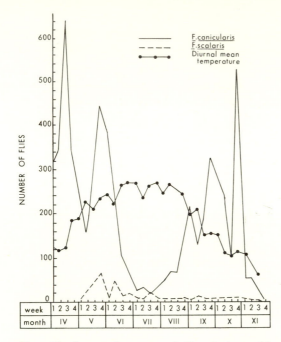

FIG. 10. Seasonal course of *Fannia* in residences in 1950 (adapted from Sychevskaĩa, 1954). The marked endophily of *F. canicularis* is noteworthy. *See* Fig. 9, which shows *F. scalaris* to be the dominant population outdoors at this time.

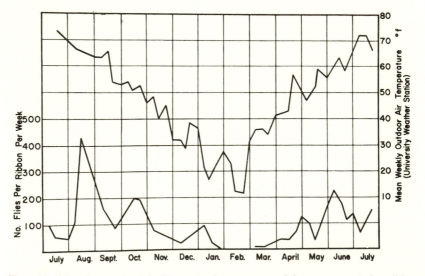

FIG. 11. Numbers of adult *F. canicularis* captured by means of fly ribbons placed over the kitchen table in a house located about 100 yards from a poultry farm heavily infested with this species. Despite its strong endophilous habit and presence in the kitchen, this fly did not readily settle on food. (From Steve, 1960.)

On Guam *F. pusio* is abundantly collected in traps baited with carrion and human excrement and is rarely found indoors. *F. canicularis* is very common in houses, and *scalaris* breeds in cesspools and latrine pits. The *Fannia* species here are commonly attracted to decaying fruits and human feces (Bohart and Gressitt, 1951).

ADULT FEEDING HABITS

Plant and fruit associations have been reported for various *Fannia* species. Both males and females are attracted to honey-dew and often collect in large numbers on aphid-infested plants (Tiensuu, 1936). Many *Fannia* species are attracted to plant sap (Tiensuu, 1938; Shillito, 1947). Several species have been taken commonly on many species of flowers, particularly the blossoms of the Umbellifera (Chillcott, 1960). Shura-Bura (1952a) reported that, in Crimea, 2.8% of the flies on spoiled grapes were *canicularis*. During July in Japan, Illingworth (1926a) found that, of 5250 flies trapped in sweetshops, 1.37% were *canicularis* and 4.88% were *scalaris*.

Although there are often reports of high percentages of *Fannia* invading houses, Chillcott (1960) and Steve (1960) observed that they did not readily settle on humans, tables, or food, which is the annoying characteristic habit of *Musca domestica* (Fig. 11). However, there are reports of direct human association. *F. benjamini*, which is well-known in western North America, has the habit of buzzing annoyingly around ears and eyes, attracted by sweat and mucous (Chillcott, 1960), and has recently been cited as a vector of the eye-worm *Thelazia* of mammals in California (Burnett et al., 1957). In addition, Haddow and Lumsden (1935) listed 22 references dealing with records of intestinal myiasis due to *F. canicularis* and *scalaris* larvae.

Breeding materials also serve as an adult food source. Anderson and Poorbaugh (1964) examined the diverticulae of both males and females of *canicularis* and *femoralis* and found chicken droppings. Dissections of the *Fannia* species of Samarkand often showed fecal matter in the mid- and hindgut and cherry or other fruit juice in the crop. In fact, the enclosed deep-pit latrines near market places served as refuges, breeding places, and food sources for these market flies. *Fannia* were particularly attracted to cherries and muskmelons (Sychevskaĩa, 1954).

OVIPOSITION AND BREEDING HABITS

Laboratory studies on *canicularis* and *leucosticta* indicate that mating occurs on the second day after emergence. However, lab studies by Tauber (1968) showed that *femoralis* females begin to mate when they are 2 to 24 hours old whereas *canicularis* females begin at 96 to 120 hours. The mating of *scalaris* has often been observed in autumn on sun-lit leaves and lasts 10-40 minutes (Sychevskaĭa, 1954). Both field and laboratory studies show a preoviposition period of 4 to 5 days (Sychev-skaĭa, 1954; Lewallen, 1954). Tauber (1968) found that *canicularis* females oviposit 72 to 96 hours after emergence and *femoralis* within 48 to 72 hours after emergence. The eggs are deposited on the surface of the substrate. *F. canicularis* oviposits both indoors and outdoors. Field observations in Samarkand showed that the *scalaris* egg stage lasts 30-40 hours in summer and 3-4 days at other times. Larval stage periods were I, 3-4 days, II and III, 3-5 days (Sychevskaĭa, 1954). Steve (1960) found similar developmental periods for *canicularis* in the laboratory as follows: egg 1½ to 2 days, and larval 8 to 10 days. He also exposed various manure samples to ovipositing females and demonstrated these preferences (percentages of the total number of eggs laid): chicken 63.5, hog 20.5, horse 8.7, cow 4.0, and sheep 3.0. The high ammonia con-tent of chicken manure was thought a possible attractive factor. In south-eastern Washington state *canicularis* was collected from mink (47%), chicken (40%), swine, cow, and human dung, and reared from chicken (92%), mink (6%), swine and cow dung (Coffey, 1966).

The most usual larval habitat for the almost 200 known species of *Fannia* around the world is mushrooms and related fungi (Chillcott, 1960). However, the larvae of the synanthropic and widely distributed species we are treating here are found in decaying animal and vegetable matter, especially excrement and occasionally in birds' nests. Highly nitrogenous materials seem to be required for *Fannia* breeding. The breeding matter of the genus in Samarkand includes human feces, chicken and pig manure, dog dung, and moist garbage; *scalaris* breeds in deep semifluid latrine material; *canicularis* in chicken manure in large shaded piles; *leucosticta* breeds in drying feces of "summer type" latrines and is rarely found in animal manures and garbage (Sychevskaĭa, 1954). *F. scalaris* is also common on human feces in the privies in Canada and Great Britain (Hewitt, 1912). The larvae are dorso-ventrally com-pressed with feather-like processes that act as a flotation device in the organic semifluid media. The larvae of *canicularis* have somewhat spi-niferous processes and usually prefer a drier medium. The eggs of *Fannia* are similarly specialized; they are elongated, fairly flat, and have narrow wing-like processes.

PUPATION AND OVERWINTERING

F. canicularis and *scalaris* usually leave their semiliquid or liquid sub-strates for somewhat drier places to pupate (Chillcott, 1960), but *pusio* pupates in the larval medium without any serious loss in viability (Illingworth, 1922). Wilhelmi (1920) reports that *canicularis* and *scalaris* may overwinter as adults in warm rooms and stables. But more commonly they overwinter as pupae (less commonly as larvae) in the earth 2 to 3 inches below the surface or under shelter on the surface (Mellor, 1920). The pupa has the appearance of a quiescent larva; it shortens and becomes robust, hardens, and darkens. In the relatively mild climate of Samarkand, Sychevskaĩa (1954) found that all *Fannia* species can overwinter as larvae, pupae, or prepupae. A long, snowy winter there in 1950-1951 killed most of the overwintering larvae, and fewer flies were evident in the spring. She also recorded seven successive generations. Mihályi (1967a) states that in Hungary the maximum population of *canicularis* is in April but it has many generations during the season. It seems likely that several generations also occur in California *canicularis* and *femoralis*. Steve (1960) reports that on Massachusetts poultry farms all stages of *canicularis*, except the egg, were found during the winter months, especially third instars. Adults in small numbers were found in cracks and crevices. The activity of the overwintering adults became noticeable in late February, and emergence from overwintering pupae occurred in mid-March. Collection data for more boreal groups indicated that they have only one or two generations a year (Chillcott, 1960).

Longevity

Steve (1960) found that the total developmental period of *canicularis* in the laboratory from egg to egg was 22-27 days. At 80°F and 65% relative humidity, 50% of the males died within 14 days after emergence, and 50% of the females within 24 days with a few surviving for 54 days. Sychevskaĩa (1954) found that some adult *scalaris* and *leucosticta* lived up to 20 days in gauze cages and some *canicularis* females lived 2 months, ovipositing 6 times. Adult longevity in nature is unknown.

MUSCA AUTUMNALIS

The most extensive study on the biology of the face fly has been reported by Hammer (1942) from Denmark. Teskey (1960) contributed an important review of its life history and habits, and more recently Ode and Matthysse (1967) reported on its bionomics. This species was first recorded in North America in 1952, and by 1967 it was distributed from coast to coast in the Canadian provinces and in the northern United States (Sabrosky, 1961; Anderson and Poorbaugh, 1968; Depner, 1969). Its habit of attacking cattle and annoying man has prompted many studies in recent years in the United States and Canada. Useful bibliographies of the face fly in North America have appeared in recent issues of *California Vector Views*, e.g. Vol. 13, No. 6; Vol. 14, No. 11; Vol. 15, No. 11; Vol. 16, No. 12; and Vol. 18, No. 4.

ADULT HABITS

In Europe and in North America, the flies are present in the field from May to October, attaining high populations in late summer (Hammer, 1942; Teskey, 1960). In Ohio, they have been observed in large numbers on cattle, horses and sheep, and occasionally on hogs, dogs, and humans (Treece, 1960). The most frequently reported feeding habit was that of sucking on various animal secretions. Large numbers of flies cluster around the eyes and nostrils of cattle and feed on the mucous and watery secretions (Treece, 1960; Teskey, 1960). Treece (1960) reports that tear production is apparently stimulated, and cattle are often seen with dark wet areas under the eyes. Face flies also feed on the blood exuding from wounds made by biting flies. Thomsen (1938) observed these flies imbibing blood left by *Stomoxys*, *Haematobia*, and *Lyperosia* spp. Tabanids are the most frequently mentioned as providing the meal, and Hammer (1942) has observed as many as thirty *autumnalis* clustering around a feeding horse fly, waiting for it to depart. He also reported that face flies left a cow for human blood from a fresh scratch. Teskey (1960) noted that his perspiration attracted females occasionally.

Only a small portion of the swarm of flies are on the animals at any time (Treece, 1964). In Nebraska, Jones (1963) placed a calf in an outdoor cage with 500 flies, and on the day of greatest fly activity only 3.8% were on the calf at any one time.

Males and females differ in their feeding and resting habits. Males were seldom noted on cattle (Hammer, 1942). Hansens and Valiela (1967) found female flies greatly outnumbering males on cattle. Treece (1960) noted that a preponderance of the flies resting on gates and fences were males, whereas collections from animals and dung tended to be predominantly female. Of 928 face flies he collected from animals in August by means of blood baits, 17% were male and 83% were female. Studies by Miller and Treece (1968) showed that the females alternately fed on blood and then completed gonadotrophic cycles and that the highest incidence of females on cattle occurred during the process of egg maturation. Males exposed to a free-choice diet showed little episodic feeding and fed primarily on malt. Hammer (1942) reported that *autumnalis* fed on the nectar of a wide variety of flowers and believed that nectar was the principal diet of the flies in the spring before cattle were pastured. While sweeping areas for face flies, when cattle were absent, Hansens and Valiela (1967) found that males predominated along the woods edge, and, there, only male flies were seen on plants in flower. Hammer (1942) observed the imbibing of dung fluids by *autumnalis*, and he did some analyses of the contents of the intestines that showed some flies in May-June with yellow matter (nectar?) and dung. Fly intestines checked in August all had dung in them. Miller and Treece (1968) demonstrated that the face fly feeds on feces to the greatest extent 1 day before and 2 days after peak oviposition. Generally the observations cited above indicate that the female is, in all likelihood, dependent on blood and dung fluids as sources of protein and nitrogenous material for promoting ovarian development and that the males may mostly feed on nectar.

Typically the fly-frequented cattle pastures are close to hedge rows, woods, swampy areas and the like, and this vegetation affords night resting sites since the flies feed only during daylight hours (Hammer, 1942). In Maryland, observations of flies marked with fluorescent dye demonstrated night resting in trees and high weeds (Killough et al., 1965). Treece (1960) observed that the largest number of face flies in the field occurred on bright, sunny days and that feeding took place during the heat of the day; the flies did not follow the animals into the barns. Yet, Benson and Wingo (1963) reported that cattle grazing or standing in the sun had fewer flies than cattle bunched in the shade. They also found higher populations during periods of low relative humidity and that wind speeds above 10 mph reduced populations on cattle. In studies of face fly dispersal, marked flies were recovered at distances up to 1500 meters in 43 and 46 hours after release (Hansens and Valiela, 1967). It is known that they can travel much farther (see Table 3).

OVIPOSITION AND LARVAL HABITS

Laboratory studies indicate that mating occurs 4-5 days after emergence; the flies copulate frequently, commonly for an hour at a time; some females copulate 2-3 times (Wang, 1964). In the field, Hammer (1942) observed that males rest on conspicuous objects and dart out to attack the females flying past. Mating is then completed on the ground. Eggs are usually laid 2-5 days after mating (Wang, 1964).

Face fly larvae are found only in fresh cattle droppings (cowpats) in pasture and range situations in North America and in Europe (Anderson and Poorbaugh, 1968; Hammer, 1942; Patton, 1933). It was the only medium used by the species in tropical Africa (Roubaud, 1911a). However, Vainshtein and Rodova (1940) also found the larva in pig dung in Tadjikistan, and Kobayashi (1919) reported oviposition in latrines in Korea. Bay et al. (1968) present data that indicate at least the possibility that the face fly may successfully propagate in nature in bison and swine feces as well as in cattle dung. The pelleted form or coarse texture of the feces of other animals was inadequate for oviposition or too low in moisture content. Changes in bovine diet are reflected by variations in the attractiveness of the resulting feces for oviposition (Treece, 1964). The flies preferred feces from animals on diets containing relatively large amounts of legumes or grass (Ruprah and Treece, 1968).

The observations of Hammer (1942) and Wang (1964) in the field and laboratory, respectively, are in general agreement on the oviposition behavior of the females. They arrive on the fresh or newly dropped dung within minutes. The flies spend several minutes creeping over the dung, imbibing dung fluids, and randomly seeking out cracks in which to oviposit if there is a thin crust or smooth areas on newly dropped dung. In the field, few flies were found on 24-hour old dung. In fact, Teskey (1960) noted that dung lost some attractiveness within an hour after deposit. Eggs are laid in a matter of seconds, singly in batches of 5-8. In the laboratory the batches most frequently comprised about 20 eggs, and females averaged 4-5 batches in their life-time (range 30-128 eggs per female). Killough and McClellan (1965) reported a maximum of 230 eggs laid by one individual. Treece (1964) found that eggs were produced at up to a maximum age of 48 days and that 80% of the eggs were laid in the first 3 weeks of adult life, with approximately equal numbers produced each week.

The eggs have a distinctive long, grayish-black mast at the anterior end (Wang, 1964). Hammer (1942) observed the survival of eggs sealed with paraffin and concluded that the mast has a respiratory function. He found average incubation periods of 10½ hours for eggs laid in the morning and 23 hours in the afternoon. In the laboratory the eggs hatch in about 16 hours at around 25°C (Treece, 1964; Wang, 1964).

In the field, the cattle droppings become progressively drier, and the larval period is about 3-4 days (Teskey, 1960). In the laboratory larval development took 5 days at 20°C and only 2½ days at 35-40°C; at 25-30°C and 50-70% relative humidity, the duration of larva I was 6½-7½ hours, that of II, 16½-17½ hours, and that of III, 55-57 hours (total range, 78-82 hours) (Wang, 1964). Wang also observed that the young larvae favored moist media and the full-grown larvae drier areas. Newly hatched larvae were sluggish, tunneling just below the surface; later instars burrowed through the whole medium; fully grown instars tended to migrate to drier crusts or even move away from the medium.

Pupation

Jones (1969) studied the dispersal of larvae from cow dung outdoors and observed masses of larvae crawling from the manure before pupation. In one test, pupae were found in a radius of 8-30 feet from the dung. The larvae pupated in crowns of grasses, under organic matter, and in cracks in the soil. Teskey (1960) found puparia under about a quarter inch of soil within a few inches of droppings. Hammer (1942) had observed larvae emigrating from droppings just before sunrise.

The pupal period is 7 days in the laboratory at 75°F and 10 days under field conditions (Teskey, 1960). In the laboratory emergence of adults from puparia occurred mainly in the early morning, and young flies did not feed until they were 2 days old (Wang, 1964). The total egg-to-adult period is about two weeks in southwestern Ontario (Teskey, 1960) and three weeks in Denmark (Hammer, 1942). It took about 12 days in the laboratory (Wang, 1964).

Overwintering

Many observations in Europe and North America indicate the adult face flies hibernate hidden in houses or other protected places in large numbers during the winter months (Graham-Smith, 1918; Hammer, 1942; Matthew et al., 1960; Teskey, 1969). Benson and Wingo (1963) found winter infestations in Missouri to have no apparent correlations with the proximity or number of cattle in the areas. The sex ratio was about 1:1 in most houses during winter. They also found that, during spring dispersal from hibernation sites, the females were fertilized and migrated to the fields, leaving the males behind. Hammer (1942) found no adults in the field during winter but suggested that they might have been well hidden in cracks and crevices in woods, because in the spring he observed large numbers on cattle grazing near woods and few flies on cattle in open pastures. Face flies generally are among the earliest to appear on cattle in the spring.

Stoffolano and Matthysse (1967) showed that both sexes of the face fly exhibit a true imaginal facultative diapause characterized by fat hypertrophy and cessation of ovarian development in females. They found that continuous light and high temperature (83°F) prevented diapause and the proportion of flies diapausing increased as the temperature was lowered and the photoperiod shortened. Almost all the flies diapaused when subjected to continuous dark at 65°F. The diapausing flies failed to feed on cow's blood and failed to mate. The face flies tended to congregate in the same sites year after year and may have been stimulated by olfactory responses to hibernate in specific places, possibly attracted to the accumulation of feces or speckings (see Barnhart and Chadwick, 1953).

LONGEVITY

Wang (1964) found that most of his laboratory reared adults lived for 4-8 weeks. Treece (1964) found the maximum longevity of adults was 55 days.

MUSCA SORBENS

This species is widely distributed in the Old World tropics and subtropics, in the Ethiopian, Oriental, and Australian regions. Its possible relation to eye diseases in Egypt has prompted Hafez and Attia (1958a, b, c) to carry out detailed investigations on the biology of this fly. Other reports come from investigators in Tadzhikistan, India, western Turkmenistan, Nyasaland, the Philippines, Guam, and China. With *sorbens*, we also include *M. vetustissima*, the Australian bush fly.

ADULT HABITS

Hafez and Attia (1958c) found this species mainly outdoors all year around in Egypt. The adults frequented bazaars, food stores, and slaughterhouses in towns and were common in areas where human defecation was left exposed. In the adjoining desert regions, the species constituted 60% of all flies collected in one study (see Table 1). Generally, two peaks of abundance occurred, in spring and in August (Fig. 12). The adults are attracted to wounds and ulcers and to the faces and eyes of humans. They are very active flyers and seem to prefer bright sunlight; at sunset, they rest on twigs and leaves in open fields or on the outside walls of dwellings. Also in Egypt, Peffly (1953) reported this fly on children's faces and attracted to the moist sweet surfaces of sugar cane, dates, oranges, and to freshly butchered meat. It fed actively on animal blood in slaughterhouses (Hafez and Attia, 1958c).

Zimin (1944) reported the species widely distributed in Tadzhikistan: in inhabited places, outdoor stands, and buildings (only during daylight). The adults first appeared in the field in May-June, and the maximum population was attained in August; the last adults were seen in December. This fly attacks man in this region; it is attracted to smelly, unsanitary persons, especially children, and feeds on sweat, mucous secretions and skin lesions. Zimin also found the adults feeding on fresh and spoiled meat, milk, fruits, sweets, garbage, animal corpses, and human feces. According to Sukhova (1954), *sorbens* is a widespread fly of the population centers of western Turkmenistan. It was most numerous from the second half of August to October. He established a relationship between the population dynamics of this fly and the incidence of acute epidemic conjunctivitis in this area. Trofimov (1963) found *sorbens* present in Azerbaijan from May to October, with maximum numbers in August-September. The adults attacked both man and animals and were commonly observed on excrement, cow dung, rotting meat, animal carcasses, and fruits and vegetables; they were a nuisance in markets and open-air restaurants.

In north China, Meng and Winfield (1938) trapped a high percentage of this fly in city foodshops, indicating its preference for indoors and for food; the flies were not particularly attracted to garbage. On Guam, it is one of the commonest flies, especially where human habitations and livestock are abundant. Bohart and Gressitt (1951) found adults commonly around garbage dumps, pig pens, native yards, privies, carrion, and various types of excrement. They fed readily on drying cycad fruits, fish fillets, jerked meat, and other sweets and processed food. They were especially attracted to human body exudations; the flies were observed to scrape sores until the skin was perforated and lymph or pus exuded. Over 50 flies were seen feeding on a yaws sore in the Solomons.

Lamborn (1935) reported *sorbens* to be the most prevalent fly collected from the backs and sores of natives in Nyasaland. In Australia, the bush fly first appears in October; the maximum abundance is attained in December-January; no flies are present from late May to late October (Norris, 1966a). It is a common house fly in India (Deoras and Jandu, 1943) and in north China, but in other localities, Australia, Egypt, Russia (Smirnov, 1940), and Morocco (Gaud et al., 1954), it is almost always an outdoor species.

BREEDING HABITS

Hafez and Attia (1958a) found a preoviposition period of 4-9 days in the laboratory. They observed that mating takes place in the superimposed position; the female carries the male and flies off to a safe place. The entire act may last from 23-48 minutes. In Egypt, no copulation was observed above 32°C or below 20°C.

Females generally show a strong oviposition preference for human feces. Workers in the eastern Mediterranean recorded the fly only from human feces and encountered great difficulty in maintaining laboratory colonies on any other media (West, 1953). When given a choice of eight kinds of dung, human feces were preferred for oviposition, but, when given no choice, small numbers of eggs were laid on cow and buffalo dung (Peffly, 1953).

On Guam, larvae were found in pig dung, horse manure, once in cow manure, and once in an isolated deposit of human excrement; they were never found in latrines or in piles of cattle droppings (Bohart and Gressitt, 1951). Meng and Winfield (1944) reported that dog manure was the important breeding material in west China, but small numbers were bred from pig and cow manure and from semiliquid human feces. On Samoa, Buxton and Hopkins (1927) usually found larvae in human feces but could not find them in any other medium. In Manila, 19.47% of the flies caught breeding in fermenting tobacco dust were *sorbens*. In Canberra, the bush fly was bred in the feces of man, cattle, sheep, and dog (Norris, 1966a). Zimin (1944) stated that, in Tadzhikistan, it breeds in human feces deposited on the ground and in distinctly smaller numbers in pig dung and in dust bins. In Azerbaijan, it reproduced exclusively in human excrement and accounted for 31% of the various species reared from samples (Trofimov, 1963).

In Egypt, Hafez and Attia (1958a) observed that *sorbens* breeds mainly in isolated human stools and to some extent in pig dung and fuel cakes (cow and buffalo dung). In laboratory preference tests, 89.4% of eggs were laid on human excrement (see Table 2). Two-day old excrement did not attract the flies. Females walked over the excrement before depositing eggs, seeking cracks and crevices. They laid an average of 3-4 batches in their lifetime. The eggs are closely packed, cemented together, and arranged in rows. A single batch may contain as many as 32 eggs; large clusters of 700 or more eggs were laid by several females. A single female may lay up to 80 eggs in her life, although Awati (1920) stated that never more than a total of 50 eggs was laid. The females feed on the excrement after ovipositing.

The eggs hatched after 6.25 hours at 31-34°C and 100% relative humidity (Hafez and Attia, 1958a). At 32°C the average larval periods were: I, 7 hours (5-8); II, 14 hours (13-15); and III (feeding and prepupal), 40 hours (38-43); total larval period, 61 hours (56-66). Zimin (1944) bred 42,000 larvae in 1 kg of human feces; development was completed in 15-16 days at a substrate temperature of 17-20°C, and at a temperature of 23.8-26.1, in 8-10 days. At a rearing temperature of 30°C and 50% relative humidity, normal pupae and adults of *M. sorbens* appeared 4-5 days and 8-9 days, respectively, after the eggs were introduced (Greenberg, 1969). Hafez and Attia (1958a) reported a pupal period of 6.21 days at 24°C and 4.56 days at 28°C. Pupation took place in the drier soil beneath the stool; the larvae had great power in penetrating hard and soft soils. In the lab, the greatest emergence took place from 4 A.M. to 8 A.M. (maximum at 6-7 A.M.) in June. In the colder season, emergence occurred later. *M. sorbens* is usually the first among the flies emerging in succession from human stools in nature.

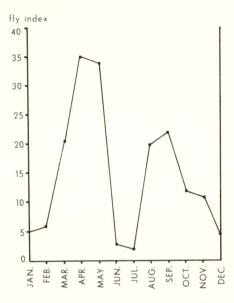

FIG. 12. Density of *Musca sorbens* in Talbia village near Cairo, Egypt, during 1954. (From Hafez and Attia, 1958b.)

TABLE 1

RELATIVE ABUNDANCE OF *M. sorbens* AND OTHER FLIES IN TRUE DESERT IN EGYPT*

SPECIES	NUMBER	PERCENTAGE
Musca sorbens	162	60.2
Musca domestica vicina	58	21.5
Musca xanthomelus	17	6.3
Sarcophaga spp.	30	11.2
Wohlfahrtia spp.	2	0.8
Total	269	100.0

*From Hafez and Attia, 1958c.

TABLE 2

ATTRACTIVITY OF VARIOUS KINDS OF EXCREMENT TO
OVIPOSITING FEMALES OF *M. sorbens**

ANIMAL DUNG	AVERAGE NUMBER OF EGGS LAID	PERCENTAGE OF EGGS LAID
Human excrement	479	89.4
Horse manure	0	0.0
Donkey dung		
Buffalo dung	0	0.0
Cow dung	0	0.0
Calf dung	20	3.7
Camel dung	0	0.0
Fowl manure	0	0.0
Pig dung	37	6.9

*From Hafez and Attia, 1958c.

The adult longevity in the lab averaged 18.5 days for the males and 21 days for the females at 24-30°C and 50-60% relative humidity (Hafez and Attia, 1958). The flies lived longer at lower temperatures than at higher temperatures. Experimentally, it appeared that sucrose was important for longevity but not the proteins that were derived from human excrement or animal dung.

Sychevskaĩa (1960) includes *sorbens* in the group of species that winters in the preadult stages in central Asia. Norris (1966) states that the bush fly may overwinter in the pupal stage around Canberra, but Hughes (1971) suggests that the species dies out in winter in the south and the region is repopulated each year by spring migrants which come down from the north.

STOMOXYS CALCITRANS

This important, worldwide, temperate and tropical biting fly is the well-known tormentor of cattle, horses, and man and a possible vector of infectious diseases. It seems to have been one of the earliest noted and most rigorously observed synanthropic flies. Observers have reported on its biology from many parts of the world.

The habits of this species vary with the climate and locality in which it occurs. The adult feeding habits restrict it to association with man's domestic animals, but, with respect to breeding places and larval development, it is difficult to generalize. It is not always associated with definite kinds of dung, and sometimes not even with dung alone. In countries with a warm climate, the adults live more in the open. Whereas in northern Europe, for example, it particularly occurs in stables (Thomsen and Hammer, 1936). Thus it is supposed that *Stomoxys calcitrans* originated in warm regions of the world, particularly central Africa where the largest number of *Stomoxys* species are found today. When animals were domesticated and stables were built, this species had the opportunity of spreading to colder parts of the world.

In order to elaborate on these worldwide variations, some of our information (especially larval bionomics) will be presented geographically: (1) the African continent; (2) southern and northern Europe; (3) the United States; and (4) Asia and the Pacific region. Then the general life history and habits will be treated.

AFRICA

Fuller (1913) describes *Stomoxys calcitrans* as a common fly distributed all over Africa. It is found in and about stables and cattle kraals and often is a serious pest. Stable and kraal manure, with which straw is well intermixed, the accumulations in cattle kraals, and the rotting remnants of haystacks are its usual breeding places. The maggots require moist decaying vegetable matter and they do not breed extensively in nor favor pure horse or cow dung. Fuller further describes an unusual outbreak along the eastern seaboard of South Africa that was traceable to excessive wet weather and abundant breeding in decaying grass straw of any sort. Horses, mules, donkeys, cattle, hogs, and sheep were viciously attacked, leaving ulcerous cavities and causing great annoyance, blood loss, and even death.

Similarly, Parr (1962) found that breeding sites in Uganda, East Africa, required the presence of rotted cattle dung, rotted straw or foliage, and shade to reduce the temperature and preserve moisture (usually urine). In the long dry season, such conditions occur only within fenced cattle enclosures and calf houses. In the wet season, suitable breeding conditions occur in open park land and banana plantations, wherever manure and vegetation occur together. *Stomoxys* populations usually increased 2-3 weeks after the rains began and fell off rapidly at the end of the rainy season. The larvae were never found breeding in the heaps of pure dung.

In the Sudan region, Roubaud (1911b) observed that in the dry season *Stomoxys* abandons the usual dung and shade of the stables and lays its eggs in bright sunlight in the sand and water-soaked dung at the edge of streams. He points out the importance of temperature and a constant preferred humidity for larval life.

Surcouf (1923) observed even more modifications of the species' usual habits under the climatic conditions of Algeria. In the northern Sahara region, cattle and horses are rare and camels pasture in the vast deserts. Aside from wild animals and some sheep, it is in goat quarters that *Stomoxys* lays its eggs. The darkest and most humid corners are preferred. The fly also oviposits in any place on the ground surface inhabited even occasionally by mammals. Pupae were encountered in a cavity of the palm *Phoenix dactylifera*, which was the home of a small mammal (*Eliomys troglodytus*). Many other animal burrows were checked, but no *Stomoxys* were found. From June onward, when temperatures reached 42°C in the shade and domestic animals moved north for food and water, eggs were laid in the very damp sand around artesian wells, on the stems of the abundant grasses and on the inflorescences of the onion. Surcouf also found *Stomoxys* larvae in great numbers in rotten trunks of palms; of these, only two females pupated and emerged. Dissections revealed palm material in the intestines of the larvae. He observed that, in Algeria, *Stomoxys* bites man frequently because there are fewer domestic animals, and he cites Roubaud's unpublished notes describing the great numbers of flies bloodying the legs of natives at Brazzaville, where there was no livestock.

In Egypt, the adults are found in the open in cultivated semiarid localities near the edge of deserts. In rural areas, Hafez and Gamal-Eddin (1959a) found that in the cold season the flies seek shelter in byres and stables and oviposit in animal beddings composed of dung, straw, food remains, fodder, and urine. In warm seasons, they oviposit in green shaded places in cattle and horse dung mixed with straw and manure. The temperature of both of these media was 22-28°C. Pure manures of cow, buffalo, donkey, or horse did not invite oviposition. Decaying vegetation on banks of water canals was thought an important fly source. In farm localities lying near the edge of the desert great numbers of flies were found to breed in decaying leaves in the soggy soil of a shady artificial forest and also sometimes in a mixture of sand and vegetable debris at the base of large *Eucalyptus* trees.

EUROPE

Diffloth (1921) states that *Stomoxys calcitrans* is found in all regions in France. Horses are the primary target from morning until dusk; cows are somewhat less attacked, and other animals not much bothered. The flies are especially abundant in August and September. He describes some particulars of oviposition and larval habits. The flies oviposited in straw piles mixed with horse manure; they did not oviposit in pure manure nor in putrid smelling material. The straw had to be piled and of sufficient depth to have enough moisture content. Oat straw was preferred over corn and rye straw. Eggs were deposited right on the straw. The larvae were observed to eat bits of straw and, if drying out occurred, to burrow into the more moist straw where they eventually pupated.

In contrast, Surcouf (1921) in France described egg laying in pure horse dung, even right under the horses, and often in the manure in the interstices of pavements, as well as on the edge of dung heaps, in debris, or in the urine-soaked ground.

In Romania, Dinulescu (1930) observed large numbers of *Stomoxys* associated with cattle and horses in river delta fishing villages. He found the humid, sandy shore line covered with fish, manure, and other debris to be favorable for larval development.

In northern Europe, *Stomoxys* is mainly a stable insect. It is never found at any great distance from farms or houses. There are few early observations of normal oviposition habits and natural breeding sites in this region. Newstead (1906) could not find larvae or pupae in the feces of domestic animals but observed the flies ovipositing in a heap of grass mowings in the field. In addition, Hindle (1914) and Thompson (1937) describe the usual larval habitats of heaps of moist, porous byre and stable manure and various plant debris. Perhaps the most thorough studies were by Thomsen and Hammer (1936) and Hammer (1942) in Denmark, where *Stomoxys* was found associated very closely with cow barns and calf stables and the larvae bred in the bedding of urine-soaked litter and dung and also in the humid remains of straw and fodder with which dung was not necessarily mixed. It is most significant that, under these climatic conditions, the only quantitatively important oviposition occurred in the stables. In oviposition experiments in the laboratory Thomsen and Hammer found that horse manure yielded the greatest number of *Stomoxys*, which supports Surcouf's (1921) observations in France. But this fact did not fit with their field observations. So the problem of whether cattle or horse manure is more attractive depends on climatic conditions of the locality. In Denmark, food and temperature preferences tend to keep *Stomoxys* in cattle sheds, and they have no tendency to seek horse manure. Whereas in France, conditions are such at times that even pure horse dung will suffice for oviposition and larval development.

UNITED STATES

The stable fly is the most numerous and irritating annoyer of livestock in the United States, most seriously in the grain belt in the central states from Texas to Canada. In 1912, Bishopp (1939) reports that many dairy men found milk production reduced from 40-60% during a severe outbreak of this fly. The breeding sites held responsible were straw stacks, especially oat straw. Other sites consisting of straw and manure in and around stables furnish breeding places especially in the early spring.

Under certain conditions, this fly may occur in large numbers from the middle of August to the middle of October in the littoral extending from the Alabama-Florida boundary east 400 miles to Cedar Keys, Florida (Simmons and Dove, 1941). Observations show that very rapid breeding occurred primarily in two species of bay grasses that wash up and ferment along the northwestern beaches. In northwestern Florida, as well as in neighboring Alabama and Georgia, the flies were found breeding extensively in piles of fermenting peanut litter left in the field after harvesting operations. Simmons and Dove (1942) reported that waste celery strippings were also an important breeding place for this fly. Dairy men reported a 25% reduction in milk output at several miles distance from the celery processing plants.

Herms (1953) and Love and Gill (1965) noted breeding sites along the sandy vegetation-strewn shores of the Great Lakes and the adults occasionally attacking bathers in late summer and fall. The studies of Love and Gill indicated that the farm populations of *Stomoxys* probably did not contribute to the beach populations. Herms also found myriads of larvae in decaying onions in the fall. We have found pure cultures of this fly breeding in lake vegetation which had been cast up on the beach.

PACIFIC AND ASIA

Bohart and Gressitt (1951) record an interesting variation in *Stomoxys* biology in the Pacific. Shortly after the U.S. invasion of Okinawa, an enormous number of *Stomoxys* bred on the unharvested cabbage crop. Since all the livestock had been killed, humans were attacked for a blood meal by the newly emerged flies. On Guam, larvae were reared from cattle droppings, piles of cattle manure, and large piles of decaying aquatic vegetation. Maggots were found only in piled horse manure, but apparently pure cattle dung is utilizable by the flies as a larval medium.

In the Federated Malay States, Pratt (1908) describes the distribution of *Stomoxys* as limited to open areas where cattle are abundant; the flies were numerous near cattle sheds from 10 A.M. to 4 P.M. There were also breeding sites in mining towns where bullocks are used for transport; he saw the flies abundant and active during the hottest hours.

In New Zealand, the flies are a nuisance to the dairy cattle. Ensilage stacks are the preferred breeding sites in rural areas; in urban areas, sources included open compost heaps, piles of grass clippings, and other decomposed organic matter (Todd, 1964).

Perraju and Tirumalaro (1956) described *Stomoxys* as occurring in small numbers all over India. They found larvae breeding extensively in rotting cake manure (containing residue plant materials), which was added as fertilizer to the melon crop. The number of flies increased when this fertilization procedure was implemented. Patton and Cragg (1913) found adults in Madras ovipositing during December and January in the green scum formed on wet sand over which sewage flowed. In Uttar Pradesh, Sehgal and Kumar (1966) found cattle-shed litter an important larval medium during the winter, garbage during the monsoon season, and human excreta especially in the spring.

It is hoped that this geographical survey of the larval bionomics of this species will aid in understanding the adaptability, variations and breeding potentials of this important synanthropic species. We have seen the fairly restricted habits of *Musca autumnalis* and *Fannia canicularis*. The tremendous opportunism of the habits of *Musca domestica* is well described in the literature. *Stomoxys calcitrans* thus seems to have habits somewhat in the middle of these extremes.

Adult habits

In Europe and in the United States, *Stomoxys* is on the wing from about May until October. In southeastern Kazakhstan, its maximum density is noted at the end of July and August (Yakunin, 1966). In Egypt, it has one annual peak of abundance in May (Hafez and Gamal-Eddin, 1958a). In the United States, its worst annoyance to cattle and man occurs in late summer and autumn. Its associations and annoyance to domestic animals have been described above.

The fly is capable of strong flight and can travel many miles. Eddy et al. (1962) recovered marked flies 5 miles from the release site in less than 2 hours (see Table 3). Flies have been found in sheltered places at some distance (1 km) from cattle (Hammer, 1942).

Studies on Riems Island, Germany, showed that below 12°C outdoor occurrence decreased; in the fall the flies sought shelter in dwellings, and by November had all disappeared outdoors (Wilhelmi, 1917). Laboratory studies by Nieschulz (1933) indicate that normal activity occurs at temperatures between 18.0 and 28.2°C and heightened activity between 28.2 and 36.8°C. Todd (1964) found daily flight activity in dairy herds in New Zealand to be diminished by low temperatures, strong winds, and heavy rains. Various authors have indicated that the flies attack less on cloudy days and that animals can obtain some protection by seeking shaded areas. The favored resting sites are sunny fences, walls of houses and farm buildings, and other painted surfaces in general. When the flies are disturbed, they usually return to the same spot. At dark, *Stomoxys* seeks sheltered areas and has been observed not to bite in stables at night (Surcouf, 1921).

Adult feeding habits

One can generalize somewhat about the feeding habits of the stable fly. Both sexes are hematophagous, primarily on warm-blooded animals. They take their first bite 6-8 hours after emergence, according to Mitzmain (1913). In France, the males have been seen to feed on the "honeydew exuded by certain plants" (Surcouf, 1921). Both sexes have been seen sucking nectar by Hammer (1942) in Denmark. Parr (1962) found that adults will feed on sugar solutions in the laboratory.

The stable fly is a vicious biter, causing a sharp pain (Fig. 19); it draws blood quickly and engorges itself in a few minutes. It attacks the least defended parts of the host animal, often puncturing the skin several times before drawing blood (Simmons, 1944). Bishopp (1939) observed flies sluggishly moving to nearby resting sites to digest a meal and occasionally feeding again the same day. Parr (1962) determined that the average blood meal (25.8 mg) taken at one time was three times the average weight of the fly (8.6 mg).

Stomoxys has been found associated with cattle, horses, man, buffalo, sheep, antelope, camels, donkeys, rhinos, elephants, dogs, cats, pigs, chickens, ducks, rats, mice, rabbits, guinea pigs, and even some large reptiles. Feeding preferences vary from locale to locale, as we have seen in the worldwide survey. Usually, either cattle or horses are preferred. But in Egypt, for example, Hafez and Gamal-Eddin (1959b) found this order of preference: donkey, horse, buffalo, cow, camel, sheep, and goat. They found that the feeding period averaged 7 minutes at 30°C and 15 minutes at 21°C. In the hot months (summer and autumn), they noted two peaks of biting activity, one in the morning and one in the late afternoon. In the colder season, there was a single peak around 2 P.M. Bishopp (1939) observed attacks on cattle in Texas from early morning until dark. Attacks on humans in most areas of the world are probably incidental, except in shoreline breeding situations.

Mating, Oviposition, and Larval Habits

Hammer (1942) in Denmark observed a swarming, mating flight about bushes and other objects in the field. Parr (1962) noted that the male was active in mating, attacking the female for some 30 minutes before copulation, which lasted 4-6 minutes. Hafez and Gamal-Eddin (1959a) described the male hovering over the female and suddenly introducing its genitalia into her vagina.

Killough and McKinstry (1965) and Harris et al. (1966) found that a few flies mated as early as 1-2 days after emergence and that most flies had mated by age 5 days. The oviposition data of the former authors indicated that the females will not lay eggs until they are 8 days old and that most females began laying eggs after the 10th day. They found 40-80 eggs laid per batch, and the largest total number laid by one female was 602. Mitzmain (1913) observed 20 egg depositions per female with a maximum number of eggs laid of 632. Parr (1962) found that females produced an average of 10-11 batches of eggs with 35.5 eggs per batch and a lifetime average of 376 eggs (see also Hummadi and Maki, 1970).

Diverse larval breeding materials have been described above in various world localities. It appears in most cases that the medium must be loose and porous, have a high moisture content and moderate temperature, be located in a shady place in the hot season, and be near domestic animals, since oviposition is preceded by blood meals (Hafez and Gamal-Eddin, 1959a). The freshness of the plant material of the medium appears to be important especially in the case of grain straw and beach grasses because the flies oviposit apparently just before or as fermentation begins to occur; the flies also usually avoid very putrid material. Todd (1964) feels that *Stomoxys* appears to have a keen sense of smell in order to detect moist, newly fermenting materials so very readily.

When the females lay their eggs, they sometimes go deep into the loose medium and prefer areas that have a high moisture content and are dark (Hafez and Gamal-Eddin, 1959a). At 80°F and 80% humidity, the eggs hatch in 24 hours (Parr, 1962). During late autumn in Texas, Bishopp (1939) found that the egg period ranges from 1-4 days. Patton and Cragg (1913) reported that, in India, the eggs hatched in 12 hours.

The newly hatched larvae bury themselves at once (Herms, 1953). In the ensilage breeding materials in New Zealand, Todd (1964) found that the maggots follow the moisture inward as the material dries out and that the larvae are negatively phototactic. Roberts (1952) states that, in the absence of adequate moisture, larval growth may be considerably retarded and maggots may even fail to mature. The following are various reported larval periods: Mitzmain (1913) 12 days in the lab at 30-31 °C; Parr (1962) 8 days in the lab at 80°F and 80% humidity; Newstead (1906) in England at 65-72°F in the field, 14-21 days to as much as 78 days under unfavorable conditions; Bishopp (1939) in the field in Texas in late autumn, 11 to over 30 days and, when conditions were very favorable, the total egg to adult period was 21-25 days.

PUPATION, OVERWINTERING, LONGEVITY

Before pupation, the third instar larva crawls into drier parts of the medium (Herms, 1953). Todd (1964) found pupae in the ensilage breeding material. Various reported pupal periods are: Mitzmain (1913) 5 days; Bishopp (1939) 6-20 days or longer in cool weather; Parr (1962) a surprisingly brief 2 days at 80°F and 80% relative humidity.

Todd (1964) states that, in New Zealand, *Stomoxys* probably overwinters as pupae mostly in partially used ensilage stacks; few adults were seen in winter. Third instars overwinter in peanut litter in northwestern Florida; slow development occurs throughout the winter and great numbers of flies emerge during warm periods (Simmons, 1944). The flies overwinter as larvae and pupae in the northern part of the United States (Bishopp, 1939). There is probably no true hibernation in tropical areas.

Bishopp (1939) put flies in cages with cattle and oviposition materials, and they lived for about 29 days. Hafez and Gamal-Eddin (1959a) found that, at room temperature (17-19°C), females lived an average of 17 days and males 26 days, reversing the usual life expectancies.

CHRYSOMYA MEGACEPHALA

This species is Oriental and Australasian in distribution. It is not known in Africa but occurs on islands in the Madagascar region. It is a common household and bazaar pest in these areas.

The adults have been reported as a nuisance in slaughterhouses and on meat, fish, sweets, fruits and other foodstuffs in market places. Patton (1930) and Illingworth (1926b) regard this fly as an important food contaminator. Roy (1938) collected it at bazaars in India. Wijesundra (1957a) states that it is commonly found in fish stalls, in slaughterhouses, and on decomposing animal bodies in Ceylon. In Manchuria, Kubo (1920) observed this fly commonly on foodstuffs, particularly fruits and especially watermelon. De la Paz (1938b) reported it as the most important fly breeding in garbage and visiting food in the Philippines. It was encountered in Manila food stores (27 of 950 flies) and occurred in larger numbers on foods in houses in April-May (Anon., 1937). Wilton (1961) found it breeding in 14 of 68 garbage cans sampled in Honolulu. Liu et al. (1957) determined that garbage and slaughterhouse offal were its important breeding sites in Taiwan.

Zumpt (1965) states that the adults are found on palm trees being tapped for toddy; they suck the juice and foul the pots and spathes with excrement. On Guam, Bohart and Gressitt (1951) observed great numbers of flies clustering on flowering trees (*Tournefortia*), which grow on the beach near breeding materials. They also noted the adults' habit of scattering feces freely while feeding and their sluggishness when on attracting materials.

The adults are strongly attracted to carrion and excrement for breeding purposes. Thomas (1951) found them in very great numbers around latrines and cesspools in southern China. In the Yangtze valley of central China, the farmers collect human excrement in liquid form and store it in large jars to be used on garden crops; the myriads of larvae cannot crawl out but the farmers pour off the top onto the ground, where the fully fed larvae pupate and emerge in a few days (Illingworth, 1926b). Of 3396 flies emerging from a privy pit on a Japan farm, 850 were *megacephala*, occurring in the greatest numbers from mid-September to mid-October (Suenaga and Fukuda, 1963). In Taiwan, Liu et al. (1957) found a high rate of infestation of larvae in pit privies and night soil tanks. Bohart and Gressitt (1951) reported that crushed crabs were the most attractive bait to these flies on Guam and that they bred in human corpses during the American occupation. Meng and Winfield (1938) stated that in north China they breed almost entirely in human excrement in places where feces cakes are dried; of 7826 flies trapped from August to December at various sites, 41.66% were *megacephala*. It was the most common species in late summer and fall. Their data also indicated that this species prefers the outdoors and tends to stay near breeding places but that it will enter dwellings.

In the laboratory the first eggs were laid 8 or 9 days after emergence, usually in the afternoon (Wijesundra, 1957a). They were deposited in mass in rows under the surface of the beef medium. In nature, this species seldom oviposits on isolated human feces but in accumulated large masses, and fresh rather than old carrion is preferred (Bohart and Gressitt, 1951). At temperatures from 24-28°C the eggs hatched in the lab in 9-10 hours (Wijesundra, 1957a). Captured wild females laid an average of 254 eggs (224-325); egg counts from ovaries of bred flies averaged 393 (214-632) (Wijesundra, 1957a).

The larvae need high humidity and high temperature for development. Meng and Winfield (1950) stated that the larvae are adapted to liquid media because their stigmal fossae are so constructed that the dorsal and ventral lips can be regulated so as to protect the spiracular slits from being blocked by liquid; however, the larvae congregate in the upper layers because of the demands of respiration. They found liquid feces (80% of all flies) the most efficient medium, especially after the rainy season in north China; city garbage (41.3%) and solid human feces (4.8%) were less productive. They observed no breeding in feces before July, an increase in August and a peak in September in this area.

The larvae breed equally well in carrion and human feces in cesspits but not in herbivore feces. However, larvae have been reported from isolated patches of cow dung. Tanada et al. (1950) found larvae in chicken manure at Oahu, Hawaii. In his lab, Wijesundra (1957a) found that 2½ ounces of beef was sufficient for the complete development of 100 larvae. The maggots are very active, voracious, and crowd out competitors. When food runs out, they writhe about in frenzied masses. Bohart and Gressitt (1951) described larvae on Guam leaving shallow concrete toilet pits by the thousands after fecal material was consumed and migrating in all directions. In camp latrines, they were observed to liquefy entire masses of fecal matter and then to consume all the contents of the pits, including paper. They migrated from several feet to several yards from the medium and burrowed in or under the surface of loose dry soil to pupate.

In the laboratory, Wijesundra (1957a) found that the larval stage lasts about 94 hours at temperatures ranging from 26-31°C; larva I, 15-18 hours; larva II, 18-33 hours; and larva III, 33-93 hours. The third instar larvae fed voraciously. The total egg-to-adult period at room temperature was 8½ days. The pupal stage lasted 100 hours. The adult longevity was shown to be dependent on temperature and humidity. At temperatures of 25-29°C and 75% relative humidity the flies lived an average of 54 days (90 maximum); at lower humidities the flies appeared to live longer (Wijesundra 1957b).

PHORMIA REGINA

This species is distributed throughout the cooler regions of the Holarctic. It is primarily a cool weather blow fly and, while it is found in the United States during the entire year, it is more abundant in the spring and fall (Hall, 1948). It is scarce in the hot months in the southern states (James, 1947). It occurs as far south as Mexico City at high altitudes and at Oahu, Hawaii, at 1000 feet elevation (Hall, 1948).

This fly is involved in wound myiasis in the New World. Knipling and Rainwater (1935) described it as the most important secondary myiasis fly in the southeastern United States. James (1947) says that it is a common sheep maggot fly in the southwest United States. Parish and Laake (1935) found that *regina* accounted for 67% of all adults reared from larvae in wounds of domestic animals in the Menard area of Texas in the spring. Deonier (1942) frequently found larvae in dehorned cattle in winter and spring in the warmer valleys of Arizona. Hall (1948) states that it is common in castration and dehorning wounds in the midwest especially from November to March.

The larvae are normally saprophagous and breed in large numbers in animal carcasses. Hall (1948) says that carcasses in the southern United States may swarm with *regina* maggots. They occur in great numbers in the paunch contents of slaughtered animals. He also states that the vast majority of these flies breeds in dead animals on the open range. Tilden (1957) found that it was the most common fly collected from dead animals in Santa Clara County, California; it was also common at chicken farms.

Paine (1912) found that 22.3% of the flies in garbage cans in Massachusetts were *regina*. Haines (1953) collected it in low numbers in animal pen wastes (9%) and in fruit and vegetable wastes (0.3%) in southwestern Georgia. Schoof et al. (1954) in Charleston, West Virginia, observed larvae in fowl excrement and in scattered and canned (most flies) garbage. Savage and Schoof (1955) found it to be one of the principal flies recovered from fish processing plants in Michigan and Kansas; they also found its peak abundance in June at a Topeka hog farm. Wilson (1968) found it to be the primary insect associated with the decomposing alewives that washed up on the southern Lake Michigan shores in 1967. Coffey (1966) in southeastern Washington found adults commonly attracted to human, mink, and swine dung and less to cow, dog, horse, and poultry dung; it occurred mostly in the summer months (May-September) and entered houses at times. Williams (1954) trapped 2,565 flies with raw fish and liver in New York City and obtained 14.2% *regina* between May and November (most flies caught with fish). Although this is said to be a cool weather (spring and fall) species, it was most common from June 16 to September 15.

Bishopp (1915) found that the preoviposition period ranged from 7-17 days. Kamal (1958) observed that the flies copulated 3-7 days after emergence. James (1947) stated that the eggs are laid in agglutinated masses of varying numbers. In his laboratory, Kamal (1958) found that the eggs were deposited in cluster form on top of the medium (beef liver) in batches of 12-80 eggs. At 80°F and 50% relative humidity, he reported an egg period of 16 hours (10-22). Melvin (1934) found that at 59°F the eggs require 52 hours to hatch; at 104°F, 8.7 hours; the optimum was 99°F, 8.13 hours; none hatched at 109°F.

Bishopp (1917) in Texas observed that the larval stage takes from 4-15 days to complete development. Kamal (1958) found these periods: larva I, 18 hours; larva II, 11 hours (8-22); larva III, 36 hours (18-54); prepupa, 84 hours (40-168); and a total immature period of 11 days (10-13) at 80°F and 50% relative humidity. He reported a pupal period of 6 days (4-9); Bishopp (1915) observed that it ranged from 3 to 13 days at Dallas, Texas. James (1947) states that cool weather favors development. The egg to adult stages takes 10-15 days (Bishopp, 1915).

Hall (1948) states that hibernating adults are found in some numbers in tunnels of wood-boring insects and on the sunny sides of buildings in winter and that females do not oviposit in winter. James (1947) states that *regina* can be found outdoors during the entire winter at least as far north as Iowa and that they apparently hibernate in the adult stage. In a cold winter in New Haven, Connecticut, Wallis (1962) found that the adults retreated to protected hiding places and emerged for brief periods regularly throughout the winter months. Mail and Schoof (1954) found this fly throughout the winter at Charleston, West Virginia; adults were taken while flying on warm days or recovered from open cellars or under porches in a torpid state during periods of freezing weather. Kamal (1958) reported an adult life span of 52 days (45-68) in his laboratory studies.

This species is common in cooler areas of the Holarctic. It occurs in considerable numbers in northern Europe, Asia, Alaska, and Canada and less abundantly in the United States (Hall, 1948). It is an early spring species in much of the United States; in northern localities, it may be more common during summer (James, 1947). It has been collected as far south as central Texas and northern Georgia. It occurs at high altitudes in temperate zones during the summer; it has been collected from the end of March to the end of June at 4000-6000 feet in Utah (Hall, 1948). Nuorteva and Vesikari (1966) found it to be the dominant species in human settlements along the coast of the Arctic Ocean. Nuorteva et al. (1964) reported that it was the first species to occur in spring at refuse depots near Kuopio, Finland. They also stated that it overwhelmingly outnumbered other species and that it had a more northern occurrence; it was more abundant in Finland than in Germany.

P. terraenovae has been involved in wound myiasis in Europe and North America (Zumpt, 1965). It has been considered the primary sheep blow fly in the early part of the season in Scotland (James, 1947).

This fly is mainly saprophagous and prefers relatively low temperatures for breeding; the larvae breed chiefly in carrion. Howard (1900) collected larvae from mammalian excrement and decaying animals and stated that, in the eastern United States, it was abundantly attracted to human feces. Coffey (1966) in southeastern Washington reared it once from cow dung and collected it rarely from May to August on horse, swine, and chicken dung. Gill (1955) trapped this species in central Alaska from April to September using beef liver, carrion, excrement, and a fruit-milk mixture as baits; he also reared larvae taken from the carcass of a dog. Havlík and Baťová (1961) found it occurring in Prague slaughterhouses. Hall (1948) collected considerably large numbers on flowers of wild parsnip in early July in northern Michigan.

Green (1951) observed the biology of this species in his studies of blow flies in slaughterhouses in England, and both he and Kamal (1958) reared this species in the laboratory. Green found a small population in slaughterhouses; other species predominated, but sometimes heavy infestations of *terraenovae* occurred, particularly in summer. When the weather was sunny, the flies flew freely and tended to rest on sunlit walls outdoors. During cold, cloudy, or wet days, they rested on refuse or vegetation and often in cracks in the walls of buildings.

Kamal observed that the eggs were deposited in cluster form on top of the medium (beef liver) in batches of 12-80 eggs; 8-15 minutes were required for deposition. Green determined the following periods at 24°C and 65% relative humidity: egg, 19-25 hours; larval feeding, 6-6½ days; larval post-feeding, 1 day; pupal, 5-7 days; total egg to adult, 15-16 days. He found that the larvae were predaceous on *Lucilia* and *Phaenicia* larvae as well as on each other. The last instar larvae began migrating at 7½ days after eclosion; the first pupation occurred at 8½ days; the main pupation at 9½ days. He stated that the larvae may complete development in 3 days under very favorable conditions and that they may pupate on the surface of the medium; he thought this species probably overwintered as larvae in the soil.

Kamal found these periods at about 80°F and 50% relative humidity: egg, 15 hours (12-23); larva I, 17 hours (12-30); larva II, 11 hours (9-20); larva III, 34 hours (20-60); prepupa, 80 hours (38-160); pupal, 6 days (4-10); total immature development, 11 days (10-13). He found that the flies copulated 3-7 days after emergence; oviposited 5-9 days after emergence (Green: 6-7 days); and the adults lived an average of 55 days (48-71) in the laboratory.

PHAENICIA SERICATA

Hall (1948) feels that *sericata* is not quite the cosmopolitan species it is sometimes considered to be; its distribution is patchy in areas outside the Holarctic. It is most common in the temperate zone of the northern hemisphere. From there it has apparently followed man to many other parts of the world, including tropical regions.

In both Africa and Australia, it has become the dominant fly in urban and suburban areas (Zumpt, 1965). In Queensland, it is the dominant blow fly in December (Johnston and Tiegs, 1923). It occurs as the most widely distributed and abundant blow fly of the Talysh region of Caucasus (Trofimov, 1965). In England, *sericata* numbers 80% of the *Phaenicia* and *Lucilia* species that make 95% of the blow fly population in slaughterhouses (Green, 1951). Mihályi (1967a) reports that it is the most abundant and dangerous fly in cities in Hungary. Tilden (1957) states that it is the dominant blow fly in Santa Clara County, California. Reyes et al. (1967) have recently reported this species involved in cases of human myiasis in Chile. This species can be collected in almost every part of the United States and southern Canada, and, in the midwest, it is the most abundant species of blow fly from late June through August.

Adult habits, associations

Before going into the details of adult biology, a brief account of surveys taken in various parts would be valuable. These were conducted to determine the possible feeding and breeding places of this fly and to obtain an index of its synanthropy.

EUROPE

In urban Hungary, Mihályi (1966) caught *sericata* on fruit (21.5% of all flies collected) and on meat (79.4%) in food marketplaces. Also 44% of the flies he trapped with meat baits in the city were *sericata* (1965), and he found adult flies frequenting human feces that were lying free or in open cesspits (1966). Havlík and Baťová (1961) reported this species common in waste meat in Prague slaughterhouses and in kitchen garbage. In Berlin, Kirchberg (1950) observed *sericata* in urban dust bins and dumps consisting of animal and vegetable debris. Morison (1937) in Scotland found that this fly was only 3% of the flies he caught in meat traps but that it was more abundant in the proximity of sheep. Thomsen and Hammer (1936) found larvae occasionally in pig manure.

NORTH AMERICA

Savage and Schoof (1955) trapped 49.2 to 71.2% *sericata* (depending on the season) in garbage dumps in New York and Michigan; 50% in a fish processing plant in Michigan; and lesser percentages in slaughterhouses, rendering plants, and hog farms in various U.S. localities. St. Germaine (1955) and Tilden (1957), in California, found larvae in dog and rabbit droppings and adults in houses, markets, dog kennels, chicken ranches, and other animal areas. Paine (1912) in Massachusetts, Quartermann et al. (1949) in Georgia, and Magy and Black (1962) in California, all reported *sericata* in surveys of garbage cans. Coffey (1966) reports it attracted to mink, swine, chicken, human, dog, turkey, horse, and cow excreta in southeastern Washington.

OTHER AREAS

De la Paz (1938b) found *sericata* visiting garbage and foods in the Philippines. Illingworth (1926a) found that between June 28 and July 28 the percentage of *sericata* caught in a Japanese sweet shop increased from 2.5% to 14.34%. Also in Japan, Fukuda (1960) found this fly emerging from pupae collected near fertilizer pits. In Egypt, Hafez (1941) states that *sericata* is mainly an outdoor fly, rarely found in houses, and is most common in farm areas where dung and human excreta are left exposed, and that it usually prefers carrion and dead animals for feeding and breeding.

ADULT FEEDING HABITS

P. sericata adults feed on some of the liquid or dissolvable solid materials described above; many of these are also larvae-breeding media. The adults feed on sweet and fermenting liquids and are frequently observed on flowering plants and on foliage in urban areas. Considerable numbers are sometimes found under the leaves of cucumbers and melons and on the leaves of vines and on the flowers of wild parsnip and wild carrot (Hall, 1948). Of several species of blow flies reared in the laboratory, Kamal (1958) noted that *sericata* consumed the most carbohydrate. Although these flies are much attracted to flowering plants and sweet materials, Mackerras (1933) and Hobson (1938) reported that females need a protein meal for maturing eggs, but this was not required for males to mature sperm.

The adults usually frequent open areas and are most active on warm sunny days. They can be collected in abundance in association with almost any kind of garbage, especially meats, slaughterhouse refuse, and damaged fruit and are attracted to carrion, open wounds, the soiled and wet fleece of sheep, and, to a lesser extent, to feces, although adults are often abundant on dog feces whose disposal, especially in larger cities, is an increasingly vexing problem.

RESTING SITES

Coffey and Schoof (1949) observed that these flies rest at night outdoors on trees, bushes, grass, and on the sides of buildings near the daytime feeding sites. During the day, they usually rest outdoors and will enter houses only in the spring and fall where they have been reported to alight on food rather than on walls or other objects. Green (1951) found 80% of *Lucilia* and *Phaenicia* species, which included mostly *sericata*, resting on vegetation near the slaughterhouse when the weather was cloudy or cold. Maier et al. (1952) found that it rested predominantly on the ground surface during the day, and MacLeod and Donnelly (1958) reported that it finds open ground exposed to the sun more favorable than ground sheltered from the wind but more shaded.

MATING AND OVIPOSITION

Johnston and Tiegs (1923) stated that copulation lasted only a few seconds when the flies were on the wing and that it took longer when they were at rest. Hobson (1938) reported that males become sexually mature in a few days after emergence. Kamal (1958) in his lab studies determined that mating activity occurred from 3-8 days after emergence.

He also found a preoviposition period of 5-14 days; Zumpt (1965) reported 5-9 days; Green (1951), 4-5 days at 24°C; and Salt (1932), about 7 days. Hobson (1938) found that 2 meat meals were essential for the development of mature eggs in the ovaries.

Kamal (1958) observed that *sericata* at first deposited its eggs in the natural cavities and crevices of the medium (beef liver); when these were filled, females tended to select moist areas under the surface of the medium. Oviposition was more frequent following the introduction of fresh media, and moist media accelerated oviposition. Smit (1928) found that eggs were laid in quick succession in crevices or under meat; they stuck together and to the meat after drying in air.

Hall (1948) states that *sericata* is attracted to media exposed to bright sunshine, but the flies oviposit in shaded areas of the medium; oviposition usually occurs between 11 A.M. and 2 P.M. at maximum sunshine. Most of the breeding materials have already been mentioned above. Most authors describe this species as normally preferring carrion and decomposing meat; meat is attacked within a few minutes after exposure. This is the first species in the ecological succession of saprophagous insects which attack dead animals, according to Hall (1948).

Green (1951) described *sericata* breeding in slaughterhouse premises on refuse and inedible offal. All combinations of garbage and even animal excrement to some extent serve as breeding media. He found that larvae can develop successfully in large numbers in fermenting pig food, which consisted only of vegetable material. Larvae will develop, albeit more slowly, in clotted milk in the laboratory. Smith (1928) in South Africa observed that the flies breed in carrion for the most part and found that flies bred from maggots on sheep will oviposit on meat and those bred from maggots on meat will oviposit on sheep.

In some parts of the world *P. sericata* is an economic problem because it causes cutaneous myiasis in sheep as described by Bishopp (1915) in the United States, Davies (1934) in Wales, Haddow and Thomson (1937) in Scotland, and Morris (1954) in Newfoundland. In other regions, e.g., Kenya (Lewis, 1933) and Australia (Mackerras and Fuller, 1937, and Norris, 1959), it has a minor role as a sheep myiasis fly; in Finland, it never causes myiasis (Nuorteva, 1959). This is possibly due to differences between biological strains; also in some areas other species are the chief myiasis producers.

The number of eggs per batch varies from 80-170 and depends on the size of the fly, according to Hobson (1938). Mackerras (1933) determined an average of 182 eggs laid at one time. Zumpt (1965) states that each female can lay 2000-3000 eggs in 9-10 batches within 3 weeks.

The rate of egg development depends on temperature and other factors. In slaughterhouse refuse in England, eggs hatch commonly in 12-16 hours. Outdoors in England on meat, eggs hatch in 10 to 52 hours in summer; at 31°C they hatch in an average of 9½ hours on the back-skin under the wool of sheep (Zumpt, 1965). In Queensland, Johnston and Tiegs (1923) reported an egg period of 16-17 hours in summer. Salt (1932) states that in high humidity and at 22°C eggs hatched in 20 hours. Smith (1928) says that in South Africa the egg period varies from 8 hours to 3 days according to the season.

LARVAL HABITS

Green (1951) observed that the larvae remain and feed near the hatching site in any light conditions and then migrate to darker areas and avoid light for the remainder of larval life. Wardle (1930) describes the first instar as a short, non-feeding period; the larvae molt in 2-3 hours. The second instar, he called the true feeding period, which lasted 1½ to 9½ days, depending on the temperature. Kamal (1958) determined the period of larva I in the lab to be 20 hours (12-28 hours) and larva II, 12 hours (9-26 hours) at about 80°F and 50% relative humidity. Under conditions of 25°C and 100% humidity in 25% animal and 75% vegetable garbage, Kirchberg (1950) stated that larval development was completed in 2-3 days. In slaughterhouse refuse, the larvae fed 4½ to 9 days (Green, 1951). Zumpt (1965) states that the feeding period on a carcass is 5-11 days (average 6 days) and that, at a constant temperature of 33°C, it was shortened to 3 days. He also states that in a sheep wound the larvae are mature in 43 hours and leave the wound for pupation. Smit (1928) in South Africa observed larvae feeding on meat for 2 days in summer and for 19 days in autumn.

Hobson (1938) says that the larvae ingest only liquid food. Michelbacher et al. (1932) found that the liquefying action of bacteria on solid food was often useful to the larvae when feeding. Like other carrion-feeding maggots, they eject digestive enzymes into the meat that partially liquefy and render it assimilable. The larvae group in closely packed masses submerged in the liquid up to the posterior spiracles, which are exposed to obtain a good supply of air. Salt (1932) determined that the larvae require at least 150 mg of meat for full development.

Wardle (1930) described the third larval instar as a nonfeeding, mobile stage, but this more accurately describes the prepupa. In his lab studies, Kamal (1958) found the larva III period to be 40 hours (range 24-72 hours) and the prepupal period, 90 hours (48-192 hours). In Salt's (1932) lab studies, the larva III left the meat in 7 days when fully fed, crawled on the sand for some hours, and then entered it and pupated. Green (1951) found that larvae could migrate 20 feet to pupation sites in the slaughterhouse situation and that this took place mainly at night.

PUPATION AND OVERWINTERING

Kamal (1958) determined an average pupal period of 7 days (5-11 days); Green (1951), 7-9 days; and Herms (1928), 4-7 days. Smit (1928) in South Africa found a minimum pupal period of 7 days in summer (January) and estimated that some flies that emerged in spring had spent 52-115 days as pupae. He believed that the flies overwintered in the larval, pupal, and adult stages. Zumpt (1965) states that in the European winter the prepupae remain inactive until the soil temperature reaches about +7°C and pupation starts at 8-11°C. He listed pupal periods as follows: 4-7 days at 32°C, 6-7 days at 27°C, and 18-24 days at 12-13°C. Green (1951) says that fully grown *sericata* larvae in the London and Home County areas are arrested in early October and winter; they pupate only a short while before emergence as adults in May. In North America, this species overwinters as larvae or prepupae (Hall, 1948). Supporting evidence is the presence of pupal fat balls in the hemolymph, nulliparous ovaries, and a full complement of bulbous setae on the antennal pedicel of flies caught in early spring (Greenberg, 1970). It breeds throughout the year in warmer countries (Zumpt, 1965). Norris (1959) states that *sericata* survives the Canberra winter as diapausing larvae that hatched from the last eggs laid in autumn.

NUMBER OF GENERATIONS AND ADULT LIFE SPAN

Wardle (1930) found that there were four generations per year in St. Paul, Minnesota, between May and October and probably eight in the southern United States. Smit (1928) found 9-10 generations per year in South Africa.

Adult longevity is 41-56 days according to Coffey and Schoof (1949). Mackerras (1933) found that a pair of flies lived for 77 days. Smit (1928) stated that the flies live about 91 days in autumn and about 35 days in summer. In his lab, Kamal (1958) found that the flies lived for an average of 46 days (40-59). In Salt's (1932) greenhouse cages at 18-28°C and with food, meat, and water provided, the males lived an average of 41 days (68 days maximum) and the females 55.6 days (104 days maximum).

CALLIPHORA VICINA

This species probably originated in the Holarctic and spread to many parts of the Australasian and Oriental regions and in the New World to the Neotropical areas (Zumpt, 1965). Hall (1948) states that it is common in North America from Mexico City to Alaska. It is especially abundant in the middle west, where it is one of the first blow flies to appear in March and early April. It becomes uncommon from May to early October, then again is found in great numbers, and is one of the last species to disappear in the fall. Norris (1959) says that it is an introduced species, which occurs throughout southern Australia and Tasmania. It is common in Europe and Great Britain. In 1948-1949 its first seasonal occurrence around Moscow was in mid-April (Sukhova, 1950). In England, it is a common species throughout most of the blow fly season, the first to appear and the last to leave (MacLeod and Donnelly, 1957a). In Hungary, Mihályi (1967a) found it mainly in inhabited areas; its numbers reached a maximum in spring and in autumn and the number of flies diminished in summer.

The adults are slow-flying and loud-buzzing and are easily caught by hand when at rest. They can be collected on foliage in cities, and the females are commonly found on carrion and other types of refuse (Hall, 1948). Zumpt (1965) states that they are attracted to any foul-smelling product of decay, especially carrion. Tilden (1957) reported it on dead animals but found it a scarce fly in California. Howard (1900) stated that it lays its eggs on meat and dead animals and has been taken on fresh human feces. Kvasnikova (1931) collected larvae in meat in Tomsk, western Siberia. MacDougall (1909) found puparia in dead birds, moles, and rats in Scotland. In studies in Hungary, Mihályi (1965, 1966) trapped *vicina* on meat (5.7% and 9.7%) and on fruit (2.1%) and reported that adults fed on feces and constantly visited houses, where they frequented uncovered meat. In Egypt, Hafez (1941) described it as an outdoor species, which commonly entered houses in search of breeding materials. He found it mostly in the country where dead animals and carrion were available and in certain areas of cities where heaps of human excrement were accumulated. Bogdanow reared more than 11 generations exclusively on human feces (Haddow and Thomson, 1937). Sukhova (1950) reported larval development in nonliquefied human feces in the Soviet Union. Larvae died in the first instar when eggs were placed on horse dung (Kvasnikova, 1931). Green (1951) seldom counted more than 10% *vicina* in slaughterhouses in England. Havlík and Baťová (1961) reported it in waste meat in Prague slaughterhouses and in kitchen garbage. Of the flies collected from meat scraps at a slaughterhouse in Moscow, 75% were *vicina* (Sukhova, 1950).

Zumpt (1965) states that *vicina* larvae have been found involved in traumatic myiasis in man and animals. In Great Britain it plays a minor role in sheep myiasis (Haddow and Thomson, 1937; MacLeod, 1937). But it is an important sheep maggot in Tasmania (Ryan, 1954).

Froggatt (1914) observed adults in gardens on decaying fruit and other vegetable matter in Australia. Sukhova (1950) saw a few adults on fruit in open stalls in the Soviet Union on sunny days in late fall. Hall (1948) collected these flies on wind-fallen fruit in his garden (U.S.A.).

In the cool early spring, Hall (1948) observed the flies resting on the sunny sides of buildings. MacLeod and Donnelly (1957) found this species in Carlisle to be most numerous in open hedge rows and estimated 50-200 flies/acre in August, 400-1000 in September and 700-1000 in October; males were more numerous in canopy cover. Sychevskaĩa (1962) net-sampled blow flies in market places in Samarkand at hourly intervals throughout the season; in cooler months diurnal activity curves were unimodal but in hotter months they were strongly bimodal. The adults avoided lethal heating in the summer by seeking shelter in shaded places; the body temperature may rise to 42.0°C in the sun's radiation (Sychevskaĩa, 1965). During frost-free winters on sunny days, she found that they can fly at temperatures of 5-6° in the shade. She also confirmed the data of E. S. Smirnow (1940) that the ovaries of this species become inactive in the hot summer months; at the end of August the inhibition ceases, eggs are laid, and the population density increases in the field.

Green (1951) determined a preoviposition period of 4-5 days at 24°C and noted that these flies commonly flew and oviposited at night in the slaughterhouses. In his lab, Kamal (1958) found this period to be 8-15 days at 80°F and the flies had copulated 5-9 days after emergence. Zumpt (1965) states that a female can lay 540-720 eggs in her lifetime in batches of up to 180 eggs. Kamal (1958) observed an egg period of 24 hours (20-28) at 80°F and 50% relative humidity. Brown (1936) found that eggs hatched in about 11 hours at 25-35°C and 40% relative humidity. Green (1951) found an egg period of 3-8 hours at 24°C and 65% relative humidity.

In the lab Kamal (1958) obtained these larval periods: I, 24 hours (18-34); II, 20 hours (16-28); III, 48 hours (30-68); prepupal, 128 hours (72-290). Green (1951) found that the larval feeding period was 4-5 days and the postfeeding period (prepupal) was 2½ to 7½ days. Bishopp (1915) reported that larval development takes 3-4 days in eastern Texas. Zumpt (1965) states that it can take as long as 9 days in cooler climates.

Bishopp (1915) reported a pupal period of 7-9 days in Texas; Kamal (1958) 11 days (9-15) in the lab. Green (1951) found that the first migration of larva III from food was at 4½ days after hatching; the first pupation at 8 days and the main pupation at 13½ days; the first emergence at 18½ days and the main emergence at 21½ days; the total egg-to-adult period, 18 to 21½ days. The larvae pupate in the soil at a depth of 1½ to 2 inches, and adults can emerge from puparia buried at a depth of 18 inches in closely packed, fine, sandy soil.

Green (1951) in England observed that the adults were active for the greater part of the year in the field; oviposition occurred until late autumn. He could still find feeding larvae in mid-November. The flies hibernated as larvae in 1 inch of ground, under light rubbish, sacks, straw, or in cracks in walls for 5-6 months. During one mild winter, he found adults emerging in mid-January at 0-2°C. Zumpt (1965) states that the winter in temperate zones is spent as prepupae (nonfeeding mature larvae). Sukhova (1950) reported overwintering in the adult stage in the Soviet Union and that the flies may breed all winter. Howard (1900) reported that thousands of *vicina* were found in a cellar in Washington, D.C., in October 1899.

Sychevskaīā (1965) found in Samarkand under laboratory conditions that there were five generations from January to June and three generations from September to December. The flies lived up to 188 days in cages. Kamal (1958) found an average adult life span of 25 days (24-35) in cages.

CALLIPHORA VOMITORIA

The species is primarily Holarctic but ranges widely. It occurs in North America from Alaska and Greenland southward to California and Virginia; it is abundant along the Canada-United States border. Gill (1955) trapped it in central Alaska from May to September on beef liver, carrion, and excrement baits. Cole and Schlinger (1969) state that it is common in Pacific Coast states, particularly California. Tilden (1957) found it the most common *Calliphora* (collected from dead animals) in Santa Clara County, California. Mihályi (1967a) rarely found it in inhabited areas in Hungary but mainly in forests. MacLeod and Donnelly (1957a) state that *vomitoria* made up 10% of the flies in the late blow fly season in Carlisle. Green (1951) seldom found it in large numbers in London area slaughterhouses.

Kamal (1958) reared this species in his lab using beef liver as the oviposition and rearing medium. In addition to slaughterhouse refuse, Green (1951) found larvae infesting tanks of blood; they floated near the surface while feeding and climbed over the side to migrate. Kamal observed that the eggs were usually laid singly and scattered on top of the medium. He found that the flies copulated 4-8 days after emergence and females oviposited 7-18 days after emergence. Green (1951) states that the flies oviposited in 4-5 days after emergence and commonly flew and oviposited at night in the slaughterhouse. Kamal recorded the following periods at 80°F and 50% relative humidity: egg, 26 hours (23-29); larva I, 24 hours (20-38); larva II, 48 hours (43-54); larva III, 60 hours (48-96); prepupal, 360 hours (240-540); pupal, 14 days. He found an adult life span of 26 days (27-36). In the slaughterhouse, hibernation took place in the ground, usually within an inch of the surface, under rubbish, straw and sacking, or in cracks in walls for a period of 5-6 months (Green, 1951). There is evidence that this species hibernates in the adult stage in the midwest (Greenberg, 1970).

CALLIPHORA URALENSIS

In Europe, this is primarily a forest species and more common in the northern regions. However, Mihályi (1967a) stated that it lives in cities and villages, and may have a maximum population in summer; it was new in the Hungarian fauna. Around Moscow, Sukhova (1950) observed the adults especially attracted to living quarters, commercial food-handling enterprises, fish-curing plants, and slaughterhouses; he estimated that 20-60% of the flies caught in such places were *uralensis*. In 1949, the flies first appeared in the field on May 6, and there was an abundance peak on May 28. This species was more prevalent than *vicina* in the summer months and the population dropped toward the end of September; adults disappeared with October frosts. He stated that, in general, it is the most numerous species of the exophilic synanthropic flies in central U.S.S.R. and more epidemiologically dangerous than other species.

Sukhova (1950) conducted bionomic observations on this species in the Soviet Union and reared these flies in his laboratory. He observed that the adults gathered in large numbers near latrines in yards and public squares on shrubbery and grasses, sometimes covering every leaf. The adults fed on human excrement; intestines of 40% of dissected flies contained fresh feces. They are also attracted to sweet and fermenting substances; open market stalls with honey, fruits and vegetables swarmed with these flies. Great numbers occurred on raspberries, plums, cherries, strawberries, grapes, melons, and on bakery goods and ice cream sold in the open. The flies frequented tables, glasses, mugs, trays, and syrup dispensers, often depositing drops of excrement and eructations.

In the laboratory at 18-25°C, Sukhova found a preoviposition period of 4-8 days. He described females ovipositing on the excrement-soiled wooden structural members of latrine pits and on the soil next to latrines which was moistened by filth; the larvae migrated into the pits upon hatching. He found that one female lays about 200-240 eggs at one time; in the lab they would oviposit on human excrement, meat, and fresh fish. He described larval breeding in nature in accumulations of human excrement in both open and enclosed latrine pits. He never observed breeding in animal dung, individual human droppings, or garbage. He obtained successful development only on human excrement in the lab; larvae developed much more poorly when fed fish, and on white mice meat there was marked mortality.

Rearing the flies on excrement at 18-25°C Sukhova found these periods: egg, 14-20 hours; larval, 10-14 days (I, 1-2 days; II, 2-3 days; III, 7-9 days); pupal, 7-10 days. He stated that the adults live up to 110 days in the lab and that this species overwinters in an arrested prepupal or pupal stage.

DISPERSAL

Fly movement is variously referred to as dispersal, dispersion, and migration—the first connoting active movement, the second, a more passive transit, and the third, directed and sustained flight, often seasonal. Each of these movements may, in fact, be important under certain conditions, and all three may operate in the life of a fly. How far, how fast, and under what conditions flies move is a subject that bears directly on their role as disseminators of pathogens. This information is also necessary for the creation of an effective *cordon sanitaire* between breeding sites and human populations.

Flies are not usually considered migratory insects in the same way as locusts, Monarchs, and pierid butterflies. Nevertheless, there is evidence that such is the nature of the southward autumn movement of *Calliphora vicina* and *C. vomitoria*, which have been observed flying through passes in the Pyrenees; this suggests that these species may be migratory in some parts of their range (Williams et al., 1956). More than a dozen species of synanthropic flies, including *C. vicina*, *M. domestica*, *M. autumnalis*, and *F. canicularis*, have been taken over a number of years under varying wind directions and velocities, aboard a lightship standing 70 km west of Flushing in the North Sea (Lempke, 1962). These records include *Eristalis* (= *Tubifera*) *tenax* and two other syrphids which Nielsen (1967) also observed in migrations over a point of land in Denmark. The fact that *M. autumnalis* has crossed the North American continent in 15 years from east to west against the prevailing winds leaves no doubt about its mobility, whatever the means. Another type of migratory movement which may be more widespread is the autumnal movement of blow flies from forests to human settlements, observed in south Finland (Nuorteva, 1966). Seasonal increases in synanthropy may have important epidemiological significance. This is a common occurrence with house flies in temperate regions.

Undoubtedly the best studied migrant is *Cochliomyia hominivorax*, the primary screwworm fly, which moves northward each spring from its overwintering sites in the southern United States. Barrett (1937) released a group of marked flies in Uvalde, Texas, which averaged 35.4 miles north and 20.3 miles northeastward per day. In another study in Texas, marked flies were recaptured 180 miles northwest of the release point in less than two weeks (Hightower et al., 1965). This underscores both the logarithmically increasing difficulties of recapture and the prodigious flight potential of the fly. The northerly movement begins as early as March and continues well into the summer, when it is halted by hot, dry weather. As a consequence, populations of this species have in the past been able to establish summer bridgeheads, far to the north of their over-wintering sites, in Kansas, Iowa, and Illinois.

The small mass of a fly, relative to its wing expanse, enables it to be carried aloft and dispersed by convection currents and wind. *Hippelates* gnats have been found at an altitude of 11,000 feet near the Texas-Mexico border. *S. calcitrans* was taken at 3,000 feet and *C. macellaria* at 5,000 feet (Glick, 1939). The fact that spiders have been taken at 15,000 feet suggests the potentialities of this means of passive transport.

There are numerous records of passive transport of blow flies and house flies by means of garbage trucks, vegetable trucks, and other such vehicles. That such traffic may aid passive dispersion was shown in a study in Egypt in which five times more marked flies were collected in a village that was on the road to Cairo than at another village that was at the end of a side road. Both villages were equidistant from the fly release site (Peffly and Labrecque, 1956).

Flies are not loathe to stow away aboard ships bound for distant ports, and they may weather these trips as well as other passengers. Illingworth (1926c) observed *Chrysomya megacephala* and *Musca convexifrons* on board a ship bound from Shanghai to Hawaii. On fine days, the former flew about the top deck near the kitchen ventilator shaft and, during inclement weather, roosted down inside the shaft. Concerning the distribution of *Sarcophaga fuscicauda*, Illingworth (1926d) says, "It is very evident that this species makes good use of the common carriers of commerce. India or southern Asia may well be considered as the home of this species, and that in extending its range it has followed the natural routes of shipping [East Indies, Australia, Japan, and Hawaii]." Saccà considers that *M. domestica* probably arose in the eastern Ethiopian region and may have spread to the Americas by way of pre-Columbian voyages (Legner and McCoy, 1966).

Air travel has lent ever more powerful wings to flies. Whitfield (1939) listed more than twenty synanthropic flies found aboard commercial aircraft flying between various countries in Africa, between Africa and the United States, and between South and Central America and the United States. Among nearly 3,000 insects taken in a three-year period, at least 850 proved to be *M. sorbens* and 550 were *M. domestica* (see Welch, 1939). More recently, cholera broke out in Bangkok and appeared a few days later in most of Thailand; this stimulated an entomologic examination of planes flying from Bangkok to Singapore. All 43 planes examined contained *M. domestica*. Although none of the flies harbored the cholera vibrio, some did have fecal organisms (Hale et al., 1960). I recently shared a flight from Rome to Chicago with a house fly and, while I sweated through customs, its entry into the New World was unimpeded. Perhaps immigrant face flies first viewed North America through the porthole of an airplane.

MARKING FLIES FOR DISPERSAL STUDIES

The marking, release, and recapture of flies for the purpose of studying their flight characteristics was initiated in the first decade of this century and a number of techniques have been used. A good label should be readily distinguishable, long lasting, innocuous to the fly, have no effect on flight or other behavior, and be inexpensive and easy to apply to large numbers of flies.

Early investigators dusted flies with finely ground blackboard chalks by gently shaking them in a paper bag. These dusts tended to be dislodged from the bodies by cleaning and agitation, but persisted at the base of the wings and on the halters. Red dust was still distinct on house flies after 20 days, whereas yellow dust was not present after 9 days under laboratory conditions (Jepson, 1908). Detection of the dust was tedious since it depended on visibility of minute particles. This was later improved by mass powdering with acetone-soluble dusts. With such dyes, a drop or two of acetone on the dead fly placed on filter paper, produced a color on the paper. Rosalic acid, aniline green, gentian violet, acid fuchsin, aqueous eosin, trypan blue, and methylene green have been applied as sprays, generally in dilute alcoholic or acetone solutions, which are not harmful to the fly. It was soon found that the addition of 2% shellac improved adhesion of the stain. More recently, primary screwworm adults were dipped into a solution of 8 ml of emulsifiable polyethylene per liter of water and then dusted with fluorescent powder to improve persistence of the dye (Hightower, 1963).

Handmarking the thorax with paints or quick drying lacquers served to trace the movements of the face fly (Ode and Matthysse, 1967). A number of marks can be applied, and this is particularly useful in multiple release studies. Its laboriousness places an obvious limit on this method.

A self-marking method developed by Norris (1957) was used by Steiner (1965) for tagging sterile fruit flies on Pacific islands. It requires dusting pupae or impregnating the substrate in which pupae are present with an oil-soluble blue dye. The ptilinum of the emerging fly collects dye, which is retained after the ptilinum is withdrawn permanently into the head (Figs. 13 and 14). The head of a captured specimen is crushed in a drop of acetone on filter paper to see the color. Flies that were more than 16 weeks old could still be identified and there could be no transfer of dye to unmarked specimens in the traps.

Dusts that fluoresce under ultraviolet light are another means of detection and can be especially useful in studies of night-resting places. Flies are readily detected with a portable ultraviolet light at a distance of six feet.

Fluorescent dyes were used by Norris (1957) in self-marking experiments with four Australian blow flies and by Murvosh and Thaggard (1966) in studying the dispersal of house flies from privies on the island of Grand Turk in the Bahamas. Dyes such as rhodamine B, uranine, auramine O, erythrosin, and tartrazine are applied in 1% aqueous or alcoholic sprays. Fluorescent dyes are also spread on the surface of privies to tag visiting and emerging adults, or fed directly to adults. *D. melanogaster* fed rhodamine B retained the dye for two months (Wave et al., 1963), and dead house flies remained fluorescent for many months after feeding on flourescein (Zaidenov, 1960). A dead fly tagged the previous summer and trapped in a spider web fluoresced brightly under ultraviolet a year later (Morris and Hansens, 1966). Many sulfide phosphors give good luminescence but are photo-labile and therefore unsuitable for out-of-doors.

We studied dispersal of flies from a rural slaughterhouse in Mexico by extensively spraying large populations with uranine solution dispensed in a 2-gallon pressure sprayer (Greenberg and Bornstein, 1964). Under ultraviolet light, marked flies showed pale yellow splotches or spots of fluorescence on various parts of their bodies. Some natural fluorescence was encountered, but this was usually in the form of a faint, diffuse glow. Eddy et al. (1962) observed natural fluorescence in the halters and wing tips of wild *Haematobia irritans* from eastern Oregon but not in those from the western part of the state.

Fig. 13. Stereoscan photograph of newly emerged adult *Phaenicia pallescens* showing ptilinum ($\times$ 25).

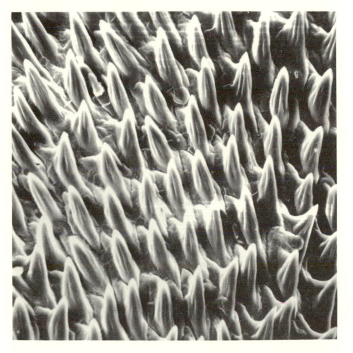

Fig. 14. Same as Fig. 13 showing ptilinum considerably magnified ($\times$ 1025).

Persistence may be unnecessary in a marker. Derbeneva-Ukhova (1947) fed a dye to house flies to test the rapidity of turnover of populations between indoors and outdoors. Flies fed a red dye were released indoors and others fed a blue dye were released at a garbage can outside a house. The color in the dissected crops of flies in the next few hours revealed a 16.4% outward migration and a 13.1% inward migration. Wolfinsohn (1953) fed finely divided ferric oxide mixed with sugar to house flies, which caused their abdomens to become red.

With the availability of radioactive isotopes came a burst of investigations utilizing radioactive phosphoric acid (^{32}P). Phosphorus is used because it is fundamental to protoplasm, has a convenient half-life, and the detection of its beta radiations is relatively easy. Radioactivity is detectable by means of a Geiger-Muller counter, which may be hooked up with an earphone or with a count rate meter to record corrected counts per minute.

Tagging flies by means of their larval food produces too great a variability in radioactivity. ^{32}P was used as a 60-second dip at a concentration of 12.5 μc/ml plus 0.1% Triton X-100 as a wetting agent; there was measurable radioactivity in several blow fly species and in the house fly for 18 days (Roth and Hoffman, 1952). Most commonly, tagging is done via the adult food.

When 1 μc per ml of milk, sugar solution, or defibrinated blood is given to flies for 24 to 48 hours, 90 to 95% of the population becomes labeled, and radioactivity around 1000 counts per minute can be measured compared with a normal background of less than 5 cpm. Counts vary with species and sex; large flies such as blow flies and females usually ingest more material than house flies and males (Jensen and Fay, 1951). But behavioral differences may reverse this rule. *Protophormia terraenovae*, for example, is lethargic in captivity, ingests less food than the smaller more active *P. sericata*, and therefore shows lower counts (MacLeod and Donnelly, 1957b). Adequate levels are reached in *Drosophila* after one day of feeding on 1 to 2 μc per gm of a banana-yeast or similar diet. In these experiments radioactivity is lost through natural decay and through fly excretion and vomiting, but, at these dosage levels, flies remain detectable for a month or more.

Two other applications of the ^{32}P method to the tagging of flies were reported by Shura-Bura (1952b), who may have been the first to use radioactive tracers to study dispersal of synanthropic flies. One application involved spraying the surface contents of pit privies with ^{32}P and subsequent detection of radioactive *M. domestica vicina* in nearby houses. This, the author feels, is superior to the visual method because it has the possibility of showing the frequency and degree of contact with excreta. Another application (Shura-Bura, 1957) involved tagging house flies by feeding them *Shigella flexneri* labeled with ^{32}P. After one day, 1/3 to 2/3 of the label was still localized in the digestive tract, but the rest was already in the tissues of the flies, presumably by digestion of bacteria and assimilation of the label. The use of a pathogen in this way has little to recommend it.

Spraying flies with innocuous microbes also has little in its favor. *M. domestica* was detected up to 84 hours after it was sprayed with a culture of the yeast, *Saccharomyces cerevisiae*, but attempts failed with coliforms, *Pseudomonas* types, or *Serratia marcescens* because they were naturally present in flies (Peppler, 1944). Love and Gill (1965) attempted to use bacteria as indicators of beach and farm populations of *Stomoxys calcitrans* in their dispersal studies in Michigan. However, their results were inconclusive; the presence of enteric bacteria in some of the beach flies may or may not have been related to their having migrated from farm areas. These results illustrate the questionable feasibility of this method; some workers have shown that streptococci, including the enterococci, may be isolated from a variety of plants and adult insects not known to be attracted to fecal matter (Eaves and Mundt, 1960; Mundt et al., 1958). Love and Gill pointed out that the success of their method required that the flies did not frequently become contaminated with bacteria from sources other than manure.

A fly tagger, which mechanically attaches an inch-long brightly colored nylon thread to the fly's thorax, has been developed for studies of fly movements, especially in enclosed spaces such as in aircraft (Klock et al., 1953).

Some mutants possess built-in markers that are convenient for dispersal studies, provided the flies behave normally. The dispersal of orange-eyed *Drosophila pseudoobscura* (Dobzhansky and Wright, 1943) and yellow-eyed *Chrysomya macellaria* (Quarterman et al., 1949) has been studied, and the former's dispersal characteristics were found to be comparable with that of wild flies.

ESTIMATION OF FLY DENSITY

The Scudder grill (Scudder, 1947) provides a rough, visual assessment of fly density, which is useful for evaluating the effectiveness of a control program. Muirhead-Thomson (1968) noted that certain errors and bias in the use of this method must be taken into account in analysis of data. In order to obtain estimates of the actual density of a fly population in an area, a known number of marked flies is released and the number recaptured in proportion to total unmarked flies trapped provides an estimation of the actual numbers. The equation is:

$$\text{Size of population} = \frac{B \times C}{D}$$

where B = total number of marked flies released;
C = total number of flies captured;
D = number of marked flies recaptured.

The difficulties inherent in this method have been thoroughly discussed by Gilmour et al. (1946):

1. The area to be censused should be fairly extensive but also coextensive with the effective flight range of the species. The use of a large number of traps reduces errors due to the variability in local population densities and in trap efficiency. Each trap should serve an equal area, which means, in practice, that more peripheral areas have to be served by larger numbers of traps than areas closer to the central release point. Despite this consideration, the equipotency of the more peripheral traps would remain questionable because the number of flies caught decreases with the distance, and at greater distances fly movement may be less randomized than at distances closer to the release point (Schoof and Siverly, 1954). Perhaps by multiple releases of smaller numbers of flies from a series of points along a circumference that is equidistant from the center and the periphery, a better mixing of the released and wild populations can occur and, therefore, a truer estimation achieved.

2. The traps should not be spaced so closely that an artificial enhancement of the attractiveness of an area occurs, thereby drawing in flies from the surrounding untrapped areas.

3. Marked flies should be released some time before trapping to allow adequate mixing. Ideally, mixing should be uniform throughout the trapping area.

4. All or nearly all of the marked flies should be in the sampling area during the trapping period.

5. The released flies should be a facsimile of the natural population with regard to such factors as age, sex, maturation, nutritional state, handling, etc.

In actual practice, the above conditions are never achieved. A serious weakness is that error increases in proportion to the square of the range of the species involved (Hocking, 1953). Since many synanthropic flies are strong fliers, they could easily move out of the study area in a day or two. There may also be independent fluctuations in numbers of both the marked and natural populations due to differential birth and death rates, which would change their proportions from day to day. Furthermore, flies are not necessarily uniformly distributed over an area but may occur in pockets as do *Drosophila*, or may be canalized by features of the environment such as hedgerows, riverbeds, tree belts, residences, open fields, hosts, and barns. Even the unexpected predation of marked flies by birds has been known to occur at the time of release. Taking into consideration possible errors of 20 to 30% due to all the above, estimates of population density by this mark-release-recapture method are interesting nonetheless because they provide an estimate of fly densities that would otherwise be impossible to obtain.

FACTORS INFLUENCING DISPERSAL AND TRAPPING

A number of factors may influence the dispersal characteristics of a fly population, and we will discuss some of them briefly here. These include possible differences related to sex, age, nutrition, density of the population at time of release, wild versus laboratory populations, placement of traps and attractivity of bait, the time and place of release, and meteorological conditions.

Few investigations have been specifically designed to evaluate the influence of these various factors. Any such studies would be overwhelmingly difficult to analyze considering so complex a behavior as dispersal in so variable an arena as the natural environment. Observations have been made however, which, although often incidental to the main study, provide worthwhile insights into the responses of flies to many of these factors.

Generally, we find no consistent differences between the sexes with regard to rates of dispersal or response to wind direction. Pickens et al. (1967) observed that three times more male house flies tended to move out from the release site than females, and that among one-, two-, and three-day old flies released together, the youngest ones dispersed most rapidly. The sex ratios of trap catches are known to be influenced by temperature, light intensity, and the nature of the baits.

We know about the effects of larval crowding on the size of flies but not about the possible effects of larval conditions on subsequent adult dispersal behavior. It has been known for some time that, in laboratory populations of *Drosophila*, supraoptimal density, despite the presence of excess food, results in reduced reproduction rates and longevity (Pearl et al., 1927). It was even found that a significant decline in the reproduction rate occurred between a mean density of 2 and 3.4 flies per half-pint bottle. It is interesting that these negative effects continued to be expressed up to 15 days after removal of the flies from this environment. Whether laboratory and natural optima are necessarily the same we do not know. We do know that natural fly populations often exist in dense pockets both as larvae and adults, due in part to the gregarious oviposition habits of females.

The carrying capacity and attraction of the area in which flies are released influences dispersal. When house flies were released in woods, 38% of the recovered flies flew 1 mile or more past farms which were situated within ½ mile of the release site; when they were released at sheep barns, only 5 to 18% were recovered beyond ½ mile; and the tendency to stay put was even greater at cattle farms, where only 1 to 3% were recovered beyond ½ mile (Pickens et al., 1967). It was also found that dispersal from a clean barn was 2 to 3 times as great as from an unsanitary barn. Thus, release in unattractive areas may stimulate a wider range of movement.

Flies are attracted by chemical cues, the composition and decomposition of the bait attracting one group of flies more than another. Indole, skatole, and trimethylamine are decomposition products of meat that attract carrion-feeding blow flies and flesh flies but do not attract the house fly. Similarly, decomposing fruit baits, which release acetic, propionic, and butyric acids, are especially attractive to *Drosophila*. Dry-ice-baited traps, so effective in trapping mosquitos, have been used for *Haematobia irritans* (Tugwell et al., 1966), but their sphere of attraction is limited.

One can obtain selective responses to meat baits from different species of blow flies. Catches of Australian blow flies were four times greater with bullock liver than with fat or lean meat; liver-baited traps proved more attractive to *Chrysomya albiceps* than to the other four blow flies (Gurney and Woodhill, 1926). Necrotic wounds in yearling steers, infected with the larvae of *Cochliomyia hominivorax*, were more attractive to adults than were meat-baited traps (Parman, 1920).

Artificial baits dry out and lose their attractiveness; the rate of deterioration depends on the microclimate. Taking these factors into account will help reduce differences in trap efficiency. More difficult to control are the minute differences between traps, their placement, and certain fly density effects, which contribute collectively to what is called trap idiosyncrasy.

The location of traps has considerable influence on the size and type of catch. Traps set in a barnyard caught about 18 times more tagged house flies than traps set in field sites (Lindquist et al., 1951). Placement in relation to the barnsite itself was important. Baited traps were effective for house flies and stable flies when they were placed within twenty feet of a barn, but lost their effectiveness beyond this distance, probably because of competition from the large amount of fly-attractive materials in and near the barn itself (Pickens et al., 1967). Placement of traps for *Hippelates* has been particularly troublesome (Dow and Hutson, 1958). So simple a device as sticky fly paper is also subject to certain variables. Fly strips were only one-fourth as effective when they were hung horizontally near the barn ceiling as when they were hung vertically from the same location. In fly-dense situations, the strips should be replaced frequently. Nets have proven a good alternative when other methods were not practical (Greenberg et al., 1963).

The question of whether laboratory-reared flies behave like the wild type when released has not been adequately studied. Orange-eyed mutants of *D. pseudoobscura* had normal vigor and dispersal characteristics comparable to a wild type (Dobzhansky and Wright, 1943). No differences were found in dispersal rates of laboratory and wild face flies (Ode and Matthysse, 1967), but Eddy et al. (1962) reported that field specimens of the stable fly were more active than their laboratory counterparts. They released 97% lab flies and 3% field flies and 20% of those caught during the first 4 hours were field flies. Inactivity in newly released laboratory insects has been often observed, but this may only be a temporary effect. The problems of acclimation and discrepancies due to patterns of daily flight activity are largely avoided by releasing flies toward evening. House flies do not travel far when thus released, whereas the movement of *H. irritans*, while not great, was found to be nocturnal (Hoelscher et al., 1968), with a possible predawn peak (Tugwell et al., 1966); some nocturnal flight and oviposition activity of *C. vicina* and *C. vomitoria* was observed in an English abattoir (Green, 1951). Most synanthropic species probably settle down for the night and are presumably acclimated by morning. Light intensity and temperature influence flight activity and, therefore, have an effect on trap efficiency, and these specific effects will be discussed in the following section. More studies on the effects of age and maturation on dispersal will have to be done before conclusive comparisons can be made between the behavior of laboratory and wild flies.

Meteorological Factors

Fly dispersal is a complex behavior subject to wind, temperature, light intensity, and possibly ionization and barometric pressure. Sufficient information exists to warrant discussion of the influence of these factors on flight patterns.

The influence of wind direction has been variously interpreted, but the evidence, in general, indicates that fly orientation to this element is plastic. House flies released from an island in a river traveled directly against or across the wind in moving toward the city of Ottawa (Hewitt, 1912b). House flies, *P. regina*, and *P. sericata* dispersed in every cardinal direction when released in open country, uninfluenced by weather or prevailing northerly winds (Lindquist et al., 1951). When released among native dwellings on the Island of Grand Turk in the Caribbean, house flies moved with, against, and perpendicular to a wind which was normally 5 to 10 mph (Murvosh and Thaggard, 1966); on a Maryland farm, house fly movements seemed to be predominantly upwind when the wind was 2 to 7 mph (Pickens et al., 1967). In Manitoba farm country, a large number of tagged house flies dispersed rapidly into a wind that averaged 8 mph (Hanec, 1956). Face flies flew two miles into a prevailing wind to locate cattle (Killough et al., 1965), and cattle-directed face fly dispersal was also noted by Ode and Matthysse (1967).

Even so frail a fly as *D. melanogaster* appears undaunted by light prevailing winds; it traveled at least 4.4 miles in 24 hours against a wind of up to 10 mph. However, these flies were grounded by a wind of 12 to 14 mph (Yerrington and Warner, 1961). Dispersal of *F. canicularis* in a Japanese village was not influenced by wind direction (Ogata and Suzuki, 1960), nor was the movement of the house fly, *L. caesar, P. terraenovae*, or *M. stabulans* from a garbage dump near Leningrad, despite a moderately strong north wind (Shura-Bura et al., 1956). In studying the movements of five sheep blow flies in Australia, Gurney and Woodhill (1926) concluded that they would follow the scent of carrion against a slight breeze but, when ranging over the country in search of feeding or breeding grounds, the majority moved with or slightly across the wind. In dry, open country in Oregon, which was devoid of livestock, the flight direction of *S. calcitrans* and *H. irritans* was also windward (Eddy et al., 1962). This again suggests that questing flights are likely to be downwind and random. On the other hand, *Hippelates collusor* moved into the wind to cross one-half mile of desert in order to reach favorable sites on the other side (Mulla and March, 1959).

The repeated observation that dispersing flies may bypass attractive sites does not negate the evidence that they follow chemical cues and will orient into the wind to do so, as we have seen. The potency of bait greatly diminishes with distance, and the random dispersal observed in some trapping studies has been interpreted as indicating little or no attraction, although it could also be interpreted as more or less equal attraction. On the other hand, barns, garbage dumps, and substandard sections of towns send out far more potent cues than baited traps, and several studies have shown a directed rather than random flight toward these foci, sometimes over a number of miles. Nevertheless, the fact that flies, presumably regardless of sex or maturation, will also disperse from a barn, dump, or slaughterhouse toward seemingly less attractive places indicates gaps in our knowledge of what attractiveness really means to the fly on the wing.

Flies may also respond passively to prevailing winds in what may be called the island effect. This was first observed with the appearance and disappearance of swarms of house flies, stable flies, and blue bottles on a crib in Lake Erie, six miles from shore, coincident with a southerly or northerly wind (Hodge, 1913). Sporadic outbreaks of the stable fly along Florida Gulf Coast beaches were also found to coincide with wind direction. Large numbers of flies appeared after a northerly (offshore) breeze blew them out of a wooded area where they had taken refuge from a previous south wind. These aggregations of flies appeared and disappeared within a few hours after the shift in the wind (Wright, 1945).

Temperature has an important influence on dispersal. On a flight mill apparatus, the flight distance and efficiency of *P. sericata* actually decreased with increasing temperatures between 15°C and 30°C, when a carbohydrate source was withheld. The calculated flight ranges at 15°C and 30°C were 32.16 km and 24.80 km, respectively (Yurkiewicz, 1968). Tethered flight, however, imposed starvation and other constraints not normally experienced by flies in nature. Under natural conditions and within normal limits, flight is directly affected by temperature. When house flies were released in two Armenian villages situated at altitudes of 882 and 1760 meters, the flies in the lower village, where the mean temperature was 18.1°C, covered the same distance faster than those in the upper village, where the weather was the same but the temperature was 13.1°C (Vashchinskaĩa, 1956). Primary screwworm flies dispersed more slowly when released during December-February in southern Texas than when released in spring and summer, although the patterns of local distribution in relation to vegetation and water remained the same (Hightower and Adams, 1969).

There may be a cessation of flight activity above certain temperatures, manifested in a diel bimodal activity curve. Activity curves for calliphorids are bimodal in those places and seasons where daytime temperatures exceed the preferenda of the flies, e.g. above 28°C for *P. terraenovae*. Moving northward, there is a lowering of the daily temperature maximum with a consequent fusing of the bimodal curve into a single broad peak. Still farther north, as in subarctic north Finland, the peak of calliphorid activity is quite narrow and is confined to the early afternoon (Nuorteva, 1965). During the rest of the day, it is too cold for calliphorids to fly even though activity patterns of other dipterans are are still bimodal.

Nuorteva notes that the highest peaks of calliphorid activity coincide with brief periods of sunshine rather than with specific temperatures. However, it is difficult to rule out the temperature effects due to insolation, especially on these dark-bodied insects.

An interesting interaction between temperature and light intensity has been widely noted and is carefully discussed by Norris (1966b). At high temperatures, a fly's light reaction inhibits flying, and this probably accounts for the bimodal activity in hot weather of *Fannia, Glossina,* and *Calliphora,* as well as *Drosophila* and *Eristalis tenax.* As might be expected, differences in thresholds occur between sexes as well as species. Males of *Calliphora stygia* and *C. augur* may have a negative light response at a lower temperature than females (Norris, 1966b). In effect, this removes the males from the available population during certain parts of the day and results in trap catches that are skewed toward females.

The adaptive nature of the light-temperature reaction is clear. When temperatures reached into the 90s in the San Joachin Valley in California, *D. melanogaster* remained in cool, shady locations until just before sunset, when lowering temperature and light intensity triggered flight (Yerrington and Warner, 1961). This phenomenon is most likely a light avoidance reaction, rather than a direct effect of temperature, as shown by the behavior of *Glossina morsitans.* When the temperature reached a certain level in an illuminated chamber, flies moved into a darkened chamber with a higher temperature even though it eventually killed them (Jack and Williams, 1937). It is not clear whether the light reaction works on the cool end of the temperature scale to stimulate flight activity, although diapause in the face fly may be controlled by this mechanism.

Temperature and humidity also interact to influence flight activity. Dispersal of Australian sheep blow flies (Gurney and Woodhill, 1926) and the primary screwworm fly (Hightower and Alley, 1963) is restricted by hot, dry weather and increases with high humidity and after rains.

Unusual fly behavior with approaching storms has long been known. Parman (1920) observed that several different muscids became active and then partially comatose with a rapidly falling barometer. Studies by Edwards (1960) and Maw (1965) seem to offer at least a partial explanation. Edwards found that a surplus of positive ions stimulated flight activity of caged *C. vicina.* Maw exposed tethered *P. sericata* to air ions at ion currents of approximately 3.4×10^{-11} amperes and noted a two- to three-fold increase in flight duration. Positive ions stimulated longer, faster flights with steeper increases and decreases in speed than did laboratory air. Negative ions resulted in fast, steady flight of generally longer duration than in laboratory air or in positive ions. As Maw suggests, it is worth considering that augmentation of flight activity before a storm, with or without barometric changes, may be due in part to the ionic content of the atmosphere. This possible effect on fly dispersal needs study.

DISPERSAL PATTERNS AND DISEASE DISSEMINATION

The restlessness and mobility of flies is one of their most important vector attributes for it helps to expose them to a vast array of pathogenic organisms and to the food we eat (see Chapters 5 and 6, Volume I). It is evident from Table 3 that flies can be relatively far-ranging animals whose movements may be channeled or restricted by features of the landscape but are not blocked by rivers, tree belts, wooded ridges, or belts of desert. The extent and rate of movement is greatly influenced by the synanthropy of the fly, which is expressed by its attraction to farms, abattoirs, and to substandard areas of villages and cities. Nash (1913) pointed out that, where houses were numerous, few house flies were to be found more than ¼ mile from their source, but, where houses were few, flies traveled farther afield. Ode (1966) has made the same observation in a Japanese village. On small collective farms, Derbeneva-Ukhova (1947) noted a striking concentration gradient of the house fly in houses surrounding a central breeding place. From an average of 110 flies in houses at less than 100 meters from the breeding source, the catch dropped to fewer than 10 flies in houses at distances of 160 to over 400 meters. Thus, the interposition of attractive sites intercepts and keeps flies in a given area, and the best way to shatter distance records is to release flies in country that offers little to detain them.

For the same reasons, the rate of fly dispersal through unattractive terrain is generally rapid. House flies moved through an Armenian village at the rate of 56 meters/day, whereas outside the village the rate was 350 meters/3 hours (Vaschinskaĩa, 1956). Released in open country, *D. melanogaster* traveled 4.4 miles in one day, but released among houses and privies its range was reduced to 500 feet (see Table 3 for other examples). Muscids and calliphorids, when released in a sparsely populated landscape, oriented to nearby towns and particularly, to a city that was 15 km away. Attractive sites in these places included cattle farms and yards, the backyards of restaurants and dwellings, where unsanitary conditions prevailed (Shura-Bura et al., 1958). "Hot spots" of a similar nature were frequented by marked house flies in Savannah, Georgia, though blow flies seemed to prefer vacant lots (Quarterman et al., 1954a). House flies have been known to bypass a relatively clean farm to reach a less clean one (Pickens et al., 1967).

The restlessness of flies is such that ample food and breeding material

TABLE 3

SUMMARY OF DATA ON FLY DISPERSAL

SPECIES	LOCATION	HABITAT	EFFECTIVE DISPERSAL RANGE	MAXIMUM DISPERSAL DISTANCE	APPROXIMATE DISPERSAL RATE	AUTHOR
CHLOROPIDAE						
Hippelates collusor	California	Desert, agricultural		4.3 mi		Mulla et al., 1959
Hippelates pusio	Georgia	Rural	~ 1 mi	> 1 mi	> 1 mi/3½ hrs	Dow, 1959
STRATIOMYIDAE						
Hermetia illucens	Canal Zone	Semiurban	2500 ft	2500 ft		Zetek, 1914
SPHAEROCERIDAE						
Borborus equinus	North Sea	Lightship		70 km		Lempke, 1962
DROSOPHILIDAE						
Drosophila pseudoobscura	California	Subboreal	~ 360 m			Dobzhansky et al., 1943
Drosophila melanogaster	Texas	Town		500 ft		Pimentel et al., 1955
	California	Open country	~ 1-2 mi	4.4 mi	4.4 mi/ 24 hrs	Yerington et al., 1961
	Maryland	Tomato fields		3 mi		Wave et al., 1963
Drosophila repleta	Texas	Town	~ 300 ft	1000 ft		Pimentel et al., 1955

Species	Location	Habitat	Distance		Rate	Reference
stercoraria	North Sea	Lightship	70 km			Lempke, 1962
MUSCIDAE						
Fannia canicularis	Leningrad	Rural	4.5 km		2 km/26 hrs	Shura-Bura et al., 1958
	Japan	Urban	480 yds (?)	115 yds		Ogata et al., 1960
	North Sea	Lightship	70 km			Lempke, 1962
Fannia pusio	Georgia	Urban	3.5 mi			Quarterman et al., 1954b
Hydrotaea dentipes	Leningrad	Rural	15 km		1 km/5½ hrs	Shura-Bura et al., 1958
Ophyra aenescens	Texas	Rural	4.4 mi			Bishopp et al, 1921
	Mexico	Semirural	3 mi		1.4 mi/24 hrs	Greenberg et al., 1964
Ophyra leucostoma	Texas	Rural	7 mi			Bishopp et al., 1921
	W. Virginia	Urban	4.5 mi			Schoof et al., 1953
Ophyra rostrata	Canberra, Australia	Pastoral	>3 mi		3 mi/24-48 hrs	Norris, 1959
Phaonia scutellaris	North Sea	Lightship	70 km			Lempke, 1962
Muscina assimilis	Leningrad	Rural	4.5 km		1 km/5½ hrs	Shura-Bura et al., 1958
Muscina stabulans	Leningrad	Garbage dump	2.1 km			Shura-Bura et al., 1956

SPECIES	LOCATION	HABITAT	EFFECTIVE DISPERSAL RANGE	MAXIMUM DISPERSAL DISTANCE	APPROXIMATE DISPERSAL RATE	AUTHOR
	Leningrad	Rural		15 km	3.5 km/1 day	Shura-Bura et al, 1958
	Kyoto	Suburb in winter		400 m		Uemoto, 1960
	Leningrad	Garbage dump		0.4 km		Shura-Bura et al., 1962
	North Sea	Lightship		70 km		Lempke, 1962
Polietes lardaria	North Sea	Lightship		70 km		Lempke, 1962
Synthesiomyia nudiseta	Texas	Rural		½ mi		Bishopp et al., 1921
Musca autumnalis	North Sea	Lightship		70 km		Lempke, 1962
	Maryland	Farms		¾ mi		Fales et al., 1964
	Maryland	Farms		4 mi	2 mi/24 hrs	Killough et al., 1965
	New York	Farms		7.3 mi	4.9 mi/2 days	Ode et al., 1967
	New Jersey	Farms		1500 m	1500 m/43 hrs	Hansens et al., 1967
Musca domestica	England	Urban		1700 yds	800 yds/35 min	Copeman et al., 1911
	Canada	City		>700 yds		Hewitt, 1912
	Lake Erie	Crib 6 mi from shore		6 mi		Hodge, 1913
	England	Urban		770 yds		Hindle et al., 1914
	Canal Zone	Semiurban	2500 ft	2500 ft		Zetek, 1914
	Montana	Small city	Throughout city	>3500 yds		Parker, 1916
	Texas	Rural		13 mi	7.1 mi/24 hrs	Bishop et al., 1921
	Russia	Rural		5.5 km		Gorodetskiĭ et al., 1937

Location	Habitat	Distance	Distance	Rate	Reference
United States	Army bivouac		3 mi		Peppler, 1944
Oregon	Farmsteads		12 mi		Lindquist et al., 1951
Arizona	Urban	< 1 mi	5 mi	4 mi/3 days	Schoof et al., 1952
Oregon	Farmsteads	½ mi	20 mi		Yates et al., 1952
Georgia	Rural	< 3 mi	10 mi	5 mi/< 24 hrs	Quarterman et al., 1954a
Georgia	Urban	½ to > 1 mi	7.6 mi	5.2 mi/24 hrs	Quarterman et al., 1954b
Arizona	Urban	½-2 mi	7.2-8.3 mi		Schoof et al., 1954
Manitoba	Farm		2 mi	2 mi/5 hrs	Hanec, 1956
India	City-rural		350-600 yds		Ranade, 1956
Leningrad	Garbage dump	2 km	3.84 km	3 km/3 days	Shura-Bura et al., 1956
Armenian SSR	Open country village			350 m/day 56 m/day	Vashchinskaiā, 1956
Leningrad	Privy pits (town)		10.7 km		Shura-Bura, 1957
Leningrad	Rural	4-6 km	10.7 km	2 km/5 hrs, 4-5 km/1 day	Shura-Bura et al., 1958
Oregon	Open, dry country		5 mi		Eddy et al., 1962
Leningrad	Garbage dump	0.7-0.8 km	3.2 km	3.2 km/24-50 hrs	Shura-Bura et al., 1962
Prague	Farm	< 500 m	700 m		Celedová et al., 1963
Mexico	Semirural	3 mi	3 mi	1.4 mi/24 hrs	Greenberg et al., 1964
Czechoslovakia	Farm		600 m		Havlík, 1964
New Jersey	Farm	0.5-2.5 mi	6.5 mi		Morris et al., 1966
Maryland	Farmsteads	½ to > 1 mi		½ mi/4 hrs	Pickens et al., 1967

SPECIES	LOCATION	HABITAT	EFFECTIVE DISPERSAL RANGE	MAXIMUM DISPERSAL DISTANCE	APPROXIMATE DISPERSAL RATE	AUTHOR
Musca domestica vicina	Soviet Central Asia	Rural		13 km		Chebotarevich, 1937
	Israel	Rural		300 m		Wolfinsohn, 1953
	Egypt	Villages	1 mi	1.5 mi		Peffly et al., 1956
	Kyoto	Suburb in winter		400 m		Uemoto, 1960
	Japan	City suburb	330 ft	1100 ft		Ogata et al., 1960
	Malaya	Villages	¼–½ mi	½ mi		Wharton et al., 1962
	Tadzhik SSR	Rural		300 m		Il'Yashenko, 1964
	Japan	Village	500 m	2 km		Oda, 1966
Musca sorbens	Egypt	Villages	1 mi	1.5 mi		Peffly et al., 1956
	Australia	Open country		3.5 mi		Norris, 1966a
Haematobia irritans	Oregon	Open, dry country		5 mi		Eddy et al., 1962
	Louisiana	Pastoral	<1000 yds	1584 yds	1584 yds/4½ hrs	Tugwell et al., 1966
	Mississippi	Pastoral	>400 yds	>400 yds		Hoelscher et al., 1968
Stomoxys calcitrans	Lake Erie	Crib 6 mi from shore		6 mi		Hodge, 1913
	Oregon	Open, dry country		5 mi	4 mi/1¾ hrs	Eddy et al., 1962
	Mexico	Semirural	1.3 mi	3 mi	1.4 mi/24 hrs	Greenberg et al., 1964

CALLIPHORIDAE

Cochliomyia | | | | >8 mi | 0.5 mi/3 hrs | Bishopp et al., 1921 |

Species	Location	Habitat				Reference
mutant)	Georgia	Urban			1.5 mi/7 hrs	Quarterman et al., 1949
	Georgia	Rural	<3 mi	10 mi	5 mi/<24 hrs	Quarterman et al., 1954a
	Georgia	Urban	1-4 mi	7.2 mi	2.8 mi/24 hrs	Quarterman et al., 1954b
	Mexico	Semirural	0.15 mi	0.15 mi		Greenberg et al., 1964
Cochliomyia hominivorax	Texas	Ranch		9 mi	9 mi/11 days	Parish, 1937
	Texas	Open country		1500 mi/season	35.4 mi/week	Barrett, 1937
	Texas	Ranch	<300 yds	1 mi		Hightower, 1963
	Texas	Rangeland		180 mi	180 mi/<2 wks	Hightower et al., 1965
Chrysomya albiceps	New So. Wales, Australia	Open range	4-8 mi	>10 mi	>10 mi/12 days	Gurney et al., 1926
Chrysomya rufifacies	Canberra, Australia	Pastoral		>4 mi	4 mi/24 hrs	Norris, 1959
Phormia regina	Texas	Rural		10.9 mi		Bishopp et al., 1921
	Oregon	Farmsteads		8 mi		Lindquist et al., 1951
	Oregon	Farmsteads		28 mi	15.5-16.5 mi/48 hrs	Yates et al., 1952
	W. Virginia	Urban	<5 mi	10.3 mi	10.3 mi/4 days	Schoof et al., 1953
	Mexico	Semirural	0.15 mi	1.4 mi	1.4 mi/24 hrs	Greenberg et al., 1964
Protophormia terraenovae	Leningrad	Garbage dump	2 km	5.6 km	2 km/1 day	Shura-Bura et al., 1956
	Leningrad	Rural		10.7 km	4 km/4 hrs	Shura-Bura et al., 1958
	Leningrad	Garbage dump		0.4-0.6 km		Shura-Bura et al., 1962

SPECIES	LOCATION	HABITAT	EFFECTIVE DISPERSAL RANGE	MAXIMUM DISPERSAL DISTANCE	APPROXIMATE DISPERSAL RATE	AUTHOR
Protocalliphora azurea	North Sea	Lightship		70 km		Lempke, 1962
Bufolucilia silvarum	North Sea	Lightship		70 km		Lempke, 1962
Lucilia caesar	Leningrad	Garbage dump		2 km		Shura-Bura et al, 1956
	Leningrad	Rural		6.2 km		Shura-Bura et al, 1958
	England	Tree belt and river		0.75 mi		MacLeod et al., 1960
Lucilia illustris	Texas	Rural		1.1 mi		Bishopp et al., 1921
Phaenicia cuprina	New So. Wales, Australia	Pastoral	2-4 mi	>4 mi		Gurney et al., 1926
	Canberra, Australia	Pastoral		4.7 mi	4.7 mi/< 30 hrs	Gilmour et al., 1946
	Canberra, Australia	Pastoral	<1 mi	>4 mi	4 mi/48 hrs	Norris, 1959
Phaenicia pallescens	Georgia	Urban		5 mi	5 mi/48 hrs	Quarterman et al., 1954b
Phaenicia sericata	Texas	Rural		1.2 mi		Bishopp et al., 1921
	Oregon	Farmsteads		4 mi		Lindquist et al., 1951
	Georgia	Urban		3.5 mi		Quarterman et al., 1954b

Species	Location	Habitat				Reference
	Australia	Pastoral		> 3.5 mi	3.5 mi/48 hrs	Norris, 1959
	Mexico	Semirural	0.15 mi	0.6 mi		Greenberg et al., 1964
Aldrichina grahami	Kyoto	Suburb in winter		400 m		Uemoto, 1960
Calliphora stygia	Canberra, Australia	Pastoral		> 4 mi	4 mi/24 hrs	Norris, 1959
Calliphora vicina	Leningrad	Rural		4.7 km	2 km/1 day	Shura-Bura et al., 1958
	England	Tree belt and river		0.75 mi		MacLeod et al., 1960
	North Sea	Lightship		70 km		Lempke, 1962
Calliphora vomitoria var. *uralensis*	Leningrad	Rural		8.6 km		Shura-Bura et al., 1958
Microcalliphora varipes	Canberra, Australia	Pastoral		> 3.5 mi	3.5 mi/48 hrs	Norris, 1959
Neopollenia stygia	New So. Wales, Australia	Open range	4-8 mi	> 8 mi	> 8 mi/17 days	Gurney et al., 1926
Anastellorhina augur	New So. Wales, Australia	Open range	4-8 mi	> 8 mi	> 8 mi/17 days	Gurney et al., 1926
	Canberra, Australia	Pastoral land		> 4 mi	4 mi/24 hrs	Norris, 1959

SARCOPHAGIDAE

Species	Location	Habitat				Reference
Ravinia sueta	Savannah, Georgia	Urban		2.1 mi		Quarterman et al., 1954b

may not suffice to make them stay put. It is frequently observed that urban flies move from substandard neighborhoods to neighborhoods with much higher levels of sanitation (Schoof and Mail, 1953; Greenberg and Bornstein, 1964). Thus, flies released in cities, criss-cross, back-track, and generally move about at random. In the country, reciprocal movements of house flies between farms occur. One restraint on free house fly movement is the severity of the intervening landscape. A study was made of insecticide resistance among relatively isolated poultry farms in California, which were separated by temperate to semiarid land and mountains rising to 2000 feet. It was found that the pattern of fly resistance was characteristic of the history of insecticide use on each farm. This suggests that there had been very little movement between farms and therefore little mixing of genes between fly populations (Georgiou, 1966).

Fly dispersal in open country may be canalized by features of terrain and microclimate, in addition to olfactory trails. In Texas, *C. hominivorax* dispersed along dry gullies, fence lines, and natural openings in brush (Hightower and Alley, 1963). At the height of the blow fly season, few *L. caesar* were caught in open situations; the majority were found among hedgerows that may have confined their movements (Cragg and Hobart, 1955). These observations merit further study since MacLeod and Donnelly (1958) released blow flies in the Scottish uplands and failed to find any canalization of dispersal by the contours of the hilly country.

When available, maximum flight and effective flight distance records have been included in Table 3. The difference between them hardly needs emphasis; it is the difference between the average performance of the "fly in the street" and the olympic performance of a few unusual individuals. We need to remember, however, that, in spite of assiduous efforts at trapping, only a very small percentage of released flies is ever recovered and that it is on the basis of this small sample that we characterize the flight potential and, indeed, the dispersal characteristics of the population. Perhaps remote sensing devices such as radar tracking will disclose an even greater propensity to disperse than has hitherto been recognized.

The first fly dispersal study was an unintentional one conducted in Cuba in 1898, during the Spanish American War. "The house flies," Vaughan (1900) remarks, "swarmed over infected fecal matter in the pits and then visited and fed on the food prepared for the soldiers at the mess tents. In some instances where lime had recently been sprinkled over the contents of the pits, flies with their feet whitened with lime were seen walking over the food." Fly traffic between kitchens and privies has since been well documented by the studies of the house fly, *Drosophila*, and other species by Shura-Bura (1952b) in the Crimea, Pimentel and Fay (1955) in Texas, Zaidenov (1960) in Chita, and Murvosh and Thaggard (1966) in the Bahamas. A rich source of pathogens was found in a rural Mexican slaughterhouse from which *Salmonella*-laden house flies and blow flies dispersed to a nearby outdoor food market, a dairy stable, and residential sections (Greenberg et al., 1963; Greenberg and Bornstein, 1964). The dispersal of house flies from a sewage pool that contained *S. enteritidis* to an army field kitchen three miles away (Peppler, 1944), and from a sewage plant (Gorodetskiĭ and Kuznetsov, 1937), documents other fly links with contaminated wastes. We have already referred to the reciprocal movement of house flies between a kitchen in an apartment building and the garbage outside (Derbeneva-Ukhova, 1947). Due to the proven flight propensity of flies, disease agents can readily circulate between well-traveled flyways among garbage dumps, animal farms, and carrion to dwellings, restaurants, kitchens, and markets.

ADDENDUM TO OVERWINTERING OF FLIES

There is a more extensive literature on overwintering of flies than our limited treatment of the subject has encompassed. For the interested reader, we include here a somewhat annotated bibliography of additional publications.

Ashworth, J. H.
  1916. A note on the hibernation of flies. Scottish Naturalist, Edinburgh, pp. 81-84.
Bishopp, F. C., Dove, W. E., and Parman, D. C.
  1915. Notes on certain points of economic importance in the biology of the house fly. Jour. Econ. Entom., *8*:466-474.
Blakitnaĭa, L. P.
  1962. On wintering and some other aspects of the biology and ecology of synanthropic flies in northern regions of the Kirghiz Republic. Med. Parazit. Moscow, *31*:424-429.
Bogoĭavlenskiĭ, N. A., and Prokopovich, K. V.
  1942. The blowfly *Calliphora erythrocephala* Mg. at Baku in winter. Med. Parazit. Moscow, *11*:133-134.
Bohart, R. M.
  1957. Five years of fly control on an agricultural university campus. Calif. Vector Views, *4*:23-26. Grigarick discovered that some house fly larvae overwinter in pockets where heat of decom-

position kept the temperature from falling below 65°F. However, no overwintering larvae were detected in poultry manure where the nature of the material allowed the temperature to drop to an observed low of 41°F.

Copeman, S. M.
    1913.   Hibernation of house-flies. Rept. Local Govt. Bd. Publ. Health Med. Subj. n.s. No. 85, pp. 14-19.
Cousin, G.
    1932.   Étude expérimentale de la diapause des insectes. Bull. Biol. France et Belgique, Suppl. 15, 341 pp.
Davies, W. M.
    1929.   Hibernation of *Lucilia sericata* Mg. Nature, *123*:759 (normally hibernate in larval stage in North Wales; others found hibernating pupae in France).
Davies, W. M.
    1930.   Parasitism in relation to pupation in *Lucilia sericata* Meig. Nature, *125*:779-780.
Fan, T. T., and Shi, T. C.
    1959.   The breeding habits of common flies in Shanghai district. Acta. Entom. Sinica, *9*:342-365 (majority of more important species hibernate as pupae or larvae and only *M. stabulans* hibernates as adult females).
Fedder, M. L., Smetlova, A. G., and Teterowskaya, T. O.
    1952.   Eclosion of flies (*Musca domestica*) indoors in towns during winter [in Russian]. Gigiene Sanit., Moscow, No. 4, p. 52.
Gorbacheva, Z. A.
    1957.   On the phenology and number of generations of several synanthropic flies in Tashkent. Med. Parazit. Moscow, Suppl. 1, p. 75.
Hallock, H. C.
    1940.   I. The Sarcophaginae and their allies in New York. Jour. New York Entom. Soc., *48*:127-153 (the pupa is the common overwintering stage, but mature larvae and younger larvae may overwinter in their hosts).
Hallock, H. C.
    1940.   II. The Sarcophaginae and their allies in New York. Jour. New York Entom. Soc., *48*:201-231.
Hanec, W.
    1956.   Investigations concerning overwintering of house flies in Manitoba. Canad. Entom., *88*:516-519.
Hewitt, C. G.
    1915.   Notes on the pupation of the house-fly (*Musca domestica*) and its mode of overwintering. Canad. Entom., *47*:73-78.

Hoelscher, C. E., Brazzel, J. R., and Combs, R. L.
   1967.   Over-wintering of the horn fly in north-east Mississippi. Jour.
           Econ. Entom., *60*:1175 (overwinters as diapausing pupa in
           soil and dung).
Hutchinson, R. H.
   1918.   Overwintering of the house-fly. Jour. Agric. Res., *13*:149-
           169.
Īatsenko, F. I., and Tishchenko, O. D.
   1932.   O zimovke v imaginal'nykh stadiīākh mukh i drugikh dvukry-
           lykh na Ukraine. [On the overwintering of the imago forms
           of flies and other Diptera in the Ukraine.] Zhur. Epidem.
           Mikrob., *3-4*:37-39.
Il'īāshenko, L. Īa.
   1963.   Hibernation of *Musca domestica vicina* in the Hissar Valley
           in Tadzhik SSR. Med. Parazit. Moscow, *32*:565-567 (hiber-
           nate as pupae, but freezing and excessive soil moisture kill
           most).
Jepson, F. P.
   1909.   The breeding of the common house fly (*Musca domestica*)
           during the winter months. Rept. Local Govt. Bd. Publ. Health
           Med. Subj. London, n.s., No. 5, p. 598; also Jour. Econ.
           Biol., *4*:78-82.
Kaver, V. A.
   1956.   Controlling hibernation of flies. Fel'd. Akush., *21*:32-33.
Kisliuck, M.
   1917.   Some winter observations of muscid flies. Ohio Jour. Sci.,
           17:285-294.
Kobayashi, H.
   1921.   Overwintering of flies. Japan Med. Wld., Tokyo, *1*:11-14.
Kobayashi, H.
   1922.   On the further notes of the overwintering of the house flies.
           Japan Med. Wld., Tokyo, *2*:193-196.
Lindquist, A. W., and Barrett, W. L.
   1945.   The overwintering of *Cochliomyia americana* at Uvalde,
           Texas. Jour. Econ. Entom., *35*:77-83.
Lobanov, A. M.
   1957.   Sites of wintering and the seasonal course of abundance of
           Hydrotaea dentipes Fil., Muscidae. Sborn. Nauch. Trudov,
           Ivanov. Gosudarstv. Med. Inst., *12*:464-466.
Lokshina, S. S., and Gorodetskiĭ, A. S.
   1956.   Presence of microbes of the intestinal group in hibernating
           flies. Dizenterii, Kiev, Gosmedizdat Ukr., SSR, pp. 242-244.

Lyon, H.
 1915. Does the housefly hibernate as a pupa? Psyche, *22*:140-141 (concludes not).

Matthew, D. L., and Dobson, R. C.
 1959. *Musca autumnalis* (DeGeer), a new livestock pest in Indiana. Proc. Indiana Acad. Sci., *65*:165-166.

Matthysse, J. G.
 1945. Observations on housefly overwintering. Jour. Econ. Entom., *38*:493-494 (most important means is continuous breeding in upper New York; barns full of flies all winter where barn temperatures are above freezing).

McDonnell, R. P., and Eastwood, T.
 1917. A note on the mode of existence of flies during winter. Jour. Roy. Army Med. Corps., *29*:98-100.

Mégnin, J. P.
 1894. La faune des cadavres. Encycl. Sci. Aide-Mémoire. Gauthier (Villars et Fils, Paris) 214 pp.

Mên, C., and Li, L.
 1950. Study of overwintering of houseflies in Chengtu. Hsinan Hsüeh [Hsinan Medical Studies], *1*:291-293.

Morgan, N. O., and Pickens, L. G.
 1967. Cold tolerance of adults and pupae of the face fly. Jour. Econ. Entom., *60*:1464-1466.

Nakata, G.
 1958. Studies on fly control in winter season, with special reference to the biology of overwintering fly pupae and their hymenopterous parasites. Sanit. Injurious Insects, Kyoto, *3* (Spec. No.):1-26.

Nroen, V. B., and Mohrig, W.
 1965. Zur Frage der Winteraktivitat von Dipteren in der Boden-streu. [On the question of the winter activity of Diptera in ground litter.] Deut. Entom., *12*:303-310.

Ogata, K.
 1959. Studies on the behavior of *Musca domestica vicina* and *Fannia canicularis* in winter season, as observed at outside and inside of a residence in Kawasaki city. Jap. Jour. Sanit. Zool., *10*:251-257.

Parrish, D. W., and Bickley, W. E.
 1966. The effect of temperature on fecundity and longevity of the black blow fly, *Phormia regina*. Jour. Econ. Entom., *59*:804-808.

Roubaud, E.
  1922.   Sommeil d'hiver cédant à l'hiver chez les larves et nymphes
          de muscides. Compt. Rend. Acad. Sci. Paris, *174*:964-966.
Roubaud, E.
  1927.   Sur l'hibernation de quelques mouches communes. Bull. Soc.
          Entom. France, No. 2, pp. 24-25.
Saccá, G.
  1954.   On the hibernation of *Musca domestica* in Central Italy.
          Rend. Ist. Superiore Sanitá, *17*:44-54 (adults in cold shelters
          survived maximum of 84 days, pre-imaginal stages, maximum
          of 90 days. Artificially heated setups e.g. stables, are most
          important in maintaining the species).
Salt, R. W.
  1963.   Delayed inoculation freezing of insects [*Cephus cinctus, Cal-
          liphora* sp., *Tenebrio molitor, Eurosta solidaginis*]. Canad.
          Entom., *95*:1190-1202.
Salt, R. W.
  1964.   Terrestrial animals in cold: Arthropods [Diptera, Homoptera,
          Lepidoptera, Orthoptera]. Handbook Physiol., *4*:349-355.
Salt, R. W.
  1964.   Trends and needs in the study of insect cold-hardiness. Canad.
          Entom., *96*:400 and on.
Sen, P.
  1938.   Overwintering of house fly *Musca domestica*. Indian Jour.
          Med. Res., *26*:535-536.
Shamsutdinov, N. K.
  1957.   On the overwintering of larvae and pupae of some synan-
          thropic flies in the conditions of Tashkent. Med. Parazit.,
          Moscow, *26*:79.
Shterngol'd, E. Ĩa.
  1939.   On the number of generations of *M. vicina* in Tashkent. In:
          Uzbeskiĭ Parazitologicheskiĭ Sbornik, *2*:377.
Skinner, H.
  1913.   How does the house-fly pass the winter? Entom. News, *24*:
          303-304 (observed wild teneral adults March 13 in Phila-
          dephia).
Skinner, H.
  1915.   How does the house-fly pass the winter? Entom. News, *26*:
          263-264.
Smetlova, A. G.
  1952.   Hibernation of *Calliphora uralensis*. Gigiene Sanit., Moscow,
          No. 8, p. 50.

Somme, L.
    1966.   Insekter i vinterdvale. [Insects in hibernation.] Fauna, *19*: 1-14.
Strong, L. A.
    1938.   Report of the Chief of the Bureau of Entomology and Plant Quarantine. U.S. Dept. of Agric., p. 60.
Tischler, W.
    1950.   Biozönotische Untersuchungen bei Hausfliegen. Zeitsch. Angew. Entom., *32*:195-207 (*Musca corvina, Muscina stabulans* and *Pollenia rudis* hibernate in houses).
Tishchenko, O. D.
    1929.   On the question of the hibernation of the house-fly (*Musca domestica*). [In Ukrainian] Sanit. Entom. Byul., No. 2, pp. 10-14 (hibernate in warm shelters; in cold shelters die before winter is over. Hibernating in cold cellars were *Muscina stabulans, Calliphora vicina, Fannia* sp. and *Drosophila* sp.).
Ushatinskaya, R. S.
    1957.   Principles of cold-hardiness in insects. [In Russian] Moscow, Inst. Morf. Zhiv. Severtsova, Akad. Nauk. SSSR, 314 pp.
Vanskaīā, R. A.
    1942.   Hibernation of *Musca domestica* L. Med. Parazit. Parazitar. Moscow, *11*:87-90.
Vladimirova, M. S.
    1941.   Sezonnoe raspredelenie i chislo generatsiĭ y mīāsnykh mukh. [Seasonal distribution and number of generations in carrion flies.] Med. Parazit. Parazitar. Moscow, *10*:548-560.
Wang, Y.-C.
    1965.   On the overwintering of house-flies (*Musca domestica vicina* Macq.) in Ch'engtu [in Chinese]. Acta. Entom. Sinica, *14*: 163-170 (survival of well-fed adults in semi-hibernation at cool temperatures for 94 days. Also active breeding in warmer shelters with breeding material).
Williston, S. W.
    1908.   Manual of North American Diptera. New Haven, Conn., 3rd edition, p. 141 (house fly hibernates as pupa, sometimes as adult).
Zhovtyi, I. F.
    1953.   Hibernation of the housefly *Musca domestica* L. [in Russian]. Med. Parazit. Parazitar. Moscow, *1*:58-62.

3
The Fly as Host

There is some soul of goodness in things evil,
Would men observingly distill it out.

SHAKESPEARE, *Henry V*

THE MOST significant transport of disease agents, in terms of numbers and persistence, usually occurs in the digestive tract. It is therefore necessary, at the outset, to familiarize ourselves with the physiology and anatomy of the tract as a basis for discussions of intake, survival, and output of pathogens. Later in the chapter we shall deal with various aspects of mechanical transport, or dissemination from the surface of the fly.

THE ADULT DIGESTIVE TRACT

PROBOSCIS

Two types of proboscis are found in higher Diptera—the lapping-sponging type of the house fly and the piercing type of the stable fly. Important contributions to our understanding of the functional anatomy of these types have been made by Kraepelin (1883), Lowne (1893-95), Hansen (1903), and Graham-Smith (1911a, 1930a, b); Patton and Cragg (1913) figure the mouthparts of intermediate forms.

The lapping proboscis is the more common type among muscoid flies, and this undoubtedly reflects the widespread habit of flower, dung, and fermentation feeding. In the higher flies the mandibles have been eliminated and the maxillae are reduced. The proboscis consists of three parts: proximally, a conical rostrum, which bears the maxillary palps; a middle piece called the haustellum, which contains the anteriorly grooved labium into which the labrum-epipharynx and hypopharynx fit to form the food channel; and distally, a pair of fleshy lobes, the labella, which taste, scrape, and filter the food. At rest, the haustellum folds onto the rostrum within a cavity of the head capsule, and only the labella and tips of the palps are exposed. Embedded in the tough, flexible membranes of the proboscis are sclerites that facilitate its extension, retraction, and other movements under the action of intrinsic muscles and blood pressure; the latter expands the labella (Fig. 15).

The tumescent labella expose a series of parallel gutters, called pseudotracheae, which traverse each lobe. Those which drain the anterior and posterior regions empty into their respective collecting channels, while those of the middle section empty directly into the prestomum. In all, there are about 50 pseudotracheae in each lobe, and they constitute a remarkable filtration system.

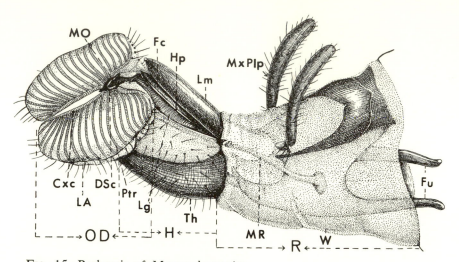

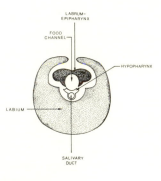

FIG. 15. Proboscis of *Musca domestica*. A. Frontolateral view. R, rostrum; H, haustellum; OD, oral disc; Cxc, main collecting channels of the pseudotracheae; DSc, discal sclerite; Fc, food channel between the hypopharynx (Hp) and the labrum (Lm); Fu, fulcrum seen in outline beneath the chitinous membrane (W); Hp, hypopharynx; LA, labella; MO, mouth opening; Lg, labial gutter; Lm, labrum; MR, maxillary rods (stipes); MxPlp, maxillary palp; Ptr, pseudotracheae; W, membrane of proboscis. (From Matheson, Medical Entomology, Comstock Publishing Co.). B. Cross-section. (Original.)

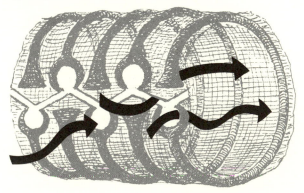

FIG. 16. Portion of a pseudotrachea of *Musca domestica*. The entire structure is invested by the labellar cuticle (cross-hatched) except for the interbifid spaces and the zigzag longitudinal fissure through which food may pass. (After Thomsen, 1938.)

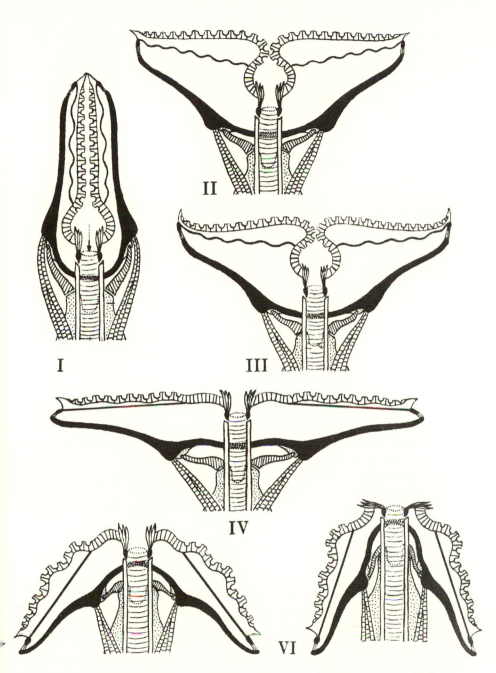

FIG. 17. Diagrammatic longitudinal sections of the labella of *Calliphora vicina* illustrating the attitudes assumed in the resting (I) and feeding positions. II. Filtering position. III. Cupping position. IV. Intermediate position with some exposure of the teeth. V. Scraping position. VI. Direct feeding position. Arrow in I shows mouth. (From Graham-Smith, 1930.)

Microscopic examination reveals that each pseudotrachea is supported by a large number of incomplete rings, which open on the oral surface of the labella. The ends of a ring face each other, a few microns apart; one end has the shape of a fish-tail, the other end forms a wish-bone. Fish-tail and wish-bone alternate along the length of the tube, giving way to simple incomplete rings near the prestomal teeth. The space between the facing elements makes a zigzag longitudinal fissure in the pseudotrachea (Fig. 16). The labellar cuticle dips into the pseudotrachea and is closely applied to the rings as well. It wraps around each fork of the bifid ring, leaving open an interbifid space to funnel fluid into the lumen of the tube. The interbifid space is 3 to 6 μ wide, varying with the location on the labellum and with the species of fly. Graham-Smith points out that effective entry into the pseudotracheae is through the interbifid spaces, since the zigzag fissures are quite narrow and are usually closed during feeding.

The rings provide a flexible "backbone" to accommodate the repertoire of labellar movements that various feeding situations elicit. For example, when feeding upon a thin film of liquid, the lobes are fully expanded and pressed against the surface; in droplet feeding, the lateral margins of the labella are cupped to embrace the contour of the drop; if scraping of the food is necessary, the lobes are curled upward by retractor muscles and the teeth are exposed; and finally, with further retraction of the lobes, the mouth is applied directly to the food. These feeding positions are illustrated in Figure 17. Larger particles, such as helminth eggs, are ingested with the direct feeding position. The largest particle that the house fly can swallow is about 40 μ; longer, narrower eggs are swallowed lengthwise, if at all. The particle size limit of *Sarcophaga* and *Calliphora* is approximately 10 μ greater, and that of *Fannia* 10 μ less, than that of the house fly; this information is summarized in Table 6. Because protozoan cysts are smaller than helminth eggs, they are ingested in larger numbers, and, when equal numbers of cysts of *Giardia intestinalis* and *Entamoeba histolytica* are present in feces, there is a selective uptake of the former by the house fly in the ratio of 50:1 (Jausion and Dekester, 1923). There also appears to be selective filtering of the vomit. When the fly regurgitates, protozoan cysts and helminth eggs are held back by the pseudotracheae, but trypanosomes and bacteria are not.

The fly's size influences the quantity of food it ingests and, theoretically, its microbial load. A single house fly which has not fed for 2 or 3 hours takes up 1 mgm of feces in half an hour (Wenyon and O'Connor, 1917b), or 0.016-0.029 ml of a 5% sucrose solution in 24 hours (Galun and Fraenkel, 1957; Greenberg, 1959a). The larger *Phormia regina* consumes slightly more (Dethier and Rhoades, 1954), and the consumption of *C. vicina*, a still larger fly, is about three times that of the house fly (Root, 1921). The intake of a fly, even under controlled laboratory conditions, is not a fixed quantity, and normally it will increase with increasing concentrations of sugar solution. Although we expect large flies to carry more microbes than small ones, so much depends on neighborhood level of contamination and habits of feeding and dispersal that generalization may be quite useless.

Food that passes into the pseudotracheae is drawn by action of the pharyngeal pump toward the collecting channels and then into the prestomal cavity, whose walls are the prestomal teeth and whose floor is the continuation of the pseudotracheal membrane. This passage continues as the food channel, formed between the labrum-epipharynx and hypopharynx, which connects with the pharynx.

In nonbiting flies, the prestomal teeth are extremely delicate and blade-like. There is a row of five such teeth arising from each arm of the discal sclerite in the muscids *M. domestica*, *F. canicularis*, and *Ophyra anthrax*. There are three rows in the calliphorids, *C. vicina* and *L. caesar*, and four rows in *Sarcophaga carnaria*. When scraping a surface, the blow fly can bring 150 teeth into action, a rather effective scarifier for the superficial inoculation of the skin, conjunctiva, or mucous membranes. Further development of the prestomal teeth was crucial to the evolution of the primary hematophage. By this term we distinguish secondary hematophages—those flies that cannot inflict a wound but profit as satellites of the others. In this country, *Fannia benjamini*, *Hydrotoea armipes*, and *H. tuberculata* have been seen feeding on blood from bites made by tabanids and stable flies, often crowding in on a horse fly and interrupting its meal. *Musca spectanda* behaves similarly in Nyasaland (Thomson and Lamborn, 1934). Such flies qualify as possible vectors of blood parasites (Garcia and Radovsky, 1962; Hammer, 1941; and Tashiro and Schwardt, 1953), though they cannot bite.

Siphunculina funicola and *Hippelates* spp., which transmit yaws and eye infections, have a lapping proboscis. The six pseudotracheae on each labellum empty directly into the prestomum and appear to remain permanently and widely open, allowing the entry of larger particles than in the blow fly. In *S. funicola*, there is no trace of prestomal teeth or an interdental armature, but their place is taken by elongated spines that rise above the inner surface of the labella; these spines appear to function as scarifiers. A similar enlargement of the proximal pseudotracheal rings in *H. pusio* provides it with a scarifying device (Graham-Smith, 1930a).

Musca crassirostris is a common pest of bovids in India. Superficially its proboscis resembles that of the house fly, yet it can cause blood to flow by scraping or jabbing its prestomal teeth into the skin. Closer inspection of its proboscis reveals a number of important modifications. The dental armature, for example, consists of four pairs of large, stout teeth with an interdental series of much finer blades—the homologues of the proximal pseudotracheal rings of the house fly (Fig. 18); the discal and other sclerites and associated muscles have been strengthened to handle the stronger thrust of the teeth, and the haustellum has been lengthened at the expense of the rostrum.

Additional modifications culminate in the true piercing proboscis of *Stomoxys* (Fig. 19). In this group, the rostrum is further shortened and the haustellum is lengthened mainly through enlargement of the mentum. The haustellum is strongly sclerotized and tapers to a point, while the labella are reduced and their pseudotracheae and labial glands are eliminated. Because of its length, the haustellum no longer fits into the head capsule and must be carried at-the-ready, bayonet fashion. In the act of feeding, the proboscis is directed downward and slightly forward, the labellar lobes are retracted, the teeth are everted and the tip of the labium is driven into the skin by thrusts of the head. The labial gutter is deeper than in *Musca* and the two elements of the food channel are more delicate. Blood is drawn directly into the channel, as in the direct feeding position of the house fly. Connection of the food channel to the pharynx is made through a short, sturdy tube, which stays open when the proboscis is flexed at the rostral-haustellar joint.

The proboscis of *Haematobia* and *Lyperosiops* is less evolved than that of *Stomoxys*. The haustellum is proportionately shorter, and pseudotracheae are still present on the medial surface of the labellar flaps to channel blood when the labella are distended (Fig. 20).

The Gut

The information in the following account is based largely on the comparative studies of Singh and Judd (1966), the monographic studies of *C. vicina* by Graham-Smith (1934) and of *M. domestica* by West (1951), and the analysis of *H. flavipes* (= *pallipes*) by Kumm (1935b).

The anatomy of the digestive tract of cyclorrhaphous flies shows little variation despite the range of feeding habits (Figs. 21 and 22). The pharynx continues dorsally in the head capsule and, as the esophagus, passes rearward through the brain into the prothorax. The crop duct arises at this level and proceeds posteriorly as a straight, slender tube until it terminates in the bilobed crop. This is a distensible, contractile sac that occupies the antero-ventral region of the abdomen. The crop,

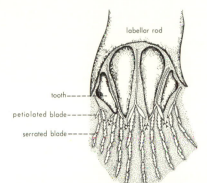

labellar rod

tooth

petiolated blade

serrated blade

F<small>IG</small>. 18. Distal portion of the proboscis of *Musca crassirostris* showing the four enlarged teeth and two types of smaller teeth, the serrated and petiolated blades. (Redrawn from Patton and Cragg, 1913.)

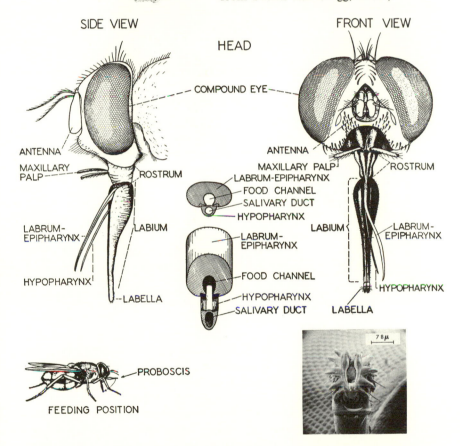

SIDE VIEW   FRONT VIEW

HEAD

COMPOUND EYE

ANTENNA
MAXILLARY PALP
ROSTRUM
LABRUM-EPIPHARYNX
FOOD CHANNEL
SALIVARY DUCT
HYPOPHARYNX
LABRUM-EPIPHARYNX
FOOD CHANNEL
HYPOPHARYNX
SALIVARY DUCT
LABRUM-EPIPHARYNX
LABIUM
HYPOPHARYNX
LABELLA

ANTENNA
MAXILLARY PALP
ROSTRUM
LABIUM
LABRUM-EPIPHARYNX
HYPOPHARYNX
LABELLA

7 8 μ

PROBOSCIS

FEEDING POSITION

F<small>IG</small>. 19. Details of head and proboscis of *Stomoxys calcitrans* (from James and Harwood, 1969). The scanning electron microscope photomicrograph shows the tip of the everted proboscis with its five pairs of teeth attached to the V-shaped discal sclerite which bounds the prestomum. (Original.)

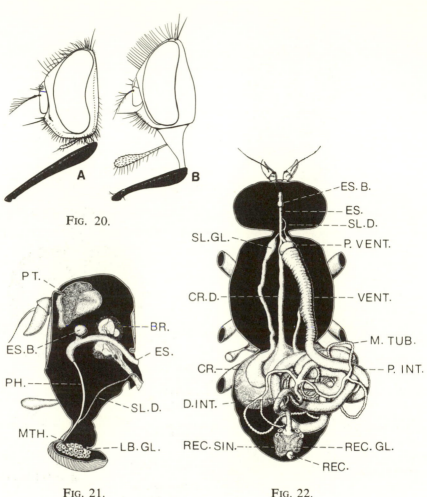

FIG. 20.

FIG. 21.   FIG. 22.

FIG. 20. Proboscis of *Stomoxys calcitrans* (A) and *Lyperosiops stimulans* (B). In the more advanced proboscis of *Stomoxys*, there is a lengthening of the haustellum at the expense of the rostrum.

FIG. 21. Sagittal view of the head of a cyclorrhaphous fly showing the relative positions of the pharynx (ph.), esophagus (es.), and salivary duct (sl. d.). Other structures shown are the mouth (mth.), labial gland (lb. gl.), brain (br.), and ptilinum (pt.).

FIG. 22. Digestive tract of a cyclorrhaphous fly showing esophagus (es.), salivary duct (sl. d.), salivary gland (sl. gl.), proventriculus (p. vent.), ventriculus (vent.), malpighian tubule (m. tub.), crop duct (cr. d.), crop (cr.), proximal intestine (p. int.), distal intestine (d. int.), rectal sinus (rec. sin.), rectal gland (rec. gl.), and rectum (rec.).

when engorged, is no longer bilobed. Graham-Smith describes four sphincters in the foregut:

1. an anterior esophageal sphincter where the esophagus arises that controls flow of regurgitated fluid into the proboscis;
2. a posterior esophageal sphincter at the level of the bifurcation of the crop duct;
3. a crop-duct sphincter; and
4. a proventricular sphincter.

When 1, 2, and 3 are open and 4 is closed, contraction of the crop forces vomit into the proboscis. When 2 is closed but 3 and 4 are open, the crop contents are directed into the ventriculus. The opening and closing of the valves have been investigated by Weismann (1964), who finds that they are under the mediation of pharyngeal receptors. When a fly feeds upon a weak sugar solution, e.g. 1%, the solution passes directly to the midgut, meaning that all valves are open except the crop-duct valve. A 5% sugar solution is shunted directly to the crop; in this case, all valves except the ventricular duct valve are open (see also Knight, 1962). The rate of crop emptying depends on blood osmotic level and is slower with a concentrated sugar meal. A meal of syrup may remain several days in the crop. Graham-Smith observed that blood does not coagulate in the crop of the blow fly *C. vicina*.

Beyond the crop duct juncture, the esophagus continues as a short tube into the proventriculus or cardia. This is the doughnut-shaped terminus of the foregut that secretes and molds the very fine peritrophic membrane that envelops the incoming meal. The proventriculus is thought to also act as a sphincter, controlling the passage of food and preventing regurgitation.

The midgut (anterior ventriculus, chyle stomach, or mesenteron) arises in the middle of the thorax and proceeds as a straight tube into the abdomen, where it coils and convolutes. The arrangement of the coils in *M. domestica*, *F. canicularis*, and *C. vicina* is similar, but there is less coiling in *Sarcophaga*. The posterior limit of the midgut is indicated by the entry of a pair of malpighian tubules that immediately branch to form a total of four long tubules. The blow fly midgut is divided into three pH regions—an anterior acid region (pH 5.0), a middle strongly acid region (pH 3.5), and a posterior alkaline part (pH 8.0). In nonbloodsucking adults, proteinase activity is weak, but carbohydrase activity is strong as compared with larvae; strong proteinases are retained in adult *Stomoxys* and other blood feeders. In addition to trypsin, common to many species, *Calliphora* females possess a proteinase with an in vitro pH optimum of 3, as do both sexes of the house fly.

No qualitative differences in proteinases appear to exist between the sexes, but quantitative differences have not been investigated. It is known that the fertilized female house fly has a greater protein intake than either the virgin female or males (Greenberg, 1959a).

The peritrophic membrane is secreted at the anterior end of the midgut. It is generally present in larval and adult Cyclorrhapha, but absent in adult *Hypoderma* and *Cuterebra*, which do not feed. In *M. domestica* and *C. vicina*, the membrane is readily permeable to amino acids, disaccharides, and smaller molecules but not to starch or proteins, e.g. albumin, casein, or gelatin. Intermediate products of digestion, e.g. dextrins and peptones, are retained by the denser internal layer, evidence that the membrane acts like a semiselective ultrafilter. It is impenetrable to the gut microflora and probably protects the insect from infection (Zhuzhikov, 1964). Stohler (1961) briefly considers the relation of the peritrophic membrane of mosquitoes, phlebotomines, and simuliids to their role as vectors of blood parasites. In these insects, the membrane restricts *Leishmania*, *Plasmodium*, *Trypanosoma*, and the microflora to the lumen, except where it is torn open in the hindgut, or where it is newly formed. It probably has much to do with variable susceptibility of tsetse to trypanosomes.

Leptomonas, which frequently infect the gut of adult *Drosophila*, are endoperitrophic and/or ectoperitrophic. They reach the latter location by escaping at the level of the rectal valve where the peritrophic membrane is destroyed and by re-invading the midgut between the membrane and the gut wall. This site seems to foster production of cysts that are better suited for the rigors of host-to-host transfer in the external environment than are the vegetative forms, which only develop endoperitrophically (Chatton and Leger, 1912).

The pyloric valve, generally not well developed, is the gateway to the hindgut. Halfway down the latter is the rectal valve mentioned above. The peritrophic membrane is broken up here by anterior and posterior groups of spined cells separated by a valve that pulls the membrane caudad across the spines. The rectal valve demarcates the anterior and posterior regions of the hindgut.

More prominent is the rectal sac, which bears four well-tracheated papillae facing the lumen. There is little question of their glandular function, and current evidence suggests that water, chloride, and probably other ions are absorbed here. Males of Culicidae and several brachycerous families are reported to have four papillae, while females have six. Sexual dimorphism in the arrangement of the papillae occurs in *Stomoxys*, *Pyrellia*, *Ophyra*, and *Lyperosiops*. Posterior to the rectal sac is the dilated rectum, which terminates in the anus.

106   THE FLY AS HOST

The length of the alimentary tract cannot be predicted from the size of the fly. In *M. domestica* it averages 36 mm, whereas in the larger *Muscina stabulans* it is 23 mm; in both *Calliphora vicina* and *Phormia regina*, it is 31 mm, but the former is a decidedly larger fly. It also is difficult to see any relation between diet and gut length in these examples. Sexual differences do exist in the ratio of gut length to body length, as in female and male *Lucilia*, where the ratios are 3.5:1 and 2.7:1, respectively. This difference presumably reflects the additional digestive activity of the female associated with egg production.

In terms of relative length, the midgut is definitely the longest segment, its percentage length ranging from 49 to 74%, compared with 9 to 22% for the foregut and 12 to 40% for the hindgut. These proportions obtain in a number of fly species.

Some flies have a phenomenal gut motility rate. Cysts of *Entamoeba histolytica* may be voided one minute (Sieyro, 1942) and five minutes (Roberts, 1947) after being swallowed by house flies. Living *Trichomonas* are passed in the house fly's feces five minutes after ingestion (Wenyon and O'Connor, 1917b), and a similar rate of passage of *Trypanosoma brucei* occurs in *Musca spectanda* (Thomson and Lamborn, 1934). However, the usual rate is somewhat longer, as it is for blow flies, and much depends on temperature and frequency of feeding. In the gnat *Hippelates flavipes*, which has fed on a blood clot, transit time is between 6 and 12 hours. In the case of such delicate organisms as *Treponema pertenue*, the agent of yaws, this sojourn is long enough to drastically reduce the number of organisms that can be fecally transmitted. But the fly maintains its transmission potential by depositing a large number of vomit droplets bearing viable treponemes. In one study a small number of flies averaged 30 spots per fly during the first four hours following a blood meal (Kumm, 1935a).

The nature of the food has an important bearing on quantity of vomit; in house flies, syrup produces a small number of droplets, sputum a large number, with milk intermediate (Fig. 23). When given one feeding of milk, house flies produced an average of 16 to 31 specks in 24 hours, most of which were vomit (Graham-Smith, 1931; also Wenyon and O'Connor, 1917b). *Musca spectanda*, an African species, averaged 27 droplets in 6 hours after feeding on an infected dog's blood; the droplets contained living trypanosomes (Thomson and Lamborn, 1934). This dangerous output can be expected to increase as the day warms up. Conversely, the house fly retains cysts of *Entamoeba histolytica* longer at lower temperatures (Sieyro, 1942).

Zimin (1944) estimates the average daily output of fecal and vomit spots by a single bazaar fly, *Musca sorbens*, at 80 to 240. Assuming a

FIG. 23. Glass from top of fly cage showing dark fecal spots and more numerous, lighter vomit spots.

sorbens population of 30 million in one Tadzhikistan market, he calculates that 7×10^9 fly specks are deposited every 24 hours on food, counters, etc. We may reduce his estimate several magnitudes without removing the spectre of a silent rain of contamination.

THE LARVAL DIGESTIVE TRACT

The gnathocephalon of cyclorrhaphous maggots has undergone extreme modification of the mouthparts with loss of such landmarks as the head capsule and antennae. The most prominent structures are the paired cephalic lobes, each bearing a dorsal, terminal, and ventral organ. Chu and Axtell (1971) describe receptors on the dorsal lobe whose morphology suggests that they mediate contact, olfactory, and mechanical stimuli. It is likely that the other lobes are also sensory. The gnathocephalon has about the same freedom of movement as the human head, with the added feature that the anterior segments can be telescoped, much like the ovipositor of the female. The mouth is located between the two cephalic lobes on the ventral surface, the center of a sunburst of fine grooves that probably channel food toward it (Fig. 24). Lacking the discal sclerites of the adult, the mouth can be stretched to engulf particles of epidemiologic importance. For instance, the house fly maggot can swallow, undamaged, the eggs of *Ascaris lumbricoides*, which measure 50-75 μ by 40-50 μ. The oft-made statement that maggots take only liquid food is not supported by the facts. Additional information relating to the transmission of helminth eggs is given in the following chapter.

The cephalopharyngeal armature differs in small details between species, but these differences together with features of the anterior and posterior spiracles, anal papillae, and cuticular spines are sufficient to provide a reliable diagnosis. The mouth armature of a number of maggots is depicted in Chapter 5 of Volume I, and a good many more are figured in the works of Hennig (1968), Zumpt (1965), Ishijima (1967), and James (1947). The cephalopharyngeal skeleton of the house fly maggot is illustrated in Figure 25.

The mouth hooks move vertically and, by grappling the substrate, aid the forward movement of the maggot. They also serve to tear or puncture the substrate, thereby hastening decomposition, as larvae of *Hylemya cilicrura* do to the surface of potatoes on which they feed. The homology of these structures remains unsettled after more than a century of controversy, some arguing that they have arisen *de novo* in cyclorrhaphous larvae, others tracing their origin to elements of the maxillae or mandibles (Menees, 1962). In *Musca* larvae, the mouth hooks are fused, with the line of fusion clearly discernible in *autumnalis*.

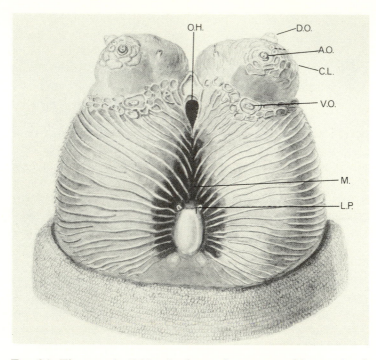

Fig. 24. The mouth (M.) of a house fly maggot with associated structures. Oral hook (O.H.), dorsal organ (D.O.), anterior organ (A.O.), ventral organ (V.O.), cephalic lobe (C.L.), and lingual process (L.P.). (Relabeled from Thomsen, 1938.)

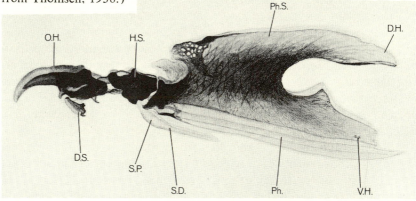

Fig. 25. Cephalopharyngeal armature of a house fly maggot. Oral hook (O.H.), dental sclerite (D.S.), hypopharyngeal sclerite (H.S.), salivary pump and duct (S.P., S.D.), pharyngeal sclerite (Ph.S.), pharynx (Ph.), and dorsal and ventral horns (D.H., V.H.). (Relabeled from Thomsen, 1938.)

110   THE FLY AS HOST

The hypostomal and pharyngeal sclerites provide structural support for the pharynx and the salivary duct, besides attachment for gnathocephalic muscles. Photoreceptors, housed within the anterior curvature of the pharyngeal sclerites, mediate the negative phototaxis of maggots (Bolwig, 1946).

The interior space of the maggot is traversed by loops and coils of gut that are bound together by tracheae. In *C. vicina*, a maggot 1 cm long has a tract about seven times longer, and in *Phaenicia* the tract is five times the maggot's length. The larval gut is much longer than the adult's, which is consistent with the maggot's primary function of food conversion for energy storage and growth. Average values for the lengths of the three major segments of the gut in *C. vicina* are: foregut, 5 mm; midgut, 45 mm; and hindgut, 23 mm.

Anatomically, little distinguishes the gut of one species from another, except for the presence of a crop in calliphorids and sarcophagids and its absence from muscids, generally. The crop is particularly useful to those maggots that liquefy their substrate, for they engorge beyond gut capacity and their crop swells with the excess. At peak of feeding the crop resembles a sausage, dominating the anterior half of the maggot and clearly visible under its skin (Fig. 26). As in the adult, the muscular crop wall squeezes fluid into the proventricular duct. We have seen no mention of crop, esophageal, or proventricular duct valves in the maggot and wonder how crop fluid is ushered toward the midgut rather than toward the mouth. In the prepupa, two days after feeding has ceased, the crop is a collapsed, inconspicuous sac.

The proventricular duct passes through the trilobed brain and Weismann's ring gland (containing the corpora allatum and cardiacum) and enters the bulbous proventriculus. The latter produces the peritrophic membrane, which is generally present in cyclorrhaphous larvae including the house fly in which it is erroneously reported to be absent. Although the membrane is difficult to see in the gut, it is easily demonstrated by puncturing the gut wall of an actively feeding specimen whereupon the contents ooze out, confined within the transparent membrane (Fig. 26).

The midgut is morphologically the most differentiated and physiologically the most active region. It has three distinct pH zones, whose average lengths in *C. vicina* are: fore-, 15 mm; mid-, 11 mm; and hind-, 19 mm. These zones seem to be strongly buffered since their pH is little influenced by the pH of the food or by the pH of the preceding zone. Thus, if one places a group of *C. vicina* maggots in sour milk overnight, and another group in a mouse cadaver, and then isolates the regions of digestive tract for micropotentiometric measurements, he will observe the following. In the group that fed on the mouse, the pH of the mouse and of the crop will be about 7.70 and the pH of the fore-midgut will be about 7.30. In the group that fed on the sour milk, the pH of the milk and of the crop will be about 5.40, while the pH of the fore-midgut will be 7.10, a lowering of only 0.2 of a pH unit. The mid-midgut of some blow fly maggots reaches an acidity of pH 2.8, but generally it is slightly higher. As Table 4 indicates, the acidity of this zone differs among species with the same diet. In general, the pH of the fore- and hind-midgut of carrion feeders hovers near neutrality, and is slightly on the acid side in the house fly. pH values of the gut are the same in germ-free maggots, showing that the gut microflora does not influence pH. The temperature at which maggots are reared also has little effect (Greenberg, 1968a). Besides the pH zones, the midgut of some blow flies is divided into five zones that are distinguished histochemically as specific ion absorption sites.

Secretion of proteinases and probably lipase takes place in the fore- and mid-midgut, while absorption occurs along the entire midgut. Carbohydrases are either weak or absent; therefore, addition of sugars to the medium does not improve larval growth. In addition to trypsin and cathepsin, the proteinases include a pepsin-like enzyme in *Musca*, *Stomoxys*, and blow flies with an in vitro optimum at pH about 3.0 (Greenberg and Paretsky, 1955; Lambremont et al., 1959; Fraser et al., 1961); the latter have shown that the source of this enzyme is the acid midgut. There is evidence that the trypsin is a complex of at least two enzymes, including chymotrypsin, which differ from mammalian trypsin (Patterson and Fisk, 1958). Perhaps the pepsin-like enzyme is also a complex, but this has not been investigated. Collagenase is secreted by carrion-feeding maggots and enables them to attack connective and other structural tissues of a carcass.

The pyloric flexure just behind the entry of the malpighian tubes marks the beginning of the hindgut. The anterior section of hindgut is distinctly narrower than either the midgut or the rectum. The hindgut re-absorbs water and probably certain ions. It has too little fluid for micropotentiometric pH analysis, but indicator dyes show that it is basic, sometimes reaching a pH of 8.5. The hindgut lacks a rectal valve for the destruction of the peritrophic membrane; instead, it is torn away as the maggot moves through the medium. The anus opens ventrally on the last or 13th segment.

Motility is an important feature of the digestive tract, for, as we shall see later, this has a direct bearing on the colonization of the gut by microorganisms. The flow of food in the foregut is rapid, as it is in man's

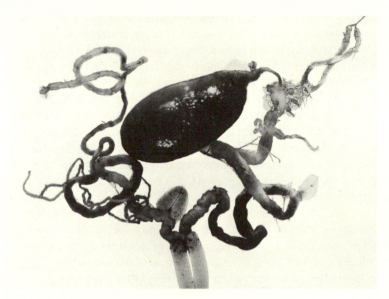

Fig. 26. Photograph of the digestive tract of third instar larva of *Calliphora vicina*.

TABLE 4

pH Values of the Maggot Midgut
Determined by Micropotentiometer*

Species	Fore	Mid	Hind
Calliphora vicina	7.33-7.37	2.80-3.20	7.08-7.48
Phormia regina	6.71-7.01	3.20-4.08	7.24-7.59
Sarcophaga bullata	6.79	3.11-3.33	7.00
Musca domestica	6.69	3.40-3.90	6.29

*From Greenberg, 1968a

esophagus, taking less than a minute to reach the midgut. Food may, however, be diverted to the crop and remain sequestered there for several days. Motility in the rest of the tract varies only slightly from one region to the next. The average flow rate is 1 to 2 mm/minute in the gut of a blow fly maggot actively feeding at 23°C. It is roughly the same in an adult blow fly, but much faster in the adult house fly, as we have already mentioned. Although ambient temperature has little effect on gut motility in homoiothermic animals, it has a pronounced effect in cold-blooded insects. For example, the transit time of carbon particles from mouth to anus in a mature maggot of *C. vicina* decreases from 65 minutes at 23°C, to 20 minutes at 31°C. Ambient temperature therefore determines the duration of exposure of microbes to various factors in the gut, and this may influence their survival and multiplication.

Sweep-out rate is also important. A population of microorganisms remains constant in a favorable segment of gut if its doubling time equals the sweep-out rate, but it is gradually washed out when its doubling time lags behind the sweep-out rate. The latter term connotes a combined effect of volume and rate of flow. One has only to compare the dissected tracts of a maggot and adult to appreciate the enormous difference in gut volume between the two, but how these factors affect microbial colonization is not known. Microecology of the gut—invertebrate and vertebrate—is a vast field still in its infancy. Modeling with the chemostat, or continuous culture method, and with gnotobiotic animals has reduced the number of variables and provided useful information. Later, we shall discuss the application of gnotobiology to the study of microbial survival in flies.

EMPUSA DISEASE OF FLIES

The first century of research on *Empusa* disease of flies previews early microbiology inasmuch as it marks the emergence of critical experimentation during a time when the causes of disease were still rooted in miasmas, humours, and other folklore. Decades before it was conceived for higher animals, the germ theory of disease was applied to the fly. In pursuit of the causes of this obscure fly disease, the first insect pathologists developed the same rigorous criteria for establishing etiology later used by Pasteur and Koch, even anticipating Koch's postulates. The obvious virtue in studying a disease of insects was that experiments could be pursued relatively free of social restraint and bias. Such study in the 19th century was well tuned to the technical means at hand, particularly to the contemporary microscope, because the pathogen is fortunately large, though the host is relatively small. Unfortunately, however intriguing these insect diseases, they could hardly have been expected to cast light on man's own afflictions. Only with the extension of Pasteur's work on diseases of wine and silkworms to those of higher animals and man could a framework develop for a unitary concept of infectious disease.

In 1782, DeGeer first noted the external symptoms of *Empusa* disease, supposing it was due to a poison ingested with the fly's food. Goethe described it as a destructive reduction-to-dust of the insect, which begins about one day after its death. The "dust" comes out of the pores on the sides of the abdomen and continues to disperse with increasing distance for four to five days, forming a white halo about an inch in diameter around the fly, which is stuck, lifelike, to the wall.

In 1827, Nees von Esenbeck first used the microscope to examine a fly dead of the disease and recorded certain accurate observations. He described the dust expelled from the fly as tiny pellets of basically dodecahedral form. He logically traced the origin of the dust (spores) to tubes (hyphae) that he found at the white rings (conjunctivae) of the fly abdomen, and attributed the distention of the fly body to fermentation. He failed to link the disease of the fly to the organism he observed.

Duméril (1835) first definitely considered the organism to be a mold that had probably caused the death of the fly. Follin and Laboulbène in 1848 described a white fungus mass they found on living flies and to which they ascribed the death of the flies. In these early observations, there is inevitably a question of the specific identity of the fungus, but the stage was now set for the monumental studies of Cohn and Brefeld.

In 1855, Cohn described in detail the course of the disease, demonstrating changes in the microscopic appearance of the blood and the development of mycelia from "innumerable free cells" before and after the fly's death. He noted that eight to ten hours after death, clubbed hyphae pierce the intersegmental membranes of the abdomen and appear on the exterior, forming white zones that continue to increase in width. Spores develop from the tips of the hyphae, and these are projected to form a dusty corona around the fly. Cohn anticipated Koch's postulates by carefully pointing out that since "It has not yet been possible to obtain germination of the spores, whether in water, in moist air, by external attachment [to flies], or by artificially inoculating them into living flies . . . it is therefore in no wise possible at present to demonstrate any influence of the *Empusa* spores on the appearance [in the fly] of this fungus and of the disease." This difficulty led Cohn to the unfortunate conclusion that the fungus is the product not the cause of the disease. He named the fungus *Empusa muscae*. Lebert, the following year, avoided Cohn's dilemma by erroneously supposing that spores of another fungus reached the blood through the tracheae or the gut and developed into the fly fungus by gradual transition.

For the next ten years, confusion was stirred by efforts to culture the fungus from dead flies. Brefeld (1873) characterized these efforts: "We would have as much right to derive the [various] fungi which develop from a dead fly from the specific fly fungus present in it as to assume a genetic connection between the whole gamut of weeds in a grain field and the rye seed with which it had been sown." Brefeld devised various media for in vitro culture of the spores, with negative results. Out of these and his other studies came the pure culture method, so important to the development of microbiology. Profiting from his experience with a related fungus disease of the imported cabbage worm *Pieris rapae*, he decided to place sick flies in a small container with healthy flies under conditions where "spores could not fail to be thrown against their abdomens, whether they walked around or over their dead comrades. After 24 hours, they were released from their close confinement and put into a larger vessel—the hospital—where they could fly about freely and enjoy a softened prune which remained free of mold for a long while. Each experiment was controlled by a like number of non-infected flies from the same source. Of the first three series, involving about 60 flies in all, two-thirds succumbed to the fungus disease on the 7th, 8th, or 9th day, the survivors being mostly small males. In the control cage, all were in good health: hence no doubt that the disease was spread by infection."

Brefeld went on to describe penetration of the spores into the fly's body, their residence in the fat body and release into the hemolymph, the development of mycelia, and the escape of the spores. He correctly attributed the higher attack rate in female flies to their larger fat body. In considering how the fungus overwinters, he ruled out transmission from diseased flies to maggots and pupae. He was able, instead, to transmit the disease from sick to healthy adults until a February cold spell terminated both flies and experiments. His final discussion is surprisingly contemporary:

"The fungus exists throughout the year in living flies. It disappears in the fall with the flies it inhabits; with these it retreats to the overwintering location. Under the influence of winter, perhaps also of the decreased life-intensity of the fly, speed of development decreases. At the beginning of spring the number of flies is markedly reduced, and from the few survivors the disease again gradually spreads. In the summer the flies live outdoors, and there the fungus escapes observation, all the more so as it is not yet very widespread. Only with the fall, with the arrival of fruit and the increasing chill of the night do the flies seek out human dwellings, and now the fungus becomes apparent, its increase over time and the increased ease of infection as the fly density increases in close quarters [combining] to produce the epidemic. Yet another and even more important circumstance explains the limitation of the disease in summer and its more rapid spread in the fall: the dryness of summer hinders infection, and the humidity of autumn facilitates it. Last summer, I found in the railway waiting rooms at Halle the fungus epidemic already widespread among the flies in July, so that it must have broken out much earlier . . . the exclusive early outbreak here, while elsewhere nothing was yet to be seen of the disease, will not appear surprising if we recall the overwintering of flies in the cellars of the building."

MAGGOTS AND MICROBES

Synanthropic flies begin life in a milieu teeming with microorganisms. Although blow fly maggots develop equally well on germ-free or contaminated meat, house fly maggots quickly die in their standard rearing medium (CSMA) if microbes are excluded (Greenberg, 1954). That microorganisms provide nutrilites essential to the maggots' development can be shown by adding a powdered concentrate of a common fecal organism, e.g. *Escherichia coli*, to a nonnutritive base and introducing disinfected fly eggs. If aseptic conditions are maintained, it can be shown that, within limits, the growth of the maggots is linear with respect to the quantity of *E. coli* "seasoning" added (Levinson, 1960). In the absence of bacteria, maggots of the house fly and *Protophormia terraenovae* do not grow on blood agar slants that contain 25% beef blood, 38.5% beef-peptone broth, 35% yeast extract, and 1.5% agar; Radvan (1960a) does not identify the limiting factor. It is still not clear whether, in a natural medium, e.g. feces or decomposing plant material, house fly maggots subsist largely on the microflora, on the vitamins that they provide, or on the decomposition products of the medium. We suspect it is probably all three, with dependence on microorganisms being more pronounced in an inadequate medium (Chang and Wang, 1958).

Other questions concerning the larval diet may not be associated with microbes. For example, in nature, maggots of *S. calcitrans* are usually limited to manure mixed with straw or hay, yet in the laboratory they grow well in CSMA. Are they excluded from manure by competing house flies and, if so, what advantage is gained when manure is mixed with straw?

Present evidence suggests that microbes play a specific role only in the nutrition of larvae. Germ-free house flies have normal longevity (Greenberg and Burkman, 1963), but studies are needed on the effect of germ-free life on a fly's reproductive capacity.

Fate of Microorganisms in the Developing Fly

To know the fate of pathogenic agents ingested by the maggot we begin with the fate of the many millions of nonpathogenic organisms swallowed in the normal course of feeding. The digestive tract of an actively feeding 3rd instar blow fly maggot contains 10^8 to 10^{10} microbes; in house flies it is about 10^7. House fly maggots breeding in manure, garbage, and privies have similar numbers of organisms. These are distributed throughout the tract, but the largest number is lodged in the crop, simply because it has the greatest capacity. As the pH of the crop reflects that of the larval medium so does its bacterial count, until the maggot stops feeding and enters the prepupal stage. Then the contents of the crop and its microbial cargo diminish rapidly. This reduction occurs throughout the tract as the last meal is absorbed or eliminated and the gut constricts. Figure 29 shows clearly that the greatest loss of microorganisms during the life of a fly occurs at this time.

The next antimicrobial mechanism is the molt, which transforms the larva into a pupa inside the puparium. The highlights of this process are depicted in Figure 28. The intima that lines the foregut and hindgut is continuous with the integument and is shown in red. The organisms remain in the intima when it is pulled out of the gut and deposited against the interior wall of the puparium at the time the pupa appears.

Fig. 27. Life stages of *Musca domestica* (Greenberg, 1965, courtesy Scientific American).

Microorganisms left in the midgut now face other adversities. The gut epithelium of the larva is destroyed by the pupal epithelium, which develops from cells of the anterior and posterior imaginal rings where the mesenteron joins the fore- and hindgut. It is noteworthy that the degenerating larval cells of these two zones are cast off into the body cavity and phagocytized, while the epithelium of the midgut is discarded into the lumen to become the "yellow body." More than a century ago, Weismann aptly described this body as floating in a honey-like mass. According to Perez (1910), the "yellow body" forms about 48 hours after the appearance of the white pupa in *C. vicina* (given a total pupal period of 18 days). The "yellow body" is only partly digested by the pupal midgut epithelium; the remainder is voided with the meconium at eclosion (Deegener, 1904; and Kowalewsky, 1887). At this time, a few drops of clear fluid are also passed.

Balzam (1937) asserts that the gut flora is released into the hemocele during metamorphosis and is phagocytized a day or two after eclosion, but the facts do not support this. Further, of 52 newly emerged house flies we tested, the dejecta from 37 flies contained no bacteria; that from another 10 flies contained less than 10 organisms per droplet; and the remainder ranged from 100 to 10,000 bacteria per droplet (Greenberg, 1959e). At the moment an emerging fly wrestles free of the puparium it may actually be sterile.

THE FLY AS HOST 119

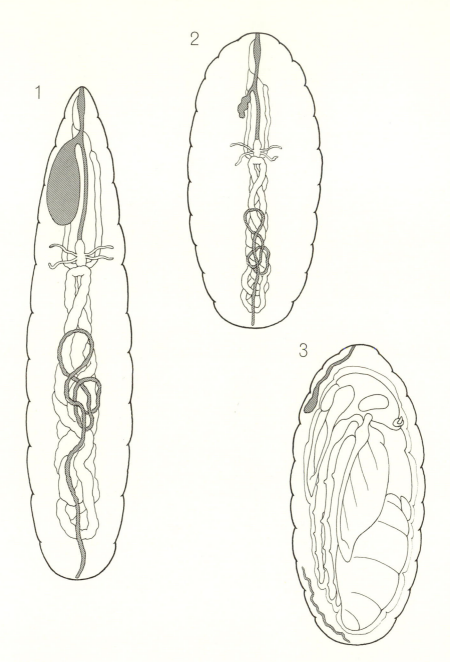

FIG. 28. The linings of the foregut and hindgut are shown (pattern area) as they occur in the maggot (1), early pupa (2), and in a pupa halfway through metamorphosis (3). (Greenberg, 1965, courtesy Scientific American.)

Autosterilization is extensive in some species, amounting to 54% of teneral stable flies, an even higher percentage of horn flies, and about 20% of house flies. These are rather impressive figures when we consider the huge cargo of microbes that has to be jettisoned in such a brief period. Pockets of survivors can usually be found on the inner wall of the vacated puparium, but these are probably beyond reach of the eclosing fly, otherwise it would routinely recontaminate itself. Autosterilization, however, is not a consistent process, and, while some flies are sterile, others from the same batch retain up to several million viable bacteria. Nothing is known that explains the emergence of saints and sinners among flies, but the phenomenon is widespread among Diptera. It has been observed by ourselves and others (Radvan, 1960b) in over a dozen muscoid flies, and it also occurs in midges and mosquitoes.

MICROBIAL INTERACTIONS

The foregoing events do not give a full story of how the fly gets rid of its juvenile load of microbes. Some 60 years ago, Graham-Smith and others made the interesting discovery that, although they virtually drenched house fly maggots with *Salmonella* or *Shigella*, adults emerged without these types, provided the surface of the puparium was first disinfected to avoid recontamination. This observation has been confirmed by a number of reliable studies involving the following bacterial pathogens and flies: *Salmonella typhi, Sal. enteritidis, Serratia marcescens* and *Vibrio comma* in *C. vicina* and *M. domestica* (Graham-Smith, 1911b, 1912); *S. typhi* and *Shigella dysenteriae* in *Sarcophaga carnaria, Lucilia caesar, Cynomya mortuorum* and *M. domestica* (Krontowski, 1913); *S. typhimurium* and *S. dysenteriae* in unspecified calliphorids and sarcophagids (Takeuchi et al., 1966), and in *P. regina* (Knuckles, 1959); *Pasteurella pestis* in the "green blow fly" (Lang, 1940); *Bacillus anthracis* (vegetative), *Mycobacterium tuberculosis, S. typhi* and *Shigella* sp. in *Calliphora* sp., *Lucilia* sp., and *M. domestica* (Wollman, 1921, 1927); and Bacot (1911a, b), Ledingham (1911), and Muzzarelli (1925), for various bacteria.

With regard to viruses, several investigators have reported the complete destruction of Coxsackie and poliovirus prior to or during metamorphosis. For details the reader is referred to the sections dealing with these diseases in Chapter 4. When Sindbis virus is injected into the hemocele of *C. vicina* larvae, thereby bypassing the gut, it will persist into the adult stage (Berkaloff et al., 1967).

Available information indicates that Protozoa are also eliminated before the adult emerges. In *Sphaerocera curvipes* and other borborines infected with *Leptomonas legerorum*, the organisms are confined mostly to the midgut within the peritrophic membrane and are destroyed in the "yellow body" during metamorphosis (Chatton, 1912). *Drosophila* rid themselves of *Octosporea* in the same manner (Chatton and Krempf, 1911).

Not unexpectedly, we find reports that contradict the above and demonstrate that bacteria and other agents occasionally cross the bridge between larva and adult. The following results appear to have been adequately controlled and are worth mentioning: *P. pestis* in *Calliphora vomitoria*, *Chrysomya macellaria*, and *M. domestica* (apparently killing the flies 15-24 hours after emergence) (Gosio, 1925); *P. pestis*, *V. comma*, *S. typhi*, and *M. tuberculosis* in *S. carnaria* (Petragnani, 1925); *S. pullorum* in *M. domestica* (in which day-old chicks became infected by eating infected flies and the flies remained infected 5 days following emergence; Gwatkin and Mitchell, 1944). In another example, anthrax spores survived two days after eclosion and for weeks in the bodies of dead blow flies and house flies (Graham-Smith, 1911b, 1912).

Trawińska and Trawińska (1958) claim that *Salmonella* sp. can be transmitted from infected adult house flies to their progeny. In one experiment, 6% infected adult offspring were obtained; in another, flies were fed swine fever virus, and larvae and pupae that developed from these flies were inoculated into two normal pigs with positive results. These interesting and unusual findings await confirmation. It is interesting to note in passing that Sindbis virus maintains a high-titered, symptomless infection when inoculated into adult *Drosophila melanogaster*, but the virus is not transmitted to the offspring (Ohanessian and Echalier, 1967a).

Protozoa of the genus *Herpetomonas*, which normally infect the gut of flies, survive metamorphosis in *C. vicina* and may even multiply in the preadult stages, according to Perez (1910). His figure 38 shows two spherical masses of organisms enclosed in the peritrophic membrane of a pharate adult.

Finally, certain worms are well adapted to life in the fly, evidenced by their ability to develop in the maggot, survive metamorphosis, and utilize the adult as a vector. The section on Helminth Diseases deals in detail with these species.

Returning to human bacterial pathogens, it has been confirmed many times that their survival through metamorphosis is assured only in the absence of other microorganisms (Tebbutt, 1913; Wollman, 1921, 1927; Ostrolenk and Welch, 1942b; Gerberich, 1951a, b, 1952; Greenberg, 1959b; and Radvan, 1960a).

Our first experiments with flies focused on this aspect of selective elimination. By the time we had confirmed the observations of our predecessors, we were convinced that the traditional qualitative approach had only limited value since it gave no inkling of numbers of bacterial survivors. For example, positive adults might harbor one *Salmonella* or ten million and we could not distinguish them. We therefore developed test strains of *Salmonella typhimurium* and *enteritidis* that were resistant to streptomycin and could grow without inhibition in streptomycin-agar media, which suppressed all other organisms. By using normal media in conjunction with streptomycin-media, we could compare counts of *Salmonella* and the normal flora. Before proceeding, we assured ourselves that the test organisms retained normal virulence for mice and guinea pigs, and were biochemically typical, despite their antibiotic resistance. A typical experiment is shown in Figure 29, in which eggs of *C. vicina* were planted on the carcass of a guinea pig that had died of infection with *S. enteritidis*. As each fly stage developed, a sample was removed, disinfected externally, and bacteriologically examined. This method, supported with adequate controls, assured us that the bacteria we reported were housed in the digestive tract.

Two observations emerged from experiments using blow flies as models: the *Salmonella* population is always considerably lower than that of the normal flora; and when feeding stops, both populations decline sharply—the normal flora from about 20 billion to 8 million, and *Salmonella* from about 4 million to 0. *Salmonella* is eliminated before onset of metamorphosis, while the normal flora persists to the adult stage. The fate of *Salmonella* and a normal flora in the life cycle of a fly are seen at a glance in Figure 29.

What kills *Salmonella*? This question led to several years of study of the larval and prepupal gut, and of gnotobiotic flies. Turning our attention first to the latter, we introduced *S. typhimurium* into a flask that contained aseptic blow fly maggots (the technique is described under gnotobiotic methods later in this chapter). Sampling the various stages as before, we found the pathogens' survival in the insect gut was much improved when other organisms were absent. Whatever inimical intrinsic gut factors may have been operating against *Salmonella*, they were not as overriding as competition between microorganisms themselves (Fig. 30).

The problem now was to single out the key participants from among the many species of microbes in the gut. Again we used the gnotobiotic approach, with the fly gut as a microarena in which candidate bacteria and *S. typhimurium* were pitted against each other. In three separate studies of dibiotic interactions, we found that *S. typhimurium* dominated *Streptococcus faecalis*, was dominated by *Proteus mirabilis* (a predominant member of the normal flora), and remained at parity with *Esche-*

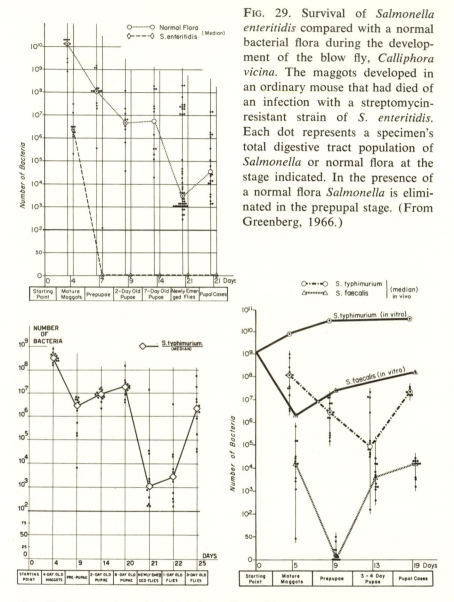

FIG. 29. Survival of *Salmonella enteritidis* compared with a normal bacterial flora during the development of the blow fly, *Calliphora vicina*. The maggots developed in an ordinary mouse that had died of an infection with a streptomycin-resistant strain of *S. enteritidis*. Each dot represents a specimen's total digestive tract population of *Salmonella* or normal flora at the stage indicated. In the presence of a normal flora *Salmonella* is eliminated in the prepupal stage. (From Greenberg, 1966.)

FIG. 30. Improved survival of *Salmonella typhimurium* in the monocontaminated blow fly. (From Greenberg and Miggiano, 1963.)

FIG. 31. Dibiotic interaction between *Salmonella typhimurium* and *Streptococcus faecalis* in the blow fly, *Calliphora vicina*, and in a static broth culture demonstrates inhibition of *S. faecalis*. (From Greenberg, 1969.)

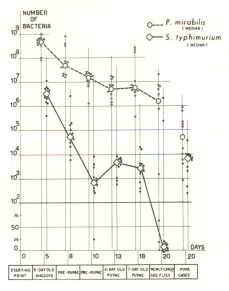

Fig. 32. Dibiotic interaction between *Salmonella typhimurium* and *Proteus mirabilis* in *C. vicina* shows a marked suppression of the former. (From Greenberg and Miggiano, 1963.)

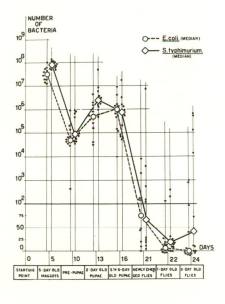

Fig. 33. Dibiotic interaction of *Salmonella typhimurium* and *Escherichia coli* in *C. vicina* shows both organisms on a par throughout the life cycle of the fly. (From Greenberg and Miggiano, 1963.)

richia coli. These experiments are summarized in Figures 31, 32, and 33. In none of these trials was *S. typhimurium* eliminated, in contrast to its routine elimination in the presence of a normal flora. Against *Proteus*, however, it fared poorly, and many flies emerged free of *Salmonella* while retaining millions of *Proteus*. The suppression ratio reached a peak in the prepupa, where we recorded 11,500 *Proteus* for each *Salmonella*. This is especially interesting because the maximum ratio of the two organisms when they are grown together in broth is only 20 to 1. This indicates that the maggot's gut does contribute to *Salmonella*'s suppression, but, before discussing this aspect, just a few more words about the response of *S. typhimurium* to other bacteria.

Dibiotically,* *S. faecalis* is overgrown by *S. typhimurium* in the fly and in a broth. The overgrowth amounts to a 4-log difference, which is maintained from the larval stage to the adult. The effect of adding *P. mirabilis* to this interaction is to suppress *typhimurium* even more than in the *mirabilis-typhimurium* association. As a result, the small numbers of *typhimurium* that persist in a few teneral adults are eliminated in the first few days of adult life. Others of this trio are also disadvantaged by the trifloral interaction—*S. faecalis* is sharply reduced and *P. mirabilis* loses ground. Triple interactions consistently brought the pathogen closer to extinction than did any of the dibiotic interactions.

Competitive exclusion of pathogens probably begins with the different microbial growth rates in the larval medium. In other words, the numbers of each organism swallowed by the maggot have much to do with that species' subsequent survival. Doubling times for some of our test organisms under optimum pure-culture conditions have been reported in minutes as follows: *E. coli*, 17; some *Proteus* species, 21; some *Salmonella* species and *S. faecalis*, about 25; and *Pseudomonas aeruginosa*, 31. Generation times in pure culture, however, are not applicable to mixed cultures. For instance, a broth of *P. mirabilis* incubated at 37°C for 24 hours has twice the viable cells as a corresponding broth of *S. typhimurium*, yet the difference is 13-fold when the two are grown together. Although *E. coli* has a shorter doubling time than *S. typhimurium*, the populations are coequal when grown together in the chick-maggot environment (Fig. 33). *Pseudomonas aeruginosa*, with a slower doubling time than *S. typhimurium*, outgrows the latter both in vivo and in vitro, whereas *S. faecalis*, with the same doubling time as *typhimurium* is suppressed. Competition for a limited resource and dependence of one species on a resource that is in limited supply or unavailable for physico-

*Mono-, di-, and tribiotic as used here refer only to the microbes, not to the fly.

chemical reasons, may be more important than growth rate *per se* in determining dominance hierarchy.

In vivo antagonisms and competition are a widespread fact of life. Their contribution to the host's defense against infectious disease has been investigated in higher animals. By modifying the intestinal flora of mice and guinea pigs, one can make these normally resistant animals succumb to cholera and dysentery. Possibly, the normal flora of dogs and cats confers on them the ability to shrug off a dose of 15 billion *Salmonella*, which would make a man severely ill. Antagonisms are known to occur between closely related strains and types of virus and between different viruses, between virus and bacteria, between different bacteria, as we have seen, and even between nematodes and between parasitic wasps. Not much is known about microbial interactions in arthropods despite the important implications for transmission of plant and animal diseases.

THE MAGGOT AS HOST

Survival of a microbe through metamorphosis is favored by a large initial input into the maggot. Let us now follow the fortunes of bacteria once they are swallowed by the maggot. The picture is assembled from experiments with gnotobiotic and conventional specimens of *C. vicina*, in which the crop, three regions of midgut, and the hindgut are assayed for bacteria. At the outset, the number of organisms in the foregut and in the larval medium are, volume for volume, the same. *Proteus* outnumbers *Salmonella* in both places by 10 or 100 to 1. The large populations of the crop conform to its volume (Fig. 34), and we find in the anterior midgut one hundredth the volume and, therefore, one hundredth the number of bacteria. The normal flora diminishes slightly as it passes down the midgut, and it multiplies in the hindgut. *Salmonella* is virtually destroyed as it passes through the acid midgut. In the prepupa, the remaining *Salmonella* are eliminated, their last stronghold being the crop. The rapid destruction in the acid midgut resembles the action of a potent bactericide.

The rate at which bacteria are destroyed in the acid midgut is remarkable. Millions of organisms continually pour into this region during feeding, but few emerge. Species of *Salmonella*, *Proteus*, and *Streptococcus* are equally vulnerable, and each type undergoes a 5-log destruction in a 5-minute transit. Since this happens equally in digestive tracts that are monocontaminated with *S. enteritidis* or *P. mirabilis*, bicontaminated with *S. typhimurium* and *S. faecalis*, or tricontaminated with combinations of these bacteria, it cannot be solely an effect of bacterial antago-

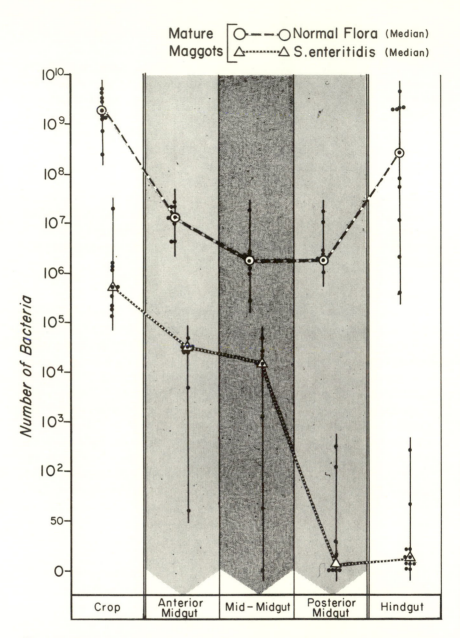

FIG. 34. Fate of *Salmonella enteritidis* and a normal flora in the digestive tract of the blow fly maggot. Both populations are highest in the crop because of its large volume. Density of the normal flora is unaffected by passage through the acid midgut, whereas *Salmonella* is virtually eliminated. (From Greenberg, 1966.)

128   THE FLY AS HOST

nism. It is noteworthy that both gram-positive and gram-negative bacteria are destroyed (Greenberg, 1966). Robinson and Norwood (1934) made similar observations by placing sterile maggots of *P. sericata* in the wounds of osteomyelitis patients and counting bacteria in certain parts of the maggot gut. They concluded that pyogenic organisms, e.g. *Staphylococcus aureus*, are largely destroyed in the hind-midgut; unfortunately they did not dissect the acid midgut.

Bactericides are of widespread occurrence in arthropods, and we shall now examine the features of the ones from flies. Duncan (1926) has demonstrated bactericidal effects in gut homogenates and feces of *S. calcitrans* and *M. domestica*. The preparations were most active against several *Bacillus* spp. and staphylococci, and had only slight activity against *Shigella dysenteriae*, and no activity against other enteric bacteria. The activity was thermostable, it was unaffected by protein precipitants and by trypsin, and was not accompanied by bacteriolysis. Picado (1935) obtained anti-*Staphylococcus* activity from gut homogenates of *Dermatobia hominis* but none from house fly larvae, nor could Maseritz (1934) find activity in similar preparations from *Phormia regina*. A burst of studies followed Baer's (1931) important disclosure of the value of maggots in the treatment of osteomyelitis wounds. Slocum et al. (1933) confirmed Baer's observation that maggots of *P. sericata* exert a bactericidal effect in the wound by demonstrating a reduction in the number of bacteria, pyogenic types disappearing first and *Proteus* remaining until the last. They noted that after 24 hours of maggot activity the wound becomes alkaline. Livingston and Prince (1932) claimed that homogenates of whole blow fly maggots were bactericidal, but this was disproved by Slocum et al., Maseritz (1934), Pavan (1949), and several other investigators. Tebbutt (1913) had already shown that a homogenate of house fly maggots is less bactericidal than normal saline solution—in fact, it makes an excellent medium for bacteria to grow in. By the end of the 1930s, allantoin, a nitrogenous excretion of blow fly maggots, was generally thought to be the actual bactericide. But Robinson pointed out in 1940 that even a concentrated 8% solution of allantoin is not bactericidal; neither are weak aqueous solutions of ammonium hydroxide (0.5 to 2%), although each brings about a reduction of bacteria in suppurating wounds. The healing agent was discovered to be surprisingly simple—alkalinization of the wound inhibits the etiologic agent *Staphylococcus aureus*. At about the same time, it became known that the washings of maggots nonetheless do contain potent bactericides.

Simmons (1935a, b) found that the washings of whole, sterile or non-sterile maggots of *P. sericata* contain an active principle that can be autoclaved or dessicated without loss of potency and that destroys suspensions of *S. aureus*, hemolytic and other streptococci, *Clostridium perfringens*, *Salmonella typhi*, and *Proteus vulgaris* without lysing cells. Five- to ten-minute exposures suffice to give 100% kill of dense suspensions. Gwatkin and Fallis (1938) extended these observations to include *C. vicina* and other *Calliphora* spp., noting that the activity was highest in recently colonized maggots and decreased with each laboratory-bred generation. They found that young larvae of *Protophormia terraenovae* excrete a thermostable bactericide that is active against various streptococci, less active against *S. aureus*, and without effect on members of the Enterobacteriaceae. Presumably, this, too, is a maggot product, although we cannot tell since the maggots were reared conventionally. Picado (1935) is convinced that maggot bactericides are of bacterial origin because maggots of *Phormia* and *Lucilia* produced anti-*Staphylococcus* activity when they were reared on putrefying meat but not when they were reared on sterile serum agar or blood agar. Recall, however, that Simmons' sterile maggots also produced bactericide. Picado further points out that vultures also have bactericides in their excreta when they are reared on putrefying meat.

With some exceptions, the general features of maggot bactericides are the following. They are active against gram-negative and gram-positive bacteria. The active substance seems to originate in the gut and is stable to heat, drying, and proteolysis.

The nature of the activity or its role in the natural history of the maggot is unknown. We also do not know whether this activity is the same as the activity we previously described as occurring in the acid midgut. It satisfies two requirements of the in vivo system—extremely rapid kill and broad-spectrum effect. But the third requirement—acid-dependence—does not seem to fit. This problem invites more study.

Shope (1927) suggested that bacteriophage has a bearing on the survival of pathogens in the gut of the fly. He showed that extracts of house flies captured in hog lots contained several phages that lysed *E. coli*, *S. paratyphi*, *S. typhi*, and *Staphylococcus muscae*. Although it is difficult to assess the importance of phages as destroyers of pathogens in wild flies, the following arguments suggest they have little importance in laboratory populations:

1. Phages are recoverable from wild flies but disappear after a few generations of laboratory breeding. Yet, the midgut destruction of bacteria can be demonstrated in flies which have been cultured for four years;

2. Bactericidal activity is not transmissible in vitro, but phage is;

3. Lysis of susceptible bacteria does not occur with maggot preparations;

4. Phage is host specific, maggot preparations are not.

Bacteriocins—bacterial products that are active against other strains or closely related species—are excluded from consideration because of their narrow spectrum and their susceptibility to heat and proteolytic enzymes.

In seeking a mechanism for the "valley of death" phenomenon, models were tried that simulated the pHs and digestive enzymes and reduced O_2 tension of the gut; but these were all too slow, taking 24 hours to destroy bacteria that the midgut destroyed in a few minutes. Further investigation led to a mechanism that seems to meet the requirements of the maggot gut and succeeds in test tube trials. The suggested scheme of how the substance might work in the maggot-microbe ecosystem follows. Needless to say, this model is purely hypothetical and, as such, is presented here with the hope of stimulating further research rather than offering a definitive solution. A substance is produced and excreted by *Proteus* and other species common in the larval medium. This substance is ingested by maggots together with myriads of microorganisms, but it has no effect on them in the anterior part of the tract, where the pH is alkaline or slightly acid. It becomes active in the acid midgut, where it rapidly destroys both gram-negative and gram-positive bacteria. This bactericide is thermostable and resists proteolysis like the bactericides obtained from maggot washings but differs from them in two important respects: its activity is inversely related to pH and ceases above pH 4.5; and it is a bacterial product, not a maggot product (Greenberg, 1968b).

Another phenomenon inviting study occurs in the immature stages of the horn fly, *Haematobia irritans*. The microbial burden is greatest in the first instar larva of this fly and undergoes a stepwise decrease with each molt until very few organisms are left in the pupa (Stirrat et al., 1955).

Bacteriostatic activity in the gut of *Hypoderma* and *Dermatobia* seems to prevent undesirable overgrowth of the maggot's environment by suppurating bacteria (Beesley, 1968; Landi, 1960; Jettmar, 1953; and Vogelsang et al., 1955). Cysts containing larvae of the chipmunk bot, *Cuterebra emasculator*, remain pus-free while larvae are present, but become heavily infected when larvae leave the cysts to pupate (Bennett, 1955). An experiment to alkalinize vacant cysts would seem indicated since this may possibly turn out to be the same bacteriostatic mechanism that operates in maggot therapy. There is bacteriostatic activity also in the hemolymph of bot flies and warble flies that is specific for pyogenic bacteria. This double line of defense indicates considerable mobilization against suppurating types of bacteria, which is probably no accident considering the maggot's need to keep a "clean" house.

The house fly, too, is capable of mustering some humoral defense. Cell-free hemolymph of laboratory-reared larvae of *M. domestica* shows bactericidal activity against *Shigella dysenteriae* but not against *Pseudomonas aeruginosa*, *Serratia marcescens*, or *Bacillus cereus* (Stephens, 1963). Enhancement of humoral protection against black widow spider venom lasts 96 hours after injection of venom into adult house flies (Bettini, 1965). The adult is thus capable of a specific, though transient, response to an antigen, but evidence of humoral responses in the maggot and of possible carry-over of immunity from maggot to adult is not yet available.

Lysozyme has been reported in the gut and fat body of *M. domestica* and *P. sericata*, but not of *Drosophila* sp. Its presence was demonstrated as a zone of lysis when the insect tissue was placed on agar seeded with *Micrococcus lysodeikticus* (Malke, 1965). According to Mohrig and Messner (1968a, b), bactericidal lysozyme in the hemolymph provides a generalized, nonspecific, and transient defense against microorganisms. It is not known at the present time how and against which organisms this system might operate in vivo, and whether or not the humoral phenomena described above are all part of this system.

In Lepidoptera, a rise in the hemolymph antibacterial principle occurs only when materials are injected into the body cavity of larvae, never by feeding the inoculum. The active principle is retained throughout life and differs from vertebrate antibody in its thermostability (Briggs, 1958.) Malke's lysozyme preparation from flies was thermolabile and transient. But his preparation was in a homogenate, and purified preparations might behave differently. It is obvious that a systematic study of the immune process in Diptera is needed. One approach to such a study might be axenic rearing in a nonantigenic medium (amino acids, purified vitamins, and cholesterol) to eliminate possible humoral responses to microbes and proteins in the larval medium.

The fly responds to invasion of the body cavity by a metazoan parasite by rallying its cellular defenses to encapsulate the agent. The parasite's fate may also hinge on the host's humoral, hormonal, and nutritional responses. For example, Stoffolano (1967) reports that the production of juveniles of the nematode, *Heterotylenchus autumnalis*, appears to be suppressed in diapausing face flies, and he suggests that hormonal imbalance and nutritional deficiency are possibly responsible.

Observations of this worm in an unnatural host, such as *M. domestica*, provide additional insights into the fly's repertoire of defenses. A successful defense reaction becomes apparent three days after the worm has invaded the maggot's hemocele. The worms become progressively melanized as they are encapsulated in a homogeneous syncytium of hemocytes variously penetrated by tracheae, fat cells, and muscle epithelium (Nappi and Stoffolano, 1971). This is the typical insect capsule formed to wall off a foreign object.

Evidently, successful host-adapted parasites can stimulate capsule formation without preventing their own development and subsequent liberation. For example, *Habronema muscae* resides in the larval fat cells that are destroyed in the pupal stage and supplanted by a cyst-like wall that is pigmented and interlaced with tracheoles (Mello and Cuocolo, 1943a, b); this cyst cannot restrain the worm larvae in the adult fly (Fig. 55).

Even in a preferred host, habronemas have not attained a refined state of peaceful coexistence. Theirs is a destructive residence, increasing both larval and adult mortality. Infested house flies bear a pendant proboscis, which they can neither retract nor extend. Such flies have a depleted fat body and do not lay eggs. Expulsion of larvae destroys the labellar membrane, and the fly dies a few days later.

H. microstoma, whose normal host is the stable fly, does not develop beyond the spiny stage larva in the house fly, but *H. muscae* and *megastoma* develop successfully in various flies, including drosophilids. The reasons for these differences are not known. Poinar (1969) has written an excellent review of arthropod immunity to worms. In Chapters 4 and 5, the reader will find additional information relating to the life cycles of *Habronema*, *Thelazia*, and other helminths.

Turning to arthropods, Streams (1968) cites instances of hymenopterous parasites encapsulated in what are considered to be their normal insect hosts. Particularly interesting is the *Pseudeucoila bochei-Drosophila* association. In the *melanica* species of *Drosophila*, the parasite does not develop despite the absence of any visible host response. In these cases it would seem that either a humoral reaction has occurred or the physico-chemical milieu of the host discourages parasite development. The host milieu cannot be totally inimical, however, because some individuals do mature. In other *Drosophila*, the parasite is able to develop in spite of encapsulization.

It has been suggested that encapsulization is the host's response to a foreign object or antigen and that worm or arthropod parasites that do not elicit this reaction may possess antigen molecules that mimic the host's own and are therefore not recognized by the host as foreign. Another possibility is that the parasite interferes with the host's reactions, mechanically or otherwise, and thus thwarts effective encapsulization.

Those interested in the general problem of arthropod immunity to animal parasites should consult Salt (1961, 1963, 1970), Weiser (1969), and Shapiro (1969).

ADULTS AND MICROBES

The average fly has a month or two to disseminate mischief. In assessing its capability, we will draw upon studies with enteric bacteria, as this group has received the most attention. Interactions involving other organisms are discussed in the next two chapters in the context of the specific diseases they cause.

MECHANICAL TRANSPORT

Flies transport agents on their integument and within their body. These are quite different habitats from the viewpoint of microbes intent on colonization. To look first at the exterior, it is common knowledge that the hairy body, the tarsi with their sticky tenent hairs and corrugated plantas, the wings, and grooved proboscis all provide hideaways for microbial stowaways.

Early in the century Buchanan (1907) showed how easily blow flies and house flies pick up pathogens from their surroundings. His photograph of an agar plate shows outgrowths of anthrax colonies arising from tracks made by a fly after it had visited a condemned carcass. This mode of transport can now be directly and rather dramatically visualized by means of the scanning electron microscope. Figure 35 presents a series of SEM photographs of the spore-laden tarsus of a blow fly after a brief stroll on moldy bread. One can readily observe the variety of types and approximate numbers of each. This instrument offers new ways to study mechanical transport.

Let us now examine the factors that oppose mechanical transport. Casual observation reveals that flies have a cleaning compulsion that partly offsets their filthy habits. They are constantly preening their wings and brushing their legs and heads, thereby dislodging organisms. As far as we know, there seems to be no place on the surface of a fly where conditions are right for microbes to multiply. Yet it would be a mistake to conclude that external phoresy is therefore of no consequence. The not infrequent recovery of pathogens from the surface of wild flies indicates the contrary.

To the cleaning activity of the fly, we now add the threat of desiccation. Flies winging from feces with their bodies covered with protozoan cysts, helminth eggs, etc., may be playing the paradoxical role of purifying agents, for their passengers may die from desiccation en route. This

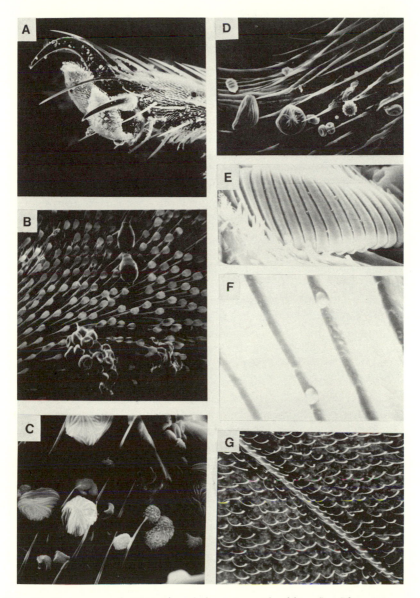

Fig. 35. Stereoscan photographs of the tarsus of a blow fly, *Phaenicia seri-cata*, which has walked on moldy bread (A-F). A. Paired claws (ungues) and pulvilli with tenent hairs (× 205). B. Two *Penicillium* spores and cluster of unidentified microorganisms adhering to tips of tenent hairs (× 1950). C. Subsegment with spiny spores of *Penicillium* sp., among others (× 2470). D. Claw showing large wrinkled spores of black bread mold, *Rhizopus nigricans* (× 1800). E. Planta (× 1800). F. Same, with coccus-like bodies, lodged in crevices (× 9000). G. Microtrichia are numerous on both wing surfaces (*M. domestica*, × 500).

is an entirely reasonable notion, which, however, still needs experimental proof. Conceivably, such an experiment might also inform us about the microclimate of the fly under various weather conditions. Some years ago, I observed that the hypopi of anoetid mites lived on the fly for at least 47 days (dying only after the flies died), but when they were kept in a petri dish in the same fly cage they all succumbed in 30 hours. Since the hypopi lack mouthparts and do not feed on the fly, it was probably the microclimate of the insect's surface that favored their survival (Greenberg, 1961).

We do not know how microbes fare on dark-bodied *Calliphora* as compared with the black- and grey-checkered *Sarcophaga* or *Musca*, but it is known that insolation can significantly raise the surface temperature of a dark-bodied fly. Under the action of solar radiation in the East Pamir, at 3,860 meters above sea level, the surface temperature of *C. vicina* reached 42.0°C when the air temperature in the shade stood at 30°C (Sychevskaĭa, 1965). Such temperatures would hasten the demise of surface populations.

The sun's ultraviolet radiation has well-known bactericidal effects to which the surface flora of the fly is susceptible, although radiation and insolation effects are not always distinguished by investigators. Exposure of house flies to a summer sun for two hours reduced the average number of bacteria from 2,540,000 to 160,000 on flies from city flats, and from 32,560,000 to 6,720,000 on flies from a manure heap (Parisot and Fernier, 1934). Analysis of flies captured in Egypt showed that *M. domestica* carried twice the number of bacteria as *M. sorbens* and that in both species there were about 20 times the number of internal bacteria as external bacteria (McGuire and Durant, 1957). Under the solar conditions of Soviet Central Asia the integument of flies was sterile or nearly so, while their intestines contained immense numbers of bacteria (Derbeneva-Ukhova, 1952; Shterngol'd, 1949). The guts of house flies captured near dysentery patients in the Crimea contained up to 30,000 shigellae, whereas their body surfaces were invariably negative for the same organisms (Shura-Bura, 1955). Ability to survive on the integument has been shown to vary with the species of microbe (Přívora et al., 1969).

If we total the sterilizing effect of self-cleaning, desiccation, and insolation and radiation, the integument appears to be a rather bleak landscape for the microflora, providing only a temporary, progeny-less, perch at best. But obviously it is not bleak enough, otherwise there would not be reports of large surface populations on wild flies. Occasionally, in the laboratory, where solar and other influences are absent, the situation is reversed, with maximum counts occurring on the surface rather than on the inside (Ostrolenk and Welch, 1942a). All such laboratory experiments have limited applicability to nature, and in the absence of sunlight should be taken with a grain of salt.

PATHOGEN PERSISTENCE

Basically, the persistence, multiplication, and output of a microbe is a measure of the host capacity of the fly. In evaluating host capacity, most investigators have obtained their information from unrestrained, conventional flies, a few have monitored input-output of mounted specimens, and still fewer have used gnotobiotic flies. The published data are so detailed that an exhaustive account would blur our purpose. Only a few representative studies will be detailed in this discussion, but we shall try to cite other studies to which the reader may turn for additional information.

When a group of flies is unrestrained, they are free to contaminate themselves and their mates with dejecta. The importance of fly-fly contamination is often overlooked in designing an experiment, and reports of exaggerated pathogen persistence have thus accrued. For example, it was found that flies remained infected with typhoid organisms for 22 days and with dysentery organisms for 13 days if they were not regularly transferred to clean vessels; otherwise they eliminated the organisms in 8 to 10 days (Wollman, 1921, 1927). Graham-Smith (1910, 1911a, b) showed that clean house flies are quickly contaminated with *Serratia marcescens* after they enter a contaminated cage. In another study (Ostrolenk and Welch, 1942b), infected house flies spread *S. enteritidis* to other flies, as well as to mice in the same cage. Under unsanitary cage conditions, flies remain infected for life. The extent of contamination can be appreciated from the following observation. About 150 house flies that had been caged with mice suffering from salmonellosis were transferred to a cage that contained pecan meats, and within 15 minutes had deposited the pathogen on the meats. Bacterial counts on the meats rose from 900 during the first 15 minutes to over 3 million after 5 hours' exposure. Fly-fly contamination with a specific microorganism continued to operate even after three successive introductions of 10 contaminated flies into a cage with 100 normal ones (Webb and Graham, 1956; also Gross and Preuss, 1951).

The following studies, carefully done in most other respects, should be viewed with this limitation in mind.

Celli and Alessi (1888) first demonstrated that viable typhoid and other pathogenic organisms fed to house flies can be recovered from their feces and intestinal tract. Simple mechanical transmission of typhoid and dysentery organisms was shown by Manning (1902), Auché

(1906), and Firth and Horrocks (1902). Teodoro (1916) demonstrated that *C. vicina*, *L. caesar*, and *M. domestica* excrete typhoid organisms for a maximum of 5 days. Graham-Smith's base-line studies are best appreciated by consulting his book, "Flies and Disease," published in 1914. Here, in summary form, are the results of feeding experiments in which he determined the maximum recovery (in hours) of test organisms from the legs (L), vomit (V), and feces (F) of house flies: *Sal. typhi*–48 (L,F); *Sal. enteritidis*–168 (L); *S. marcescens*–168 (L,F), 48 (V); *Mycobacterium tuberculosis*–312 (F); *Bacillus anthracis*, spores–480 (L), 336 (F), 144 (V); *B. anthracis*, vegetative–18 (L), 48 (F); and *Vibrio comma*–30 (L,F).

Differences in the survival times of different species of *Salmonella* appear real and have been corroborated by work from the laboratory of Gross and Preuss (1951). These investigators found that, on the exterior of the house fly, *S. typhi* and *S. paratyphi* B persist for 10 and 11 days, respectively, while *S. paratyphi* A persists for 7 days. Again, when flies are given an infective meal of 200,000 typhoid organisms in a sugar solution, their vomit and feces are infective for 3 days and 7 days, respectively; under the same conditions, *S. paratyphi* A is found in the vomit for 3 days and in the feces for 12 days.

Typically, pathogenic bacteria on the surface of a fly are more quickly eliminated than their enteric counterparts. In one study, *Salmonella pullorum*, agent of white diarrhea of chickens, remained on the legs and wings only six hours but survived at least 5 days in the house fly gut (Gwatkin and Mitchell, 1944). In another study, dysentery and typhoid organisms persisted for 5 to 10 days in the tract, while they were wiped out on the surface in one day (Shterngol'd, 1949). When house flies are exposed to the feces of a typhoid patient, they remain surface-contaminated for about 5 days, but Ara (1933) claims that the effective period for mechanical transmission is 24 hours or less. In fact, 40% of the flies lost all recoverable *typhi* within the first hour (also Ara and Marengo, 1932). Krontowski (1913), Alcivar and Campos (1946), Ara (1933), and Evtodienko (1968) also report that the digestive tract is a better refuge.

Studies of shigellae generally indicate survival of less than one week in flies. Reinstorf (1923) provides additional data on the average persistence (in days) of three *Shigella* types, as follows: proboscis–3; legs–4.8; and gut–8. Viable *S. dysenteriae* have been recovered from the guts of house flies for 5 days (Bahr, 1912) and for 6 days (Aradi, 1956). According to Gromashevskiĭ, shigellae remain viable on the legs for 2 days and in the tract for 3 days (Elkin, 1958, p. 233). Turning to blow flies, we find pretty much the same picture. *S. flexneri* survived two days in *C. uralensis* (Sukhova, 1950), and *C. megacephala* hosted *S. dysenteriae* superficially and internally for 5 days (Chow, 1940). Near-total elimination of enteric pathogens in one day (Thomson, 1912; Dudgeon, 1919) is probably as atypical as reported persistence of 2 to 3 weeks or more (Ficker, 1903; Faichnie, 1909a, b; R. M., 1937; Ara, 1933; Stewart, 1944). Extended survival of *Salmonella* and *Shigella* in flies is probably an artifact due to recontamination.

Some of the above records of maximum recovery exaggerate transmission potential in another way by equating mere presence of viable organisms in the gut with transmissibility. Pathogens are dangerous only when they leave the fly.

Extended pathogen survival may occur in hibernating flies or when the insects are subjected to periods of cool weather. This is suggested by Bychkov's (1932) study, in which *S. marcescens* was retained in house flies for 17 days when the flies were kept at a temperature that ranged from 3°C to 15°C. Even on the 17th day, the rectum contained up to 115,000 *Serratia* and the crop had up to 27,000 organisms. Such experiments should be repeated with pathogens and designed to include diurnal temperature swings, as well as uniformly cooler temperatures than those normally used in laboratory experiments.

Biting flies have received little attention as potential participants in the transmission of enteric diseases. It is therefore interesting to note that *S. calcitrans* transmitted *Sal. breslaviensis* for 14 days following an infective meal and that 6 out of 9 mice became infected from the bite of a single fly (Birk, 1932).

MULTIPLICATION OF MICROBES IN FLIES

When carmine, a biologically inert material, is fed to flies, the particles are completely eliminated in 3 days from the house fly tract (Shterngol'd, 1949) and in 6 days from the tracts of most *Phormia regina*, *Phaenicia sericata*, and *Sarcophaga bullata*, particles sometimes remaining in a few flies for as long as 13 days (Melnick and Penner, 1947). On the basis of these average sweep-out rates, it has been assumed, with doubtful validity, that a population that remains in the gut beyond these periods has probably multiplied. More persuasive evidence of multiplication can be obtained from experiments in which the input and output of a microorganism are quantitated. Only a few such studies have been reported, and in each of them mounted flies were used to minimize recontamination.

In one study (Hawley et al., 1951), the output of test bacteria from conventional house flies was compared as the infective dose was increased. Feces were collected in physiological saline solution or on a nonnutrient agar block. When inputs were below 10^3 (*Escherichia coli*, *Sal. schottmulleri*, or *Sh. dysenteriae*), the organisms were not excreted in the feces. The threshold for colonization of the fly's gut, evidenced by decided multiplication, was an initial feed of 10^4; above this number, output increased with input, and it also increased during the 6-day observation period. Haines' unpublished study (Lindsay and Scudder, 1956) has failed to confirm these findings and suggests a higher threshold, as does a study by Ingram et al. (1956). The latter fed about 10^6 *E. coli* to house flies, and the bacteria behaved like carmine—the greatest recovery in the feces occurring within the first 2 or 3 days. In only a few cases was there possible multiplication in the fly. Unless Hawley's work is substantiated, it would appear that conventional house flies are resistant to colonization. In some of these experiments, different concentrations of sucrose and milk were used to feed the mounted flies, with no effect on colonization or output of *E. coli*. Likewise, Shterngol'd (1949) reports that diets of syrup or of mixed garbage do not influence the persistence of *Sh. flexneri* and *dysenteriae* in house flies.

The so-called normal gut flora of the flies was not controlled in the above studies, and they could have harbored a complex or simple flora in their digestive tracts. We have already seen the important effects of microbial interactions on pathogen survival in preadults. It is also possible to reproduce these effects in adults. Thus, *S. typhimurium* multiplies in monocontaminated *P. regina* when the fly is fed a minimum dose of 15,000 organisms, but the pathogen requires a larger input when *Proteus morganii* or *Aerobacter* sp. are present. *S. typhimurium* also seems to be less tolerant of competition than *S. schottmulleri* (Knuckles, 1959).

The host capacity of blow flies and house flies has been scrutinized and compared in so many ways we wondered whether there might also be detectable differences between species in their innate resistance to colonization. Our method was to mount germfree *M. domestica* and *P. sericata* on paraffin and to feed different groups of flies graded doses of *S. typhimurium*. Each day, a fly's fecal output of bacteria was monitored to give data on persistence and possible multiplication of the organism. Gnotobiotic and other techniques used in this type of study are described in detail in a separate section of this chapter. With the house fly, we learned that a high input is not required to produce a high proportion of excreters. Among 22 flies given an average of 90 *Salmonella* cells, multiplication occurred in 16, and 18 became excreters. In fact, the organism multiplied—up to 1.4×10^7 in one case—in more than half the flies fed as few as 20 cells. Despite these low inputs, the flies excreted the pathogen for about ten days. The green bottle fly appeared to be a poorer host than the house fly. When both flies were fed 100 *Salmonella* cells, the percentage of green bottles that excreted the organism was 45% compared with 81% in the house fly; also the percentage showing multiplication was slightly less than half that of the house fly. In both species, there was significant multiplication of the pathogen at very low inputs, provided there was no competition.

The antagonism of *S. typhimurium* by *P. mirabilis* that can be staged in the digestive tract of the blow fly maggot also occurred in the adult house fly. When flies were fed 10^3 *Salmonella* only, the organism was excreted for at least 8 days and the percentage of excreters was 87% on the first day and 62% on the second day. When a second group received 10^3 cells each of *Salmonella* and *Proteus*, there was a reduction of *Salmonella* excreters to 27% on the first day and to 0% from the second day on, while excretion of *Proteus* continued at high levels for the life of the flies. The same number of *Salmonella* mixed with a mouse's fecal flora (which had few *Proteus* spp.) and fed to a third group of flies had an intermediate survival time (Greenberg et al., 1970).

There are few observations on the localization of organisms in the gut of the adult. Three days after an infective meal, we found that *Salmonella* cells are localized in the midgut and hindgut of the house fly, with few, if any, organisms left in the crop. In Knuckles' experiments, the greatest multiplication occurred in the crop, and Vanni (1946) found that certain nonpathogenic bacteria are more numerous in the crop than in the hindgut of the house fly. Some of these discrepancies may be due to the nature and richness of the food, frequency of feeding, and time of sampling following the infective dose.

It is clear that the fate of *Salmonella*—and probably other pathogens—in an adult fly is subject to the following factors:

1. species of fly—both house fly and green bottle fly are synanthropic and both are natural carriers of salmonellae and other pathogens, but the house fly appears to be a distinctly superior host;

2. size of input—low inputs can result in massive multiplication, at least in gnotobiotic flies, but the percentage of successful implantations increases with dose;

3. microbial condition of the fly gut—interspecies antagonism may lead to rapid elimination of *Salmonella*, or there may be a more gradual suppression by a mixed flora. Either condition reduces the natural vector capacity of the fly.

In no way is our discussion of persistence meant to negate the medical importance of transiently infected flies. In all cases, we are concerned with the important question of a fly's dosage delivery. A dose of 10^5 or fewer *Salmonella* cells may provoke illness in a baby or an undernourished child, whereas an adult ordinarily requires at least 10^7 cells. From available information and our own experiments (1964), it appears unlikely that a single fly normally delivers such a dose. But this is no problem where bacteria are deposited on food under conditions favoring multiplication. It is a real problem, however, with viruses that cannot multiply outside a suitable cell. Some of the constraints of virus transmission will become clearer from the following example with poliovirus. The numbers of poliovirus per gram of feces are variously reported at $10^{2.8}$ TC and $10^{3.2}$ TCID$_{50}$ for paralytic cases, and $10^{2.7}$ TC for subclinical cases (Brown, 1955). *Phormia regina* retains virus for more than two weeks and excretes virus for ten days when fed a stool suspension with a minimum titer of 1:100 per ml. During the first six days, virus recovery from whole flies is 28 times greater than the original virus input, and from the excreta, 11 times greater (Melnick and Penner, 1952). Therefore, the average titer of virus in an infected person's feces is probably adequate to "infect" flies. What is unresolved is the fly-man link in the epidemic chain. It is known that a dose of 10^5 TCD$_{50}$ of attenuated virus will infect over 80% of susceptible persons, and the infective dose for virulent strains is probably less. It is not known whether flies can deliver such a dose.

Selection may alter the properties of a pathogen population that has passed through a fly, provided extensive multiplication has occurred. There has been little work on this aspect of the fly as host except for one report, which found no significant alteration of properties of shigellae after passage through the fly intestine (Shura-Bura, 1955). The problem is worthy of study, particularly since vaccine strains of poliovirus are readily picked up by flies in poorly sanitated environments, and possible virus mutation in the fly has not been ruled out.

FLY DANGER-INDEX

By recognizing the key variables, one can formulate a tentative danger-index (D) of synanthropic flies, as Mihályi (1967) has done. The three parameters basic to such an index are: A. infection potential, or the proportion of flies that carry the pathogen; B. infective potential, or the proportion of flies that can transmit; and C. fly-pathogen factors. The following formulation is a modification of Mihályi's, taking into account additional factors discussed in this chapter, as they concern adult transmission only:

$$D = \underbrace{(a + b + c + d)}_{A.} \quad \underbrace{(e + f)}_{B.} \quad \underbrace{(m + n + o)}_{C.}$$

A. where

    a = the fly visits feces for mating, oviposition, etc. (a positive case is designated as 1, a negative as 0);

    b = female feeds on feces;

    c = male feeds on feces;

    d = flies feed on other infectious matter, e.g. yaws sores, TB sputum, trachomatous, and conjunctival secretions.

B. contact by such flies with food, sores, mouth, eyes:

    e = indoors, in the home, restaurant, food store, food processing plant, hospital, etc.

    f = outdoors, in the market, slaughterhouse, farm, encampment, etc.

C.

    m = size of fly;

    n = transmissible pathogen density on or in fly;

    o = duration of fly's infective state.

Mihályi determines a fly's size volumetrically and scores it in relation to *M. domestica*, which is 1. An index based on the average weight of a species would be equally valid and simpler to obtain.

REARING GNOTOBIOTIC FLIES AND OTHER TECHNIQUES

Gnotobiology is the study of organisms that have a known biota. Insect gnotobiology deals mainly with three types of microbiologically modified hosts: aposymbiotic (free of cellular symbiotes, but not intestinal microbes; aseptic (free of intestinal microbes, but not of symbiotes); and axenic (free of demonstrable cellular and intestinal microbes). Fly gnotobiology deals mainly with the latter.

Achieving and maintaining the gnotobiotic state requires familiarity with aseptic techniques and microbiological routines that are not a usual part of entomological training. Since the gnotobiotic route is being increasingly traveled for nutrition studies, infectivity studies, and in vitro tissue and organ culture, possession of these skills is an asset in research.

PHYSICAL STERILIZING AGENTS

1. *Moist Heat.* The autoclave produces superheated steam under pressure (15 lb. at 121°C). High pressure provides effective penetration, and moist heat coagulates protein more readily than does dry heat. Sterilization time varies upward from 10 to 15 minutes, depending on type and volume of material. The autoclave can be used for solutions,

rubber tubing, instruments, sawdust, sand, cages, etc., but not for most plastics.

2. *Boiling*. Five minutes destroys vegetative bacteria, but some spores survive ordinary boiling for 16 hours. The addition of 2 to 5% carbolic acid to boiling water destroys spores in 15 minutes.

3. *Dry Heat*. This penetrates slowly and denatures protein less rapidly than steam under pressure. Canisters of pipettes should be allowed 4 hours at 150°C. (Cotton plugs turn brown at 200°C.) Ninety-five percent alcohol is recommended for flaming of instruments during aseptic procedures. We have used it to flame insects, pupal cases, and oothecae to render their surfaces sterile. Although convenient, this technique requires experience to avoid cooking the internal organs.

4. *Irradiation*. Ultraviolet, X-rays, and gamma rays are bactericidal and viricidal, with UV at 2650 Å most effective. "Sterilamps" are mercury arc lamps, with a small amount of neon or argon for visibility, which transmit at 2537 Å. UV is useful for sterilization of paraffin blocks and plastic that cannot be heated, and for the interior of sterile transfer chambers. Gamma rays (400 R/min.) took 400 minutes to inactivate 90% of *Bacillus thuringiensis* spores, whereas UV inactivated 99.997% in 10 minutes (Cantwell and Franklin, 1966). However, this wavelength is harmful to insects (Steinhaus and Dineen, 1960), and caution should be exercised to avoid exposing them needlessly.

CHEMICAL STERILIZING AGENTS

1. *Disinfectants*. Those commonly used in the laboratory such as Roccal, Hyamine, Zephiran, Detergicide, and Surgi-Bac (1% hexachlorophene) generally act to destroy the permeability of the bacterial membrane. Phenol (5%) has been widely used, and its toxicity is enhanced with the addition of NaCl. Lysol, a phenol derivative, is four times more potent.

2. *Oxidizing agents*. Free chlorine in water above pH 2.0 becomes germicidal hydrochlorous acid. Sodium hypochlorite (0.1-2%) is widely used for disinfecting eggs and tissues and as a laboratory disinfectant. $KMnO_4$ (1:1000) is sometimes used in insect work.

3. *Ethylene oxide*. This gas is employed in sterilizing plastics and synthetic diets, including tissue culture media. Bacterial spores are hundreds of times more resistant to heat and chemical disinfectants than are vegetative cells, but both are almost equally sensitive to this gas. After disinfecting an enclosed space (e.g. sterile transfer box) with any volatile agent, time must be allowed for vapors to disperse.

4. *Alcohol* (70%). This is a generally poor disinfectant. Ethyl alco-

hol, however, enhances the potency of HgCl$_2$ but reduces the potency of phenol and formaldehyde.

5. *Heavy metals.* Soluble salts of Hg, Ag, and Cu are bactericidal. HgCl$_2$ (1:500 or 1:1000) is used in White's solution to sterilize insect eggs. White's solution consists of HgCl$_2$ (0.25 gm); NaCl (6.5 gm); HCl (1.25 ml); ethyl alcohol (250 ml); distilled water (750 ml).

6. *Antibiotics.* Antibiotics are a mixed blessing. In some cases they act deleteriously to delay development and increase larval and pupal mortality (Singh and House, 1970). They have their place with chemical agents in suppressing microbial populations in synthetic media, but their use should always be warranted. They are definitely not a substitute for aseptic technique.

DISINFECTING THE EGG

Fortunately, the insect egg is able to withstand rigorous chemical treatment due to the chorion with its waxy layer, and plastron, which is a complex meshwork common in fly eggs (Hinton, 1960, 1970). An effective disinfectant must be surfactive to penetrate the plastron and wet the waxy layer, destroying microorganisms without damaging the egg. In some cases the age of eggs bears on the success of disinfection. Young eggs of the onion maggot, *Hylemya antiqua*, can be disinfected by immersion in 6.5% formalin for 30 minutes, but older eggs are more contaminated and are not disinfected by this treatment (Friend, 1955). Older blow fly eggs, on the other hand, are more resistant to the toxic action of a disinfectant such as HgCl$_2$ and yield a higher egg hatch than freshly laid eggs (Mackerras and Freney, 1933). House fly and stable fly eggs can be stored for several days on moist toweling at 4°C, but blow fly eggs will hatch at this temperature. If hatching occurs, the larvae should be discarded since their guts may have become contaminated, and no amount of surface sterilization will alter this.

Blow flies, e.g. *Phaenicia sericata*, *P. cuprina*, *Chrysomya rufifacies*, and *Phormia regina*, lay eggs glued together with a cement that can be dissolved with a 1% Na$_2$SO$_3$ solution. The egg cement of *M. sorbens* does not dissolve in this solution but dissolves fairly well in 0.25% trypsin (pH 7.2), after 35 minutes at 33°C. Trypsinization, followed by disinfection in 1% NaOCl, may reduce egg hatch (Greenberg, 1969).

No elaborate apparatus is required to disinfect eggs; simple items such as fine brushes, spatulas, and wire loops facilitate handling. Eggs may be sterilized inside a piece of glass tubing covered at both ends with nylon organdy. The tube is transferred from one solution to the next by means of a string. More elaborate setups have used Gooch crucibles attached by tubes to reservoirs of disinfectant and sterile water (Child and Roberts, 1931). The simplest method exploits the nonbuoyancy of the viable, undamaged egg. Eggs are swirled in the disinfecting solution in a sterile flask, allowed to settle, and the solution is carefully decanted. Fresh solutions are introduced and decanted aseptically.

Aqueous $HgCl_2$ was first used by Bogdanow in 1906 to disinfect the eggs of a blue bottle fly, but mortality and contamination both ran high. Similarly, eggs of *Drosophila melanogaster* failed to produce sterile larvae when treated with $HgCl_2$ (5:1000) or $KMnO_4$ (1:1000). Only by heroic resort to raising flies on acid media with frequent transfers to fresh media did some colonies finally achieve sterility (Delcourt and Guyenot, 1912). Immersion for 6 or 7 minutes in a saturated alcoholic solution of $HgCl_2$ (70%) improved survival (Loeb and Northrop, 1916); addition of 0.2% HCl to $HgCl_2$ prevents formation of a protective layer of mercury albuminate around the egg, and substitution of alcohol for water increases wettability. This is the basis for White's solution, used extensively for fly eggs (Kadner and LaFleur, 1951; Begg and Sang, 1950; Greenberg, 1954; Wallis and Lite, 1970).

White's solution, used in the following sequence, has given us almost perfect success with the eggs of a dozen species of flies. Eggs are thoroughly agitated in detergent in a test tube to clean and separate them. After several rinses with distilled water, the aseptic sequence is: 5 minutes in 5% formalin; 2 rinses in distilled water; 2 minutes in 1:1000 NaOCl (fresh laundry bleach); 2 rinses; 1 hour in White's solution; 5-7 rinses. A 5-ml pipette with opened tip is useful for aseptically transferring measured volumes of eggs. Roughly, 500 house fly eggs equal 0.1 ml.

Other methods have proven equally effective. Merthiolate (1:20,000) for 20 minutes disinfected the eggs of various blow flies (Causey, 1932); 2 to 3 minutes in 3% Lysol also worked well for blow fly and stable fly eggs (Gingrich, 1960). A 20-minute immersion of house fly eggs in 10% neutral formalin containing 0.05% Tween 80, or 30 minutes in a mixture of 5% formalin and 1 % KOH were also effective (Perry and Miller, 1965). House fly and blow fly eggs may also be disinfected with 0.1% benzalkonium chloride, a quaternary ammonium compound, for 25 minutes (Brookes and Fraenkel, 1958; Brust and Fraenkel, 1955).

Compounds that liberate chlorine may severely reduce egg hatch, therefore careful regard must be given to immersion time. Blow fly and house fly eggs have been treated by soaking them for 1 minute in a bleach solution (which yielded 0.5% available chlorine), rinsing, and then immersing in 4% formaldehyde for 3 minutes (Child and Roberts, 1931; Henry and Cotty, 1957). Monroe (1962) uses 0.1% NaOCl for

20 minutes to disinfect house fly eggs, and we have had good success in disinfecting the eggs of house flies and *M. sorbens* with a 1% solution for 10 minutes (Greenberg, 1969).

THE LARVAL MEDIUM

The larvae of carrion flies develop rapidly in fish, liver, or other meat placed in a jar and covered with sawdust. Pupation occurs in the upper layers of dry sawdust, which is introduced before the maggots stop feeding. Odor is minimized when the quantity of meat and the amount of moisture are not excessive in relation to the number of maggots. When the combination is out of balance, relatively innocuous decomposition shifts to offensive putrefaction. Moistened dog biscuit is said to provide a relatively odor-free substrate and incorporation of activated charcoal is recommended by Berndt (1969).

In the early days of house fly research, maggots were reared in manure for want of a better substrate. An esthetic milestone was attained with the introduction of a medium made up of two parts of wheat bran and one part of alfalfa meal, moistened with a yeast suspension and a solution of malt extract. The proprietary medium called CSMA (Chemical Specialties Manufacturers' Association) has further promoted the rearing of these flies. As put up in this country, the medium is a standardized mixture of brewer's grain (40.0%), wheat bran (33.3%), and alfalfa meal (26.7%). Other fermentation feeders, e.g. *Stomoxys* and *Fannia*, grow readily in this medium. There is no need to dwell further on conventional rearing procedures since West (1951) covers the subject very well. Additional sources of information on laboratory rearing of house flies and stable flies are Spiller (1966), Jones (1966), and Needham et al. (1937).

Numerous sterilized diets have been devised for fly larvae, which range from completely undefined to synthetic. The former are easy to prepare and usually promote faster development, but the latter provide known nutrients. The investigator selects the menu to fit his goal. Here, we can make only the briefest mention of suitable diets, because the volume of detail would take us far afield.

For calliphorids, sarcophagids, *Muscina*, and other carrion feeders the simplest undefined sterile diet, and the one least subject to risk of contamination, is the nearly mature embryonated chick. In removing it aseptically, the shell is disinfected with a 5% tincture of iodine, a cap is cut away with sterile scissors, and the chick is dropped through the mouth of a liter flask onto an inch of sawdust. Flasks can be stored in the freezer indefinitely. One embryo supports about 100 blow fly larvae, which develop normally. Additional sterile sawdust is introduced prior to pupation, otherwise migrating prepupae will burrow into and through the cotton plug. Causey (1932) reared normal blow fly larvae on a diet of rat fetuses aseptically removed from full-term mothers. The head was mashed to expose brain tissue and the fetus was placed in nutrient agar. Young larvae do poorly on autoclaved fish or beef because their external digestive powers are not yet adequate to cope with coagulated proteins. Introduction of proteolytic bacteria or Tyndallization of the meat enable young larvae to surmount the critical period.

Important nutritional differences exist between different organs of the same animal and between the same organs of different animals. Thus, blow fly larvae grow well when fed sterile guinea pig liver or brain, but their rate of growth is one-fifth the normal rate on rump and back muscles from the same animal (Hobson, 1932). Sterile beef muscle, on the other hand, will permit normal development (Fletcher and Haub, 1933). The problem is circumvented by using a whole animal, such as the chick embryo or rat fetus.

An axenic crop of house flies, *Musca sorbens*, etc., can be grown in modified CSMA, that is, CSMA that has been routinely prepared, placed in Erlenmeyer flasks, and incubated for 2 or 3 days at 37°C to maximize microbial proliferation. The medium is then broken up and thoroughly mixed, a 2- to 3-inch overlay of sawdust is added, the flasks are plugged with cotton and autoclaved for half an hour. When cool, the flasks are seeded with about 200 disinfected eggs/100 ml of medium. Maggots develop normally and pupate in the sawdust. Zhuzhikov (1963) recommends the following medium for sterile house fly maggots.

A. 500 gms minced pig liver boiled in 1 liter of H_2O for 30 minutes—autoclaved and refrigerated.

The nutrient medium consists of:

A	250 gms
powdered brewer's yeast	25 gms
bran	75 gms

Thoroughly mix, dispense about 50 ml into a 100-ml beaker, place beaker on sand in flask, and autoclave 30 minutes at 121°C.

Maggots of *Agria affinis* thrive on liver brei sterilized with liquid ethylene oxide (Barlow and House, 1956).

Synthetic diets have been developed for the house fly (Perry and Miller, 1965; Monroe, 1962; Brookes and Fraenkel, 1958; Barlow and House, 1956; Levinson and Bergmann, 1959) and for blow flies (Kadner and LaFleur, 1951; Brust and Fraenkel, 1955; Sedee, 1953). A synthetic diet for the stable fly has also been developed, but growth is retarded, even with addition of yeast extract (Gingrich, 1960).

Pupae

The pupa is immobile and requires little attention except to be strategically deployed in preparation for adult emergence. Should a question of contamination arise, pupae may be disinfected in 5% carbolic acid for 10 minutes. White's solution is not recommended because it reduces emergence. Two-day-old *Drosophila* pupae may be sterilized in 80-85% alcohol for about 10 minutes, provided contamination is light.

Adults

Flies may be maintained gnotobiotically in several ways, as shown in Figure 36. A diet of autoclaved powdered milk solution or a mixture of this with 5% sucrose solution will support normal fecundity and longevity in blow flies and house flies. Teneral flies to be fixed to paraffin should be allowed to harden before they are immobilized with filtered CO_2, ether, or cold. Survival time of mounted flies is usually less than two weeks when they are maintained at a temperature of 25°C and a relative humidity of 70%; this is reduced to 2 days at 32°C and 35% R.H. (Ingram et al., 1956). We found equal survival times for mounted *Phaenicia sericata* and *M. domestica* (Greenberg et al., 1970), but others report greater endurance among blow flies (Penner and Melnick, 1952). Flies are fed once or twice a day; under normal conditions, the second feeding seems to be gratuitous.

In our recent studies, flies are fastened by their wings to paraffin on strips of aluminum foil; these strips are first sterilized by exposure to ultraviolet light for 3 days. The mounted flies are then placed in vials containing 2 ml of physiological saline and are positioned vertically with their abdomens just above the saline. The day after they are mounted, dead flies are culled and the rest are fed a known dose of organisms by means of a 10-μliter Hamilton syringe calibrated in tenths. Infective meals of 2 to 10 μliters can be delivered with an accuracy of ±5% and are quickly swallowed by the fly (Fig. 37). Quantitative feeding can also be done with pipettes of 0.1 or 0.2 ml capacity graduated in thousandths. The tip of the pipette is flamed shut, and a hole is blown out just above the tip, big enough to accommodate a fly's labellum (Frings and Frings, 1946; Hawley et al., 1951). Daily output of the test organism in the fly's feces is sampled from the saline solution. At the start of a sampling period the fly is aseptically transferred—aluminum strip and all—to a vial that contains fresh saline solution. The sampling period is best kept to 4-6 hours to avoid possible errors due to multiplication or death of organisms in the saline. Needless to say, each experimenter will tailor a technique to fit his own requirements.

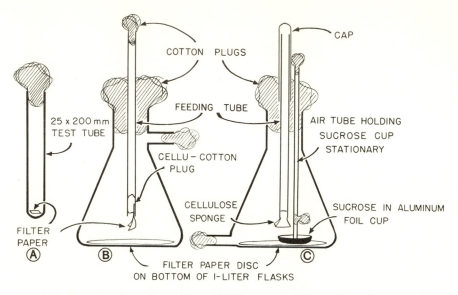

FIG. 36. Housing for a few (A) to 30 (B, C) gnotobiotic flies. Food is delivered by syringe and needle. B and C provide side-arms for removal of specimens with minimum risk of contamination or escape. Filter paper on bottom of flasks absorbs sticky feces and vomit. B and C are suitable for studies of nutrition, fecundity, and longevity of one or more generations.

FIG. 37. A. Flies are fixed to sterile paraffin and fed once or twice daily from tip of sterile disposable swab. B. Microbial output of fly is monitored by sampling saline solution. Possible error due to multiplication or death of organisms in solution is avoided by limiting exposure of the solution to the fly to 4-6 hours. C. Immobilized germfree specimen of *Phaenicia sericata* dining on 2 μliters of a solution containing *Salmonella typhimurium* delivered by micro-syringe.

150   THE FLY AS HOST

Arthropod organ and cell culture offers new ways of studying virus infection, pathology, biochemistry, hormones and development, and somatic cell genetics. Arthropod cells in vitro have proved surprisingly fastidious, and progress has been slow. Most successes have been achieved with ticks, lepidopterous pests, mosquitoes, and *Drosophila*, relatively few with muscoid flies. Over 100 papers have been published on the cultivation of *Drosophila* tissues alone.

Turning first to a brief overview of accomplishment with *Drosophila*, we find that organ culture received a big impetus from the work of Kuroda (1954), Horikawa (1958), and Schneider (1964). The latter devised a medium in which imaginal eye discs developed into ommatidia with cornea and lens that contained the eye pigments, ommochrome and drosopterin; antennal discs differentiated to the tripartite stage. The explants survived a maximum of 48 days. A novel and highly productive in vivo approach is the implantation of isolated imaginal discs into the abdomens of adult flies. Male genital imaginal discs, for example, have been cultivated for more than six years by transferring a piece of the proliferating disc into the abdomens of successive hosts. In this case, proliferation is greatest in fertilized females, less in virgin females, and least in males (Hadorn and Garcia-Bellido, 1964; Hadorn, 1968).

In vitro culture of *Drosophila* cells owes much to the work of Horikawa and Kuroda (1959) on hemocytes; Castiglioni and Raimondi (1961) on ganglia; Schneider (1963) on ovaries; and Horikawa and Fox (1964), Echalier et al. (1965), Seecof and Unanue (1968), and Echalier and Ohanessian (1970) on embryos. The last investigators succeeded in maintaining several cell lines for more than two years. Ohanessian (1967) infected *Drosophila* cell cultures with σ virus, and demonstrated that multiplication of the virus occurs without visible cytopathic effect. The reader may recall that this virus is transmitted to the next generation through the egg; infected flies are sensitive to CO_2 and are killed by a normally innocuous exposure to the gas.

Differentiation of imaginal discs of muscoid flies in vitro has received only slight attention. Gonadal discs from prepupae of *C. vicina* differentiated a little (Leloup and Gianfelici 1966), but eye discs of pupal *Glossina palpalis* developed distinct ommatidia with brown pigment (Trager, 1959). Explants of imaginal discs, midgut and other tissues from 6-day pupae of the tsetse produced cellular outgrowths that survived about 4 weeks. When *Trypanosoma brucei* and *T. congolense* were inoculated into these cultures, they grew readily and showed a variety of forms, but none was infective. Likewise, *T. rhodesiense* proliferated without producing metacyclic infective forms when injected into pupae of *G. fuscipes quanzensis* (Nicoli and Vattier, 1964).

Cell lines from dispersed house fly embryonic tissues have been maintained up to a year, with mitotic activity highest in early cultures and declining as the cultures aged. Both fibroblast and epithelioid cell types were present, and it is especially noteworthy that the dissociated cells reaggregated to form muscle and nerve tissue (Eide and Chang, 1969). These cellular recognition patterns are similar to what Hadorn observed when he implanted mixtures of cells from different discs into *Drosophila* hosts. The cells apparently moved around until they found their proper associates, so that cells from a leg disc never joined those from a wing disc.

Primary monolayers from house fly embryo homogenates have been obtained by Greenberg and Archetti (1969), and Wallis and Lite (1970). The latter maintained cell sheets from various preadult stages for up to 6 weeks, but the heaviest outgrowths developed from day-old pupal explants. Prof. Archetti and I also obtained primary cultures of axenic *Musca sorbens*—from embryos, larval salivary glands, 1- and 2-day pupae, and the cut end of adult fore-midgut. Success with embryonic tissue of *Musca* seems to correlate with use of eggs that are less than 6 hours old; pH of successful cultures has ranged from 6.85 to 7.4.

Fluid expressed from conventionally reared young pupae of *Sarcophaga bullata* has also been a source of primary cultures. Cells aggregated during the first 24 hours, and by the third day formed a confluent layer. The cultures were maintained for nine weeks, and mitosis was observed in 8-week old cultures (Shinedling and Greenberg, 1971).

It is lamentable that there are so few publications on tissue culture of muscoid flies, considering the virus problems that are now waiting to be resolved. Flies are naturally infected with polio, Coxsackie, ECHO, and, in all likelihood, hepatitis viruses, but it is still not known whether they (and cockroaches) are a viable link or a dead end. It was mentioned earlier that multiplication in the insect is crucial to virus transmission via this route, and studies of fly-virus compatibility at the cellular level would provide the same sensitive assay of vector potential now being developed for other viruses and arthropods. The interested reader is urged to consult the book by Smith (1967) and a recent published symposium (Weiss, 1971), which deal broadly and specifically with the field.

4

Flies and Human Diseases

Fly in April, dead child in July.

RUSSIAN PROVERB

THE DISEASES discussed in this and the next chapter are arranged according to the systematic position of the etiologic agent, starting with viruses and ending with helminths. In the vast majority of the more than sixty diseases presented, the evidence for fly involvement is unmistakable. A few other diseases are included that are not fly-borne but were once thought to be occasionally fly-borne; these are mentioned to provide the reader with a retrospective. Thus, the likelihood of fly involvement in smallpox and diphtheria is remote but the existing literature should be dealt with, if only to formally inter the question. Diseases transmitted by mosquitoes, tabanids, *Culicoides*, *Phlebotomus*, *Simulium*, and *Glossina* are not included except when synanthropic flies are possibly involved. In these cases, the evidence has been sorted with a view to assessing the relative importance of synanthropic flies; otherwise, it is beyond our purpose to deal with the medical entomology of these other Diptera. Zoonoses are treated in Chapter 5, with the exception of nontyphoid salmonellosis, which is included in this chapter to round out our discussion of enteric infections.

POLIOMYELITIS

Even before the introduction of polio vaccine, the vast bulk of polio infections (90 to 95%) were asymptomatic. Some few persons experienced fever, malaise, and muscle paralysis following attack of the central nervous system.

The poliomyelitis viruses, types 1, 2 and 3, are readily distinguished immunologically. The usual period between contact and onset of paralysis is 10 to 15 days. Virus may be isolated from feces or throat secretions; after an individual becomes infected with virus, replication initially occurs within the naso-pharynx and gastrointestinal mucosa. Man is infected mainly by fecal-oral transmission, though transmission by way of pharyngeal secretions may occasionally occur. Asymptomatic cases serve as reservoirs of virus, and close association with infected persons is an important means of spread. The greatest communicability is apparently in late incubation and the first few days of overt illness, though the virus persists in feces for 3 to 6 weeks or more.

Polio prevails throughout most of the world, occurring both sporadically and in epidemics at irregular intervals, with the highest incidence in summer and early fall. The rate of transmission in urban communities is closely related to the level of sanitation. In tropical areas where sanitation is poor, the infection is acquired in infancy or early childhood and paralytic disease is uncommon. In developed areas, more people mature without contacting the virus, but adults are more severely affected and run a higher risk of paralysis and death.

FLIES AND POLIO

There have been two widely separated peaks of interest in flies and polio, first in the 1910s and again in the 1940s and 1950s. Probably the earliest report came from Flexner and Clark (1911), who showed that house flies, after feeding on infected monkey spinal cord, retained the virus superficially or in their gut for at least 48 hours, shown by intracerebral injection of filtrates of crushed flies into rhesus monkeys, which contracted the disease. Howard and Clark (1912) found that house flies carried viable virus for several days externally and for several hours within the gut. Mosquitoes and lice were refractory, but one bedbug harbored virus for seven days.

In the same year, Brues and Sheppard drew attention to *Stomoxys calcitrans*, which were abundant in the vicinity and rooms of polio victims. Monkeys in all stages of the disease were bitten for several weeks by *S. calcitrans*, which transmitted the disease to 6 of 12 healthy monkeys (Rosenau and Brues, 1912). This was quickly confirmed by Anderson and Frost (1912), who used several hundred stable flies to transmit polio between two infected and three well monkeys. However, these same investigators (1913) could not confirm their own work, nor could a number of others (Sawyer and Herms, 1913; Rosenau, 1914; Levaditi and Kling, 1914; Boudreau et al., 1914; Clark et al., 1914; Campbell et al., 1918); Francis' (1914) attempts with *Lyperosia irritans* were also unsuccessful.

Josefson (1912) demonstrated that the virus adheres to objects and can remain virulent, but he failed to show this in the case of flies. Noguchi and Kudo (1917) obtained no transmissions with the bite of *Culex pipiens*.

A waning of research interest in insects followed. It was noted that even large quantities of blood from an infected monkey injected into a healthy monkey did not produce the disease, that *Stomoxys* only occasionally bites humans, and that polio epidemics also occur in winter. These facts and numerous failures with mosquitoes seemed to exclude biting insects from any probable role in the transmission of polio (Dick, 1925). In 1927, Dr. Byrd Hunter (Lyon and Price, 1941) drew attention to the similar distribution of dysentery and polio cases in Huntington, West Virginia, but the idea that polio was fecal-borne was not ready to be accepted.

POLIOVIRUS FROM WILD FLIES

Research interest was re-kindled when, in 1937, Rosenow et al. demonstrated poliovirus in filtrates of wild house flies. This set the direction for subsequent research. Toomey et al. (1941) placed fly traps near the mouth of a brook that emptied raw sewage into Lake Erie from an area where a large number of polio cases were found; mostly large blow flies and an occasional house fly were collected. A suspension of 2000 ground flies injected intracerebrally and intraperitoneally and instilled intranasally into two monkeys produced polio, which was confirmed by additional passage.

Paul et al. (1941) described two instances in which virus was recovered from flies trapped during epidemics of this disease. One sample of about 1000 to 1200 flies collected from a summer camp in Connecticut included mostly *Lucilia illustris*, *Bufolucilia silvarum*, *Phaenicia sericata*, and *Phormia regina*, and lesser numbers of house flies, *Muscina stabulans*, *Sarcophaga haemorrhoidalis*, *Ophyra leucostoma*, and *Protocalliphora*. Inocula prepared from pools of these flies produced polio in one monkey. A second sample of flies collected near a Jasper, Alabama, epidemic area also produced the disease. Trask and Paul (1943) noted later that in most instances in which virus was isolated, the flies had had a chance to feed on fresh human feces that might have contained virus. Repeated isolations from wild flies have been made from urban, rural, and suburban areas of widely scattered sections of the country (Sabin and Ward, 1941, 1942; Trask et al., 1943; Melnick et al., 1949; Melnick, 1950; Melnick and Ward, 1945; Horstmann et al., 1959; Brygoo et al., 1962; Asahina et al., 1963; Downey, 1963; Paul et al., 1962; Riordan et al., 1961). Such frequency of virus isolations from muscoid flies still leaves the question of their possible importance in polio transmission.

PERSISTENCE OF VIRUS IN FLIES

In attempts to answer this question, many laboratory investigations have been carried out. Bang and Glaser (1943) found that Theiler's virus persisted in adult *M. domestica* for at least 12 days, but the Lansing strain survived only 2 days. It appeared that rapid disappearance of virus was due to excretion by the insect; Lansing virus was recovered from fecal and vomit spots; but there was a sharp decrease in the amount of detectable virus in the abdomen of the fly from 2 to 7 hours after ingestion. Lansing virus was not recovered from *L. illustris*, *M. stabu-*

lans, C. vicina, or *S. haemorrhoidalis,* but Theiler's strain behaved in these flies as it did in *M. domestica.* Rendtorff and Francis (1943) obtained similar results with the Lansing strain in *M. domestica* and concluded there had been no multiplication of virus in the fly. *Drosophila melanogaster* was refractory even as a temporary mechanical transmitter when it was exposed to a 1% suspension of Lansing virus; and, even when grossly contaminated with a 5% suspension, the flies lost infectivity within 24 hours (Toomey et al., 1947). A mouse-adapted Lansing strain inoculated into the hemocele of the German cockroach and the house fly survived for 15 and 12 days, respectively (Hurlbut, 1950).

Given these data that limit the length of time infected flies appear to retain and excrete virus, the next two reports were very significant. Ward et al. (1945) reported that food exposed to flies in homes of polio patients in an epidemic area acquired sufficient virus to produce a non-paralytic infection or asymptomatic carrier state in chimpanzees. Melnick and Penner (1947) found that human poliovirus naturally present in patients' stools fed to *Phormia regina* could be detected in these flies for about 2 weeks and in their excreta for about 3 weeks, but, when murine-adapted strains of poliovirus and Theiler's TO strain of mouse encephalomyelitis virus were ingested by *P. regina, P. sericata,* and *Sarcophaga bullata,* they behaved like inert material and were found in gradually decreasing quantities for 5 days. This work suggested a biological association between poliovirus and fly, and it was the first controlled study indicating that human poliovirus can be excreted by flies for a considerable time. Levkovich and Sukhova (1957) showed that poliovirus is excreted for 10 days in experimentally infected *M. domestica, M. stabulans, O. leucostoma, P. terraenovae, L. illustris, C. uralensis,* and *C. vicina.*

FATE OF VIRUS DURING METAMORPHOSIS

Noguchi and Kudo (1917) were the first to report destruction of virus during metamorphosis of flies. Filtrates of adults and pupae of house flies and blue bottles reared on infected monkey brain produced no infection when injected intracerebrally into a normal monkey. In 1943, Bang and Glaser found that the Theiler's mouse strain and the Lansing mouse-adapted strain of human poliovirus could not be recovered from *M. domestica, L. illustris,* or *M. stabulans* adults that had developed from infected larvae. This was also observed by Gordon (1943) in *M. domestica.* Melnick and Penner (1952) confirmed these observations with a Lansing strain in *P. sericata.* Minimal recoveries of virus were made from maggots only 1 and 3 days after the feeding of virus.

DOES POLIOVIRUS MULTIPLY IN FLIES?

Melnick and Penner (1952) fed virus quantitatively to *P. regina*, *P. sericata*, and *M. domestica* from naturally infected human stools. Though virus was recovered from flies and their excreta between 5 and 17 days and between 4 and 10 days, respectively, no conclusive evidence of virus multiplication in these insects was obtained. An indication of virus decline was that virus could not be carried for more than two serial passages in adult flies. Virus persisted in the dried excreta of flies for at least 1 or 2 days at room temperature and 3 or 4 days at 4°C.

Gudnadóttir (1960) studied the fate of type 1 polio viruses in *P. regina*. The amount of virus in infected carcasses fed to flies and in fly excreta was determined by the plaque assay method. Flies and their excreta remained infective for 11 days when kept at room temperature or when incubated at 36°C for 2 hours a day. There was no significant drop in virus titer in flies kept in hibernation for 3 months. A relative increase in titer occurred between 9 and 18 hours after the feeding if the flies were incubated at 36°C for 5 to 15 hours a day; the peak occurred later, at 40 to 52 hours after feeding, if less incubation was used. Titers in excreta were parallel to titers in carcasses. The author stated that these results showed that these flies are not solely mechanical carriers of type 1 virus but are capable of supporting multiplication of this virus or retaining it in a masked form, unmasking it after a few hours. The experiments of Davé and Wallis (1965) gave no evidence of multiplication of type 1 or type 3 attenuated vaccine strains (Sabin) in *P. sericata*; poliovirus was recovered in gradually diminishing quantities for as long as 15 days from flies fed on type 1 virus and for at least 13 days from those fed on type 3.

FLY SURVEYS CORRELATED WITH EPIDEMIOLOGIC DATA

In systematic studies of the fly population of New Haven, Connecticut, Power and Melnick (1945) found that the peak of fly incidence preceded the peak of the polio epidemic by 4 to 5 weeks. During the height of the epidemic, *Phaenicia*, *Lucilia*, *Phormia*, and *Sarcophaga* were the only genera present in the trap collection; *P. sericata* was the dominant fly, representing 80-90% of the total catch. The species fluctuated in numbers during different parts of the season. Using fresh normal adult human stool as bait, *Phaenicia*, *Phormia*, *Sarcophaga*, *Muscina*, and *Calliphora* were the dominant genera trapped (Francis et al., 1948). Zumpt and Patterson (1952) conducted similar studies in South Africa. Melnick (1949), in an attempt to determine possible correlation between single species of flies and polio, examined over 93,000 flies trapped at random during an epidemic. At the peak of the epidemic, only *P. regina*, and *P. sericata* yielded virus. Two weeks later the positive species were *P. regina*, *M. domestica*, and *Sarcophaga* spp. In the last 2-week period, when the epidemic had practically ceased, all species were negative except *Cynomyopsis cadaverina*, which became prominent at this time. It is important to note that the occurrence of virus in a particular fly was not merely a reflection of its greater prevalence, but also of its habits; thus, both *Sarcophaga* spp. (4% of the total) and *M. domestica* (3%) were found to be positive at a time when *O. leucostoma* (20%) and *P. sericata* (24%) were both negative.

Francis et al. (1953) reported a marked correlation between presence of poliovirus in privy stool samples in Texas and in flies tested in a corollary study there by Melnick and Dow (1953). Poliovirus was found to be more prevalent in flies during the year of the epidemic than later; in April-May, 27% of fly samples yielded virus, only 9% in June, and none in July-November. In the same locale, fly control was instituted 6 months before and was maintained during the outbreak, but this failed to reduce the number of polio cases or to affect the time-course of the epidemic (Paffenbarger and Watt, 1953). Though flies could not be ruled in or out, a significant number of paralytic cases in this epidemic had a history of contact with a preceding paralytic case. Past attempts to measure the impact of fly abatement on prevalence of polio have been fraught with difficulty, largely centered on the need to achieve fly suppression in advance of an epidemic that may never have come, possibly for other reasons (Melnick et al., 1947).

Nuorteva's (1963) excellent contributions to the ecology of synanthropic flies in Finland unfortunately contain several unsubstantiated speculations about polio. He discusses fly transmission from infected carcasses, but it is not clear what these might be, since man is the only known natural host for poliovirus. No studies have shown diseased carcasses to be infective for adult flies, and several studies have shown that adults do not acquire virus from infected maggots. Nuorteva also considers that blow flies in temperate zones may infect man by feeding on his wounds, but none of the species he discusses frequents human wounds. There is evidence that wounds and trauma, e.g. tonsillectomy, adenoidectomy, and extraction of teeth, may tip the host-virus balance toward frank infection (Paul, 1955), but this has nothing to do with flies.

The introduction of vaccine strains with genetic markers opened new possibilities for tracing the spread of poliovirus in human and fly populations. A study among Yaqui Indians in Guadalupe, Arizona, during an oral poliovirus vaccine trial was the first of this kind (Riordan et al.,

1961). Unfortunately, little evidence was found that local flies picked up the LSc attenuated type 1 vaccine strain, since in only a single instance was an LSc-like strain isolated from flies. This was probably due to lack of opportunity, since only 3 children were excreting the vaccine strain and the number of flies present at the time was relatively small. It has been shown that flies collected in nonepidemic areas and periods are less likely to harbor enteroviruses (Trask and Paul, 1943; Trask et al., 1943; Downey, 1963), and this is also true of flies collected from sanitary environments (Paul et al., 1962). Nevertheless, a striking feature of the study was the frequency and ease with which non-vaccine strains of polioviruses and other enteroviruses were found in or on flies. The correlation between the various wild strains isolated from flies and from children was fairly good.

Further observations were made by the Yale group (Paul et al., 1962) during an oral poliovirus type 3 vaccine trial in Costa Rica. Again, viruses recovered from flies mirrored those recovered from children, and there seemed to be a rough parallel in the frequency with which poliovirus type 3, at least, could be recovered from the two populations. Many of the type 3 poliovirus strains in flies were indeed wild strains, but some of the type 3 strains and all of the type 1 strains from flies had genetic markers characteristic of the vaccine strains. A significant finding was the speed with which vaccine strains spread within a household. In most instances the virus was acquired by the sibling of the index child within five days after the latter was infected. It suggests that in this study virus was spread by intrafamilial contamination. Nonetheless, the authors again expressed surprise at the ease with which poliovirus and enteroviruses were isolated from flies.

Data obtained by Sheremet'ev (1964) also show a close correspondence between vaccine strains in flies and people. Studies carried out in 1961 and 1962, after mass vaccination of children, showed all three types circulating in fly populations. The attenuated strains were not found in flies during the periods between vaccinations, although Coxsackie and ECHO viruses were recovered all year long, particularly in June and July.

As we have seen, there is ample evidence that human populations readily infect flies with poliovirus. But we are woefully ignorant whether and to what extent flies return the favor. Further ecologic studies of flies and vaccine strains have been strongly urged by investigators. The circulation of vaccine strains in nature now raises two new considerations. What are the chances that the multiplication of avirulent virus in flies will produce virulent mutants? Granted the chances of such an occurrence are slight, we may have to grudgingly accept our traditional villain as an ally since it may now be active in vaccinating populations with beneficial strains. But let us study flies further before we praise them. For example, enteroviruses and flies abound in the same communities, and we still know very little about the outcome of competition between wild and vaccine strains in human and fly populations.

Other problems cloud the picture. Aside from winter epidemics, which definitely rule out flies, there are the frequent recoveries of virus in sewage and less frequent but most interesting recoveries in cockroaches. Despite the many brilliant investigations that have been conducted in the field of polio epidemiology, no reliable evidence unequivocally establishes insects, water, food (other than milk), or sewage in transmission of the infection.

COXSACKIE DISEASE

The first member of this group of viruses was recovered in 1948 in Coxsackie, New York, during screening of stools for poliovirus from patients who had developed muscular weakness during a polio outbreak that year. Extracts of these stools produced a fatal disease when injected into suckling mice. The Coxsackie viruses are worldwide in distribution. The incidence of antibodies in the general adult population is high— the result, as in polio, of inapparent infection. Frequent outbreaks involve all members of the family, and the spread of infection involves fecal and possibly droplet contamination.

The Coxsackie viruses are divided into subgroup A (24 types) and subgroup B (6 types), which differ in their tissue pathology in infant mice; the A viruses produce degeneration of skeletal muscle, whereas B viruses generally spare the skeletal muscle and produce inflammatory lesions in the brain, heart (smooth muscle degeneration), liver, and pancreas. In man, these viruses produce diseases associated with the central nervous system, heart, pleura, and respiratory tract. They are, in fact, the causative agents of several different clinical entities. It is sufficient for our purpose to briefly describe two such syndromes.

Herpangia is characterized by fever and the appearance of vesicles or ulcers in the mouth. The disease is not fatal and, in most cases, lasts 2 or 3 days. At least 6 different immunologic types of C virus have been isolated from throat swabs and feces. The usual source and reservoir of infection are the pharyngeal discharges and feces of subclinically infected persons. The disease is communicable during the acute stage of illness and perhaps longer since the virus persists in stools for several weeks. It most often affects children 5 years of age or younger; age incidence varies, however, and persons over 18 have contracted the disease. It occurs throughout the world, both sporadically and in epidemics, with the greatest incidence in summer and early autumn. Contaminated flies have been found, but no reliable evidence exists for dissemination by insects, water, food, or sewage.

Group B Coxsackie viruses are responsible for pleurodynia, an acute febrile disease, characterized by severe pain, headache, anorexia, and malaise. This may last a few days or several weeks, but there are no sequelae. The disease is not common and usually occurs in epidemics, with outbreaks reported in Europe, England, Australia, New Zealand, and North America. It is a summer and early autumn disease, occurring in all age groups but most commonly in children and young adults. The viruses are present in the throat or feces and have been found in sewage, flies, and mosquitoes.

Melnick (1950) made collections of fecal material from hospitalized polio patients, of sewage, and of flies during a polio epidemic in May through the end of August in three cities in North Carolina. Coxsackie virus was recovered from stool samples in both Greensboro and Winston-Salem. It was detected in flies collected throughout Greensboro until the end of July; it was also found in sewage samples from all three cities. The finding of an increase of C virus antibodies in normal individuals during the summer when the agent was readily detected in sewage and flies indicated that subclinical infection was common. Strains of C virus were also recovered from flies and sewage in Connecticut during the summer of 1948, and a Texas strain was recovered from flies collected in the Rio Grande Valley in the same year (Melnick et al., 1949). Other similar studies were reported by Melnick and Dow (1953), Francis et al. (1953), and Horstmann et al. (1959).

With warmer weather, an increase in both fly numbers and activity improves the opportunity for flies to contaminate themselves with virus. Thus, Riordan et al. (1961) observed a gradually increasing incidence of Coxsackie B5 virus in flies of the Guadalupe, Arizona, study area from February to May.

This was previously borne out in a study by Melnick et al. (1954), who carried out tests for up to 4 years for virus in urban sewage and flies in Arizona, Connecticut, Kansas, Michigan, New York, and West Virginia. In almost every urban area studied, virus appeared in some of the specimens collected in the summer and fall and then disappeared during the winter and spring. The recovery of C viruses was more regular from sewage than from flies. Virus was also found in flies in residential areas where there was no obvious source of contamination.

It was not evident from this study whether the presence of C virus in sewage and in flies is a direct or even an indirect link in the chain that leads these viruses from one infected person to another.

The finding of virus in flies raised the question of whether the agent multiplies within the insects or whether it is merely carried on the surface or within the gut. However, from experiments with laboratory-infected *P. regina*, *P. sericata*, and *M. domestica* via naturally infectious human stools, no evidence was obtained for virus multiplication; nevertheless, it was found to persist within flies or to be passed in their excreta for > 12 days in *P. regina*, 14 days in *P. sericata*, and < 2 days in *M. domestica* (Melnick and Penner, 1952). Duca et al. (1958) reported that virus survived 16 days in the house fly, under conditions that tended to minimize but not preclude self-contamination. Though multiplication could not be confirmed, they did note a 3-fold increase of virus in the flies compared to the original inoculum. The virus did not survive within the metamorphosing insect but was present on the exterior of the puparium and re-contaminated the emerging fly. Of particular interest were their epidemiologic findings. Dalldorf A-type 10 virus was isolated from house flies captured in five homes in different neighborhoods where people lived under crowded conditions and sanitation was poor. The virus was found to be strongly neutralized by the patients' own antisera. In a few instances, these virus isolations from flies occurred 34, 37, and 100 days after the illness. Besides disseminating virus into other neighborhoods, flies may serve as temporary reservoirs during periods when people may not be excreting virus in large numbers.

Thus, the Coxsackie viruses fit generally into a similar scheme with flies as do the polio and other enteroviruses. We have too little evidence to decide whether the presence of C virus in flies and sewage signifies a dead end or a link in the cycle of human infection.

ECHO VIRUSES

These viruses were also first isolated during the screening of human stool samples for polioviruses, and they did not appear to be associated with any specific disease. But, many viruses in this group are now known to cause aseptic meningitis and other infections such as rash, fever, and respiratory disorders. They occur with some frequency in the intestinal tract, especially of children, in the summer and under poor socio-economic conditions, and most isolations are made from stools or rectal swabs and some from the throat and cerebrospinal fluid.

The viruses are distributed worldwide, but all reported epidemics have occurred in the temperate zones. Paul et al. (1962) in Costa Rica found in most cases that the ECHO viruses isolated from trapped flies were the same as those harbored by the local children. Horstmann et al. (1959), who trapped mostly *M. domestica* and *P. sericata* as well as other feces- or flesh-eating flies, isolated ECHO as well as Coxsackie and polioviruses. Riordan et al. (1961) obtained similar results, and Downey (1963) isolated types 6 and 7 from pools in which *P. sericata* and *L. illustris* were most prevalent, with *P. regina* and *Calliphora* present but less numerous. The group appears ecologically along with the polio and Coxsackie viruses in the same situations.

SMALLPOX

This account is given with the understanding that fly involvement in the spread of smallpox might occur under conditions approaching total societal disorganization and then only as a minor means.

The virus of smallpox is quite stable and can be dried under relatively unfavorable conditions and still retain its viability for some time. It is transmitted from man to man by contact with a patient or his immediate surrounding. The lesions of the skin and mucous membranes and respiratory discharges of a patient are rich in virus, and a patient thoroughly contaminates the area he occupies. Transmission by contaminated bed clothes, dust from a patient's room, and via the respiratory route have been reported, and the persistence of the virus makes the handling of a patient's body after death a matter of considerable danger.

Smallpox is found in all regions and climates of the world, but it occurs more frequently during the winter, which argues against fly transmission. Historically a few observers have remarked on the possibility of flies spreading the disease, but no experiments were ever conducted.

Wawrinsky (1888), in discussing an epidemic of this disease in Stockholm in 1884, stated that flies and other insects might carry the infection; he remarked that the disease had spread in the warmest season, when flies were in abundance.

Armstrong (1905) in England observed that house flies were a pest in smallpox hospitals, settling on the faces and other exposed parts of the patient when the eruptions were pustular or crusting. He reasoned that their feet and proboscis became contaminated and that flies might be carried by the wind to any distance to infect persons, food, and clothing; for years he had protected his patients from flies with muslin netting.

Hunziker and Reese (1922) in Basel described the smallpox epidemic there in 1921 and observed many flies in the hospital rooms of smallpox victims; three weeks after fly prevention procedures were instituted, the epidemic subsided. In their opinion there was a strong possibility that flies may transmit this disease.

INFECTIOUS HEPATITIS

Our understanding of the epidemiology of this disease has awaited the discovery of a suitable experimental animal in which infection with the virus is demonstable. The outlook has improved with the recent finding by F. Deinhardt and his group at Presbyterian-St. Luke's hospital in Chicago that the marmoset is a suitable host. In man, the virus has an enteric entry and exit, which qualifies the usual contaminative factors, including flies and cockroaches, as possible disseminators. The literature on flies is understandably small, but it is already conflicting.

Kirk (1945) made an analysis of the circumstances surrounding an epidemic of infectious hepatitis among New Zealand soldiers at El Alamein during the North African campaign in 1942. The disease broke out among 1059 soldiers out of a total of 7500 men of Group I, 35 to 40 days after they occupied a 5-mile square area on the front line; it lasted until 35 to 40 days after the group withdrew. The area was a piece of barren desert, with no scrub or water or any natural feature to distinguish it from the rest of the Western Desert. It was possible to eliminate such factors as food, water, hardship, living conditions, and climate since these were similar among all the units in the line. The circumstances which strongly implicated flies were the following:

1. The ground had just been recaptured from Italian and German troops who had suffered an epidemic of hepatitis. The area was heavily contaminated with feces and inadequately buried dead bodies.

2. Flies were present in "incredible numbers"—the men "were unable to protect their food, mess-tins, mouths or hands from the hordes which swarmed over everything." Field conditions made it difficult to obtain a satisfactory level of sanitation.

3. A unit of Australian troops that occupied a similar area suffered similarly.

4. A unit that replaced the New Zealand unit in the same area also suffered an epidemic at the end of the incubation period.

Kirk's conclusion that flies were responsible for the epidemic seems entirely reasonable.

A different kind of campaign, namely fly control, was waged in the Russian city of Tbilisi during 1959-1962 and 1963-1966 and led to the opposite conclusion. Spotarenko (1969) and his colleagues analyzed the effect of fly suppression on the incidence of dysentery and infectious hepatitis morbidity. They found that a marked reduction in the number of flies during 1963-1966 led to a reduction in the seasonal elevation of dysentery among those who were over seven years of age. However, there was no reduction in the incidence of infectious hepatitis, and the Russian scientists concluded that the role of flies is therefore insignificant. Since flies may transmit subclinical, possibly immunizing doses of virus, and since most hepatitis infections are subclinical, ruling them out may be of questionable validity under the circumstances.

CHOLERA

The ancient home of cholera seems to have been Bangladesh and the West Bengal Province of India. Here it remained endemic for centuries until 1817, when the first known pandemic occurred. During a seven-year course, the disease struck populations in Africa, Europe, and even Australia. At least five large-scale pandemics have occurred since that time, and, at the moment of this writing, an outbreak in Egypt has spread to Russia, despite quarantine and other international sanitary conventions. The disease has frequently appeared in China, Japan, the Philippines, and other Asiatic regions but is considered to have been introduced, and not truly endemic in these places.

Cholera generally occurs between the latitudes of 50° north and 30° south and is most prevalent in warm, humid weather. The evidence for water-borne cholera has stood on firm ground ever since John Snow's classic, epidemiologic analysis of a cholera outbreak in 1854, which he traced to a sewage-polluted part of the Thames used by part of London for its water supply. The Hamburg epidemic of 1892 provided similar evidence on an even larger scale.

Intermittent rains promote the diffusion of cholera in places with inadequate sanitation by washing contaminated human feces into streams and water supplies. On the other hand, disease incidence drops markedly shortly after onset of the rainy season (Flu, 1915). This is attributed to a mechanical flushing of the ground and waters, a reduction of fly breeding due to washing away of feces, and the cesspool contents becoming too liquid for domestic flies to breed. It has been shown that algae in water tanks favor the pathogen's survival by raising the pH of the water through photosynthetic activity. Heavy rainfall prevents this rise.

Man is the only known natural host of *Vibrio comma*, the cholera organism. Since the role of flies in the general mode of dissemination stems from their association with man, it is well to begin with him. The organism is neither invasive like salmonellae nor destructive of the intestinal mucosa like shigellae. Though it may be found in the gall bladder and in the urine, its primary residence is the intestinal mucosa, where it multiplies abundantly and releases an exotoxin called choleragen. This toxin affects the permeability of the gut, causing excessive loss of fluid and electrolytes, which, in severe cases, reaches a stool volume of 9 liters a day. If unchecked, the consequences are hemoconcentration with acid shift in blood pH, coma, and rapid death. Cholera symptoms have been reproduced in rabbits and human volunteers with purified choleragen or sterile filtrates of patients' stools demonstrating the essentially toxic nature of the disease. Now it is known that the choleragen, and probably similar products of other diarrhea-inducing organisms, stimulates production of cyclicAMP (a form of adenosine monophosphate) in the epithelium of the small intestine, which causes the copious release of salt and water (Hirschhorn and Greenough, 1971).

Convalescing patients usually cease to excrete the organism after a few days, and persistent carriers are infrequent. Nevertheless, some individuals may continue to shed the organism in the feces from inapparent infections of the gall bladder, or even in the urine. Healthy carriers may also serve as sources for contact spread of the disease. Outside the human host, desiccation, high temperature, low pH, and microbial competition are some of the factors inimical to the survival of *V. comma*. Nevertheless, the organism persists one or two weeks on fabrics and other surfaces when the air is moist and the temperature is moderate.

Cholera may remain sporadic and endemic, or, given a concatenation of circumstances—crowding of susceptible individuals, meteorologic factors, societal disruption as in Odessa in 1922 (Elkin, 1961), possible changes in the pathogen, fly contamination of uncooked food—the disease may flare into epidemic proportions. Explosive outbreaks, including the most recent one in Pakistan in November 1970, which followed a devastating tidal wave, are the result of contaminated water. But authorities argue that the installation of safe water supplies in endemic areas has not reduced the death rate due to protracted or diffuse epidemics. Some vehicle other than water has been held responsible for these slight and sporadic flare-ups.

Explosive epidemics have no seasonality, whereas protracted epidemics, at least in temperate regions, are confined to summer and fall, casting suspicion on the fly. The first public writings of Sir William Moore dealing with the subject date from 1858 (1893). Based on a lifetime of experience in India, Moore cautioned against flies as conveyors of isolated cases of cholera. Opportunities for fly transmission must have been numerous. He recollects that the baskets of boiled rice that had been placed outside the regimental kitchen for the water to drain away were covered with a black coating, which, on closer inspection, turned out to be a myriad of flies.

In 1850, Nicolas, a British naval doctor, wrote: "The 'Superb' had been at sea for six months, with the rest of the Mediterranean fleet; cholera broke out on board, the flies were very numerous on the ship; but little by little, they diminished and with it, the epidemic. Returning to Malta, but without making any contact with the port, the flies returned and cholera reappeared."

In 1855, a Pennsylvania physician, J. F. Reigert, implicated "Plague flies" (*Musca ochrapesus = Chyromya flava*) as the cause of cholera. He observed vast clouds of the small yellow flies immediately before a cholera epidemic in 1852 and again in 1854 and 1855. Reigert also noted dead flies of this species around a water pump and thereby erroneously established both the cause of cholera (flies) and the cure (limewater). Contemporaneously, Snow studied a water pump in Broad Street, London, and correctly deduced the means for the spread of cholera, thereby establishing the first classic epidemiologic model.

An isolated outbreak of cholera among children in a French hospital was assumed, because of the controlled water and food, to have been carried to them by an abnormally large population of flies from other children dying of the disease in a nearby section (Lesage, 1921). Earlier, Buchanan (1897) described a cholera episode among a small group of inmates of an Indian prison where the drinking water, milk, and other vehicles were logically ruled out. Unusual swarms of flies were evident at this time, and it was presumed that many of them flew from nearby huts, which were the site of recent cholera cases and deaths, to the exposed food of this prison group, which happened to be in the direct line of flight. MacKaig (1902) described outbreaks in the Bombay area, in which the water was ruled safe and contamination of food by flies was strongly ruled in. At one point he remarks, "One receives a great shock in visiting a native stricken with cholera. The patient is usually lying on a floor, and only partially covered with a few rags. The ground round about the patient is in a state too awful to be described. A cloud of flies is present, which rise with a loud hum on a motion being made near the patient, but at once settle again."

VIBRIO COMMA IN WILD FLIES

Credit for the first isolation of cholera vibrios from wild flies belongs to Cattani (1886) and Tizzoni and Cattani (1886), who isolated characteristic organisms from flies captured in the vicinity of a cholera ward in Bologna. Since the cholera-red test is not specific and agglutinations were not performed in early studies, we cannot be certain of the specific identity of the organism.

MaCrae (1894) was the first in India to demonstrate that flies carry cholera organisms. Milk was boiled and then set out in several open areas of Gaya jail during a period when flies swarmed and cases of cholera existed among the prisoners. Vibrios (presumed to be of cholera on microscopic appearance) were found in the milk after incubation. Tsuzuki (1904) succeeded in isolating cholera vibrios from flies captured in Tientsin, China, on the premises of a house whose residents were stricken with cholera. As Flu (1915) points out, it is natural to suspect flies in outbreaks among Europeans careful of their drinking water. He isolated the pathogen from flies taken in homes in the kampongs of Batavia where deaths from cholera had occurred. In one such effort, 10 out of 21 houses contained flies that yielded cholera vibrios agglutinable at a dilution of 1:10,000. Similar, but nonagglutinable, isolates were reported as negative. This impressive result led to a study of 2572 flies from 124 houses in four different kampongs. A positive result was obtained from one group of 15 flies on 7 January, though the last reported case in this kampong had been 6 December, and from neighboring kampongs, 29 December. Since it was known that the organism does not persist in the fly, it was felt that here was an instance of flies becoming infected from an unreported or undiagnosed case. Twenty-three years of experience in India led Ross (1927-28) to conclude that the prevalence of epidemic cholera in Bihar and Orissa is associated with an extraordinary prevalence of flies. They swarmed on the cold food that the laborers customarily left over in order to have something to eat for breakfast. The sacred water tanks at Puri did not yield *V. comma*, but a large proportion of the flies from the temple area in the center of town did. The proportion of positive flies diminished to zero as the distance from the temple increased. In Albania, during May to October 1927, a cholera epidemic resulted in 288 fatalities among 435 cases. The organism was recovered from flies (no species given in this or the previous accounts) and from cereals to which these insects had access (Anon., 1928). These several findings throw light on flies as possible extra-human sources of the pathogen during endemic periods.

THE FLY AS HOST FOR VIBRIO COMMA

A year before the first isolation of the vibrio from wild flies during a cholera epidemic in Bologna, Maddox (1885) reported that *Eristalis tenax* and *C. vomitoria* excrete viable vibrios after feeding upon a culture of the organisms. In the next few years, a number of investigators addressed themselves to the questions: can cholera be transmitted to people via the fly, and how long does the organism live in the insect? Three investigators published their work in 1892, two (Simmonds and Uffelmann) briefly corroborating Maddox's work. The third, Sawtschenko, used the cholera outbreak in Kiev as a source of materials for experiments with the house fly and an unidentified blow fly, probably *Calliphora*. He found that the flies began to excrete viable cholera organisms 2 hours after an infective meal and continued to do so, in increasing proportion to the saprophytes, for 24 hours. He was careful to show that vibrios were absent from controls. Particularly important was his demonstration that flies, especially blow flies, became infective by feeding on patients' excrement and the entrails of a corpse. The fly excrement remained infective after repeated "washing out" of infected flies with sterile feedings, and this suggested to Sawtschenko that multiplication had occurred. But he admitted that such an important question required more exact experiments.

Craig (1894) recovered viable cholera organisms from the excreta of house flies after they had fed for three days on contaminated food. Chantemesse and Borel (1905) showed that the pathogen remains alive for 17 hours, but not for 24 hours, on the legs, proboscis, and in the gut of the house fly. In a series of feeding experiments, Ganon (1908) demonstrated that the house fly could contaminate food with agglutinable *V. comma* up to 20 hours after an infective meal. The following account of the Russian investigator Passek's (1911) work is taken from Pollitzer (1955). Passek found cholera vibrios to be most abundant in the gut of house flies and blue bottle flies 21 to 24 hours after infection and to be absent after 72 hours. In the intestines of adult flies, *V. comma* was usually present in association with *Proteus vulgaris*, a symbiosis that, in the experience of Passek, greatly enhanced the virulence of the cholera vibrios. A corollary to this was that the vibrios isolated from the intestines of flies just emerged, in which the symbiont was still absent, had a lessened virulence. Highly virulent vibrios associated with *Proteus* could be isolated from the feces of adult flies up to 3 hours after voiding.

The transitory existence of the vibrio in flies was further confirmed by the work of Alessandrini (1912). Experiments with five species, *M. domestica*, *C. vomitoria*, *C. vicina*, *L. caesar*, and *S. carnaria*, demonstrated that most flies completely rid themselves of the organism within 24 hours, and all flies by 36 hours. The organism was not carried through metamorphosis. Similar results were obtained with an entirely unrelated fly, the phorid *Aphiochaeta ferruginea*, which commonly breeds in human feces and has a wide distribution in the tropics. Its minute size enables it to pass readily through ordinary window screens. In studying the possibility of cholera transmission by this fly in the Philippines, Roberg (1915) found that the organism does not survive metamorphosis but may be harbored in the adult's gut for 26 hours and on the exterior for 10 hours.

The most definitive study of flies and *V. comma* was reported in 1930 by Gill and Lal. Cholera organisms were fed to flies (presumably *Musca*) and were recovered in their excreta for a maximum of five days. This was confirmed by agglutination and cholera-red tests. It is noteworthy that residence in the gut of the fly caused morphological changes in the vibrios. They were thinner, more variable in size and shape, and had a more open curvature. The authors state that the vibrios disappear from the bodies of flies after about 24 hours, but re-appear on or about the fifth day. This has led some to the interpretation that the vibrio undergoes a cyclic development in the fly that includes an occult phase. The data presented by the authors are too scanty to justify such a conclusion, and no subsequent work has borne on this interesting idea. In fact, a few years later, Lal et al. (1939) reported that vibrios could not be recovered from near-axenic flies after 48 hours, except in one instance of five days. Following up on possible mutability of the vibrio in the fly, the authors studied the effects of serial passage through as many as ten fly hosts. They found no changes in fermentation or agglutination reactions of the vibrios, but changes in respiratory and glycolytic rates were common. The authors were unable to draw any general conclusions as to the relationship of flies to the possible occurrence of variant strains in nature. This work, like the other, is most interesting but unfortunately was never developed further. There are also the reports of Shortt (1937) and Soparker (1938), who, with few exceptions, could not obtain survival of the vibrios in flies (presumably *M. domestica*) for more than a few hours and found that gut homogenates were vibriocidal. These observations appeared in institutional reports not published elsewhere. No further work has appeared on flies in three decades. Perhaps the efficacy of vaccines and antibiotic and electrolyte replacement therapy has dulled interest in fly-mediated cholera. Further studies such as quantitative input-output are needed to answer the questions raised above if we are to have a better understanding of the fly as a host. Experiments should include commensal cockroaches, which may also serve to circulate the vibrio under certain conditions.

The League of Nation's Epidemic Commission to China, reported that the most important vehicle of cholera in central China was river water that became contaminated from infected human excrement and returned the infection to its human users. Vendors were observed to moisten their fruits and vegetables with polluted river water. The Commission also noted that flies abounded and had ample opportunity to contaminate sliced and whole fruits and vegetables (Robertson and Pollitzer, 1938).

Epidemiologists are in general agreement that flies are second-echelon disseminators of cholera. Given an outbreak in which the water proves good, the focus on flies is sharpened if cases are scattered but concentrated in poorly sanitated quarters, large numbers of flies are present, and food is exposed to them. Under such circumstances, fly control together with vaccination and other control measures should be instituted, as in the El Kurein outbreak in Egypt in 1947-1948. Unfortunately, in this instance, it was not possible to assess the effect of fly control nor of the other measures taken to localize the epidemic. A recent study by Sinha et al. (1967) in Calcutta emphasizes the importance of the asymptomatic human carrier in maintaining endemic cholera either by person-to-person contact or via water, flies, food, etc. In assessing the carrier level of flies, they collected 90 batches of fly samples (4 to 6 each) from 55 suspected houses and found that 4 batches were positive for *V. comma* and 3 for nonagglutinable vibrios.

ENTERIC INFECTIONS

Some of the diseases included under this rubric are vaguely defined, while others, such as typhoid fever, have well-defined symptoms and a specific etiology and host. Diarrheal disease, summer diarrhea, enteritis, bacillary dysentery, infectious colitis, gastroenteritis, and certain kinds of food poisoning are included here. These diseases are caused by various members of the family Enterobacteriaceae, and for simplicity the term "enteric infection" will be used as the umbrella term. This account excludes staphylococcal and clostridial food poisoning, polio, Coxsackie, enterovirus, and hepatitis infections, cholera and amebic dysentery, which have each been treated elsewhere in Chapters 4 and 5.

The most important pathogens in the Enterobacteriaceae are shigellae, enteropathogenic serotypes of *Escherichia coli* and salmonellae. Shigellae have the simplest epidemiologic cycle and parasitize the human intestine almost exclusively. Salmonellae are spread through complex zoonotic cycles and frequently invade the body of the host. Pathogenic *coli* seem to be intermediate.

SALMONELLOSIS

In 1885, Salmon and Smith isolated a bacillus from a case of hog cholera that, as *Salmonella choleraesuis*, later became the first species in a new genus named in honor of Salmon. There are now more than 1000 antigenic types of salmonellae, and it is doubtful whether all are indeed distinct species. But the epidemiologic value of precise serotyping has completely overshadowed academic restraints imposed by the question of what constitutes a *Salmonella* species.

With the exception of *S. typhi*, which is specific for man, and *S. pullorum* and *S. gallinarum*, which are generally restricted to fowl, most salmonellae enjoy a wide range of hosts. They are worldwide in distribution and have been found in the following animals: sheep, horse, bovids, biltong, swine, dog, cat, rabbit, chipmunk, rat, house mouse, deermice, ground squirrel, fox, muskrat, mink, moose, caribou, chinchilla, guinea pig, porcupine, nutria, lion cub, neotropical bat (*Glossophaga soricina*), Brewer's blackbird, horned lark, rufus-sided towhee, turkey, goose, duck, chicken, canary, pheasant, peafowl, chukar partridge, black-headed gull, green finch, mountain finch, bullfinch, sparrow, starling, thrush, jay, tanager, parakeet, secretary bird, green mamba snake, garter snake, gopher snake, Cuban snake, Galapagos turtle, Gila monster, iguana, Pacific fence lizard, chameleon, painted turtle, laboratory frogs, whitefish, cockroaches, ticks, flies, and shellfish.

The ability of salmonellae to infect mammals, birds, reptiles, and possibly amphibians and fish is a remarkable attribute shared by few other groups of microbes. It is the basis for the complex epidemiology and epizootiology about which we still know relatively little.

Despite this catholic taste, there is some evidence of host adaptation. In Australia, *S. bovis-morbificans* seems to occur most frequently in livestock and man. In New Guinea, *S. weltevreden* is the predominant type in the human population, while elsewhere in the world it is quite rare. *S. dublin* is endemic in cattle in many parts of the world. *S. typhimurium* was originally named as a specific disease agent of the mouse, but it is more common in birds and man.

Animals and their products may contain salmonellae as a result of a true infection (endogenous) or contamination (exogenous). Infections may be symptomless, mild, or fatal. In true infections, the organisms may be confined to the gut, producing diarrhea or constipation, but often there is systemic invasion with septicemia and pyemic lesions of internal organs. This accounts for isolations of salmonellae from hematophagous arthropods, e.g. ticks and lice. *Salmonella* infection in the gut may be associated with symptoms of food poisoning resulting from the release of endotoxin. This is different from food intoxications caused by *Clostridium* or *Staphylococcus*, which produce their toxin in the food. Among approximately 28,000 incidents of food poisoning in England and Wales in 1953-1956, *Salmonella* accounted for 16,214, *Staphylococcus* for 626, with about 11,000 not accounted for. The highest *Salmonella* attack rates are in young children, and the highest fatality rates are among the very young and very old. Asymptomatic cases and carriers (especially among animals) may shed and spread the pathogen as contaminators of food. Other means of dissemination are flies, cockroaches, and rodents, and preparation of food in contaminated areas (Fig. 38). Water-borne salmonellosis occurs rarely (Anon., 1971).

A listing of foods found to be contaminated with *Salmonella* is as impressive as the host list. It includes the following items: animal by-products, blood meal, bone meal, calf milk replacer, complete feeds, dog food, dried buttermilk and whey, milk, egg products (egg concentrate, frozen whites, dried whole eggs, frozen yolk, dried yolk), fat, feather meal, fish meal, laboratory animal feeds, livers, meat scraps, poultry feeds, swine supplements, tankage, pigeon meat, curdled milk, cream pastries, beef, pork, poultry, frog legs, dehydrated soups (made with contaminated noodles), gelatine, lettuce, raw ice, candy bars, cake mixes, cookie doughs, dinner rolls, corn bread mixes, and pizza dough.

The most frequent sources of common vehicle epidemics of salmonellosis are poultry and poultry products, beef, and pork, in that order. Naturally, this depends on local patterns of food consumption.

The organisms are generally resistant and may survive in the external environment for considerable periods of time. Hatcher chick fluff from commercial hatcheries in Japan was kept one to four years at room temperature and still yielded 10^4 to 10^6 viable *Salmonella* per gram (Miura et al., 1964). *Salmonella* has been isolated from stuffed natural poultry toys. *S. oranienburg* has survived in cockroach feces for more than four years (Rueger and Olson, 1969)! *S. typhimurium* endured 251 days in garden soil exposed to ordinary English weather. In these investigations, viability, not virulence, was tested, but it is known that *S. typhimurium* will retain its pathogenicity for mice after an 8-week exposure on fabrics (Wilkoff et al., 1969).

Given their widespread distribution in the biosphere, it is not surprising that salmonellae occur in sewage, river, lake, and sea water, corrals, abbatoirs, and in the environs of a variety of food processing plants. All of which makes for increasingly complex and enlarging pathways of dissemination that finally reach global proportions, for international commerce now binds the world in a giant food chain. For example, shipments of *Salmonella*-contaminated fish meal destined as a livestock feed additive for Dutch cattle originated in Peru, Chile, Argentina, South Africa, Angola, Morocco, Norway, and Iceland (Jacobs et al., 1963). In another case, contaminated bone meal from India ended up in England (Harvey and Price, 1962). It would be a mistake to conclude that countries with high standards of sanitation are free of guilt. In a large shipment of bone meat-meal imported into Bulgaria from the United

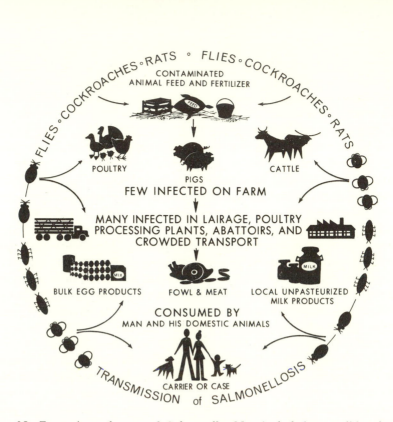

FIG. 38. Zoonotic pathways of *Salmonella*. Not included are wild animal cycles, about which little is known. (Modified and redrawn from Bowmer, 1964.)

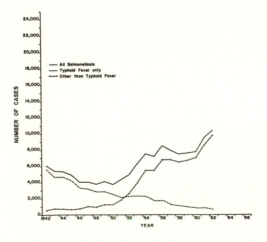

FIG. 39. Reported incidence of human salmonellosis in the United States from 1942 to 1962. Data taken largely from U. S. Public Health Service. (From Sanders et al., 1965.)

States, 69% of the samples were contaminated with *Salmonella* (Wessel-nikoff and Toneff, 1968). Almost a third of the legendary pigeons of Venice were recently discovered to be excreting *Salmonella!*

As can be seen from the accompanying graph (Fig. 39), nontyphoid salmonellosis in the United States has been on the increase in the past several decades. Part of this is due to increased surveillance and report-ing, but authorities are in general agreement that the upward trend is significant and alarming (Peluffo, 1964). During the same period, typhoid fever incidence has steadily declined. Some authors peg the annual incidence of salmonellosis during 1963-1967 at 20,000 cases a year* and consider the true incidence to be much higher (Aserkoff et al., 1970). A report of a WHO Expert Committee states that the salmo-nellosis morbidity rate generally is estimated to be 10-1000 times greater than the figures given in official reports (Anon., 1968).

The many measures that caused the decline of typhoid fever, e.g. pas-teurized milk, chlorinated water, improved personal hygiene and general sanitation, as well as detection and treatment, have not reduced the incidence of salmonellosis. The reasons for this have already been sug-gested and stem from the increased use of ready-cooked foods in the home, and bulk preparation and widespread distribution of human food and animal feed on a national and international scale. Many of these foods contain ingredients from a variety of sources. In 1970, one U.S. company had to recall 13.9 million servings of soup suspected of being contaminated with *Salmonella*. The organism was traced to the manu-facturer of noodles used in the soup mixes. There is also increasing importation of salmonellae by tourists (Ringertz and Mentzing, 1968).

The volume of publications on salmonellosis has paralleled its rising incidence and is truly staggering. For a comprehensive review of the problem, the reader is referred to *The World Problem of Salmonellosis*, edited by van Oye (1964).

SHIGELLOSIS

Another notorious group of species belongs to the genus *Shigella*. They are comparatively few in number, about 30 serotypes in all, and occur almost exclusively in man. In rare cases, they have been found in monkeys, dogs, and in a neotropical insectivorous bat. *Shigella* species have no invasive ability, and infection is confined to the gut. Dissemina-tion is therefore exclusively through human excreta. Despite this restric-tion, members of the group have a worldwide distribution and are often the most important agents of diarrheas and dysentery. This is certainly

* The U.S. Public Health Service figure for 1970 is 24,216.

the case in older children and adults; in young children shigellae vie with *E. coli* as an important cause of diarrheas. *Salmonella, Entamoeba histolytica, Vibrio comma, Proteus,* and various viruses can also cause clinical dysentery. This needs to be emphasized because dysentery is not synonymous with shigellosis.

The Pan American Sanitary Bureau's report in 1956 indicated that intestinal diseases, e.g. diarrhea and gastroenteritis, were the leading cause of death in 8 of 17 countries in the Western Hemisphere. The first decade of life is the period of greatest risk. Hormaeche et al. (1943) point out that, in Uruguay, infant mortality in the first year of life is as high as 10%, yet this is one of the lowest rates in South America! It decreased very little from 1895 to 1935, and, in the latter year, it was estimated that 28% of the deaths of babies resulted from enteritis and diarrhea, and that the total figure was probably no less than 40% if undiagnosed and unreported cases were taken into account. They believe that the diarrheal death rates are 2 or 3 times greater elsewhere in South America. As Hardy (1959) points out, the accuracy of such data is determined by the reliability of the original diagnosis; and, in some countries from which vital statistics are available, it is evident that a significant proportion of the births and deaths of infants goes unrecorded. From 1930 to 1959, in one rapidly developing progressive South American country, 40 to 60% of death certificates had no medical diagnosis.

Another factor is that bacteriological techniques seem to be less sensitive for detection of *Shigella* than for *Salmonella*. Isolation of *Shigella* from a case only once out of several tries reflects this difficulty and possibly other problems that are not well understood. This variation in the effectiveness of the same bacteriological procedures may be related to differences in the composition of the enteric flora, which, in turn, are influenced by an individual's nutrition, age, diet, and general condition.

Studies in Guatemala and Egypt show that *Shigella* are not prevalent in infants despite their high diarrheal death rate, but, with weaning, the incidence of shigellosis increases. *Shigella* has been isolated in 35% of children with diarrheas in an Egyptian village (Higgins et al., 1955). In a study of Mexican children under 2 years of age who suffered from diarrhea, only 21.7% yielded *Shigella, Salmonella,* or pathogenic *E. coli,* with percentages of 5.1, 6.1, and 10.5, respectively (Sandoval et al., 1965). Hardy and Watt (1948) reported 75% of children who died from diarrheal diseases in New Mexico and Georgia were positive for *Shigella,* while, in Great Britain, *Shigella* infections seem not to be prevalent in diarrheas of children. In general, shigellae play a variable role in provoking disease for they are often present in symptomless children (Ghai et al., 1969); malnutrition incites diarrheal disease and this is linked with poverty and low socio-economic level. This aspect has been summarized by Verhoestraete and Puffer (1958), and there have been a number of more recent studies especially by the Institute of Nutrition of Central America and Panama.

Bacillary dysentery is classically associated with diarrheal symptoms but not invariably, as shown by a study of 168 *Shigella* infections in Egyptian children in which 85 had no diarrhea (Higgins et al., 1955). When diarrhea occurs, the stools may contain blood, pus, and mucous threads. *S. flexneri* and *dysenteriae* generally attack the lower part of the tract, producing the symptoms just described, while *sonnei* may favor the upper gastrointestinal tract, producing vomiting and nausea. In infants, *Shigella* and *Salmonella* generally attack the ileum and colon with greater frequency than more anterior regions of the gut.

Shigellae probably do not survive as well as salmonellae in the external environment, but viable cells have been recovered from flour and milk kept at ordinary room temperatures for more than 170 days, and in eggs, shrimp, and shellfish for more than 50 days (Taylor and Nakamura, 1964). In naturally contaminated feces, shigellae survive one month at room temperature and two months at 3-6°C (Idina, 1959). Hardy considers food-, water-, and milk-borne epidemics of *Shigella* unusual. It is generally agreed that the two most important vehicles are flies and fingers.

COLI ENTERITIS

Prakash (1962) reviews progress with enteropathogenic *Escherichia coli* and mentions that Adam in 1927 first seriously considered that infantile diarrhea and gastroenteritis could be caused by certain biotypes of *E. coli*. The hypothesis was substantiated in the 1940s, when serological techniques were developed that distinguished pathogenic from normal gut types on the basis of their somatic (O), surface (K), and flagellar (H) antigens. The frequency of isolations of certain *coli* serotypes from cases of infantile diarrhea, as well as experimental studies, has confirmed their pathogenicity (Taylor, 1961; Guinée, 1963).

Pathogenic *E. coli* strains have been isolated from monkeys, calves and lambs ill with scours, pigs, cats, dogs, baby zoo wildcats, chickens, budgerigars, and mourning doves. There is a general predilection for young hosts. Pathogenic serotypes from animals are infrequently found associated with human diarrheal disease and the reverse is also true. However, certain O serotypes are more common to man and primates, and human types have been isolated from cats, ground squirrels, marmots, deermice, and chipmunks.

Human serotypes are confined to the gut and colonize both small and

FLIES AND HUMAN DISEASES   179

large intestines; they have the ability to cause dilatation of the ligated rabbit gut. Similar serotypes have been isolated from water, hens, and other animals but none of these affects the ligated rabbit gut; animal types typically produce a septicemia; there is some host specificity among animal types. Human types may further be divided into those that can cause keratoconjunctivitis in guinea pigs, dilatation of the rabbit gut, and can multiply in Hela cells, and other types that cannot. The former group is responsible for dysentery-like symptoms, while the latter group produces a gastroenteritis (Ogawa et al., 1968).

Pathogenic *E. coli* are often the most prevalent pathogens in cases of infantile diarrhea, and they are an important cause of nursery epidemics in Europe and America. There is also strong evidence of their involvement in "turistas," or travelers' diarrhea, in adults (Rowe et al., 1970).

Proteus species also deserve mention as occasional pathogens, particularly of infants (Burke et al., 1971).

FLIES AND THE MILITARY

Enteric disease was undoubtedly widespread in the ancient world. The Ebers Papyrus (1550 B.C.) repeatedly refers to the "bad sickness," which probably included diseases associated with diarrhea such as amebic dysentery, bloody enteritis, and cholera. The tiled lavatories and sanitary drains of Ur and Kish, dating from about 3000 B.C., suggest Mesopotamia had already come to grips with environmental health problems.

People and armies on the move, however, have always been scourged by enteric disease. The following Mosaic ordinance concerns aspects of personal hygiene in the field that remain pertinent to this day. "When thou goest forth in camp against thine enemies, then thou shalt keep thee from every evil thing. If there be among you any man, that is not clean by reason that chanceth him by night, then shall he go abroad out of the camp. He shall not come within the camp. But it shall be, when evening cometh on, he shall bathe himself in water; and when the sun is down, he may come within the camp. Thou shalt have a place also without the camp. Whither thou shalt go forth abroad. And thou shalt have a paddle among thy weapons; and it shall be, when thou sittest down abroad, thou shalt dig therewith, and shalt turn back and cover that which cometh from thee" (Deuteronomy 23:10-14). Mosaic sanitation was strict in its requirement of adequate bathing and proper disposal of feces, two precautions we now know to be of prime importance in curtailing diarrheal diseases.

The first record of the effect of fly suppression on disease incidence is found in an account by Aldrovandi (1602). He relates the following incident, which occurred during the expulsion of the Moors from Spain. Alfonso X, King of Castille and Leon, laid siege in 1257 to the fortress of Miebba in Vandalia. A tremendous fly plague broke out during the siege, and the flies not only molested the soldiers endlessly but were swallowed with their food. Dysentery broke out in the camp, which, Aldrovandi says, was spread by feces. Following destruction of the flies by public edict, the soldiers were also rid of the disease.

Enteric diseases seem to cling to old campsites, and stops of even a few days are sufficient to incite an outbreak. It is believed that more Union soldiers were killed by diarrhea than by the enemy during the American Civil War.

In comparing battle deaths with deaths from disease among soldiers, Skinner (1917) points out that, in the Spanish-American War, 454 soldiers were killed, but 5277 died from disease, mostly typhoid fever. In many of the army camps in this country and in Cuba, men were compelled to walk through human excrement to get to the latrines, and the food in the mess tents was black with flies. In some instances where lime had been sprinkled in the slit trenches, flies with white feet were seen walking on food (Reed et al., 1904). Skinner recalls "that the mouths of sick soldiers in the hospitals and hospital tents could not be seen for houseflies and these insects were sucking the juices from the lips of unconscious soldiers." Similar experiences were recorded by military doctors serving in South Africa during the Boer War. Some were struck by the affinity of flies for the enteric patients in a tent full of variously ill men. The moment an enteric patient put out his tongue, one or more flies would settle on it (Tooth and Calverly, 1901; see also Chmelick, 1899).

In 1906, Baron Takaki, the Director General of the Imperial Japanese Navy Medical Department, reported a remarkable reduction in the incidence of typhoid and dysentery among his men during the Russo-Japanese War, compared with the Sino-Japanese War just a few years earlier. Typhoid cases decreased from 37.14 to 9.26 per 1000; dysentery cases decreased from 108.96 to 10.52. This success was attributed to an improved sanitary organization, better control over food and drink, and an intensive campaign against flies.

In reviewing his military experiences in the Boer War, World War I, and in India, Faichnie (1921, 1929) recalls the paradox that, in army camps in India, enteric disease always broke out about 6 weeks after the camps had been made sanitary and had been "inspected by all the experts." The reason for this was that inadequate privies or poorly managed slit trenches allowed the proliferation of fecal-bred flies, and, in due course, cases of enteric disease began to appear. For example, in

Benares, India, the introduction of a 3-inch system of shallow trenching caused the attack rate to rise from 20 to 120 per 1000, where it stayed for three years. With the introduction of a better system of trenching the rate immediately returned to 20 per 1000. The report of the Sanitary Commissioner with the Government of India (Anon., 1909) describes a similar amelioration of typhoid fever among troops in an infantry division. Dunne (1902) and Jones (1907) in India, and Dansaur (1907) in southwest Africa were also convinced that few flies meant few or no cases of typhoid among the soldiers.

At a post in India, Faichnie (1909a) found flies contaminated with *S. typhi* in an officers' mess, a coffee shop, and around kitchens and barracks. By flaming the exterior of the flies, he demonstrated that the pathogens were being carried internally.

The Indian Field Service Hygiene Notes (Anon., 1945) points out that the incidence of diarrheal disease and bacillary dysentery in India is greatest in spring and autumn, when flies are most numerous, although outbreaks can occur at any time of the year. When troops are on active service, these diseases are more likely to occur in epidemic form during the fly seasons, whereas amebic dysentery tends to be more insidious and to produce a steady stream of cases throughout the year. Flies are regarded as the major vehicle, and a plague of flies is considered the inevitable prelude to an epidemic of dysentery and diarrhea among the troops.

S. typhi was isolated from house flies in the kitchens of typhoid patients in a British Army installation in Bermuda. In the early phases of the outbreak, patients' feces were not disinfected or adequately disposed of, facilitating fly transmission (Cochrane, 1912).

During the Salonika campaign in World War I, British troops were seriously hampered by enteric disease. The extent of the dysentery (bacillary and amebic) problem can be seen from the ratio, per 1000, of hospital admissions for 1916-1918: 1916–63.89; 1917–28.89; and 1918–52.23. Dysentery was second only to malaria in decimating the army. These were the days of horse and mule transport when house flies still "enjoyed an era of uninhibited fertility." With the arrival of warm weather and flies, dysentery erupted among staff and patients to such an extent that patients had to be sent elsewhere and the hospital was closed. Flies captured in wards, kitchens, and latrines of two hospitals yielded agglutinable *S. dysenteriae*, *flexneri*, and *paradysenteriae* from their legs and guts (Boyd, 1957; Dudgeon, 1919; Taylor, 1919; also Woodcock, 1919 in Egypt).

In the Near East, Manson-Bahr (1919) notes that the house fly was rare from June to August. After that, when the intense heat lessened, the fly returned and dysentery began to appear among the troops. He could find no evidence that food or water was involved. It is interesting that he was able to isolate *S. dysenteriae* from the intestines of house flies caught in the open desert, about 2 miles from the nearest camp and far from any known human fecal deposits.

Epidemics of dysentery in military barracks have been described in Aarhus, Denmark (Tulinius, 1943), and in Cherbourg (Maille, 1908), where food and water were above suspicion and conditions for fly transmission were ideal. At Aarhus, the latrines functioned poorly and at Cherbourg, they adjoined the kitchens. At Cherbourg, the epidemic subsided with the disappearance of flies after the first frost.

An exceptionally high frequency of dysentery carriage by flies was found in an outbreak during the summer of 1918 among civilians, returning hospitalized soldiers, and prisoners of war in a French town. The entire country and the sick rooms, in particular, swarmed with flies, which lit on dysentery stools left in patients' rooms or deposited in the woods nearby. Of 30 house flies examined, 12 contained *S. dysenteriae*, which was also the only pathogen isolated from the feces of patients. Boiled milk was exposed to flies in the surgery rooms with the result that 6 out of 26 samples became contaminated with *S. dysenteriae*.

American soldiers in both World Wars were not immune to the torments of fly-borne dysentery. Simmons (1923) describes an outbreak among an encampment of soldiers in an isolated dunes area along the Bay of Biscay in France. All went well from January until the fly season commenced in June. Poor feces disposal in camp led to a proliferation of flies such that "it was necessary to eat with one hand and to use the other to keep the flies away." What followed was a dysentery epidemic that attacked practically the entire command. Figure 40 and its caption and Figure 41 epitomize the timelessness of the problem.

An investigation of an outbreak of gastroenteritis among U.S. troops on bivouac during World War II disclosed heavy house fly breeding 3 miles away at a sewage treatment plant. The digestion chambers of the plant were being repaired and flies were breeding in raw sewage. Flies that were sprayed with a suspension of brewer's yeast as a marker were recovered 36 hours later at the camp. *S. enteritidis* was found on flies at the sewage pond and later on flies in the field kitchens (Peppler, 1944).

A massive epidemic of bacillary dysentery in an army division on bivouac in the southern United States occasioned one of the most thorough investigations of fly-borne dysentery in a military establishment. A total of 1557 cases of dysentery was recorded, of which 383 were bacteriologically positive, the agent in most cases being *S. paradysenteriae* Boyd 88. The onset of the epidemic occurred two weeks after the troops

FIG. 40. "It is in man's company as long as it lives, and takes the freedom to taste of all his food, oil only excepted, because it is poisonous to him. For him are goats milked, and the bees make honey for the flies as well as for man. For him do the confectioners make their sweetmeats, and he tastes them before the kings themselves, with whom he feasts . . ." (Lucian of Samosata, ca. 150 A.D.). Woodcut from *Hortus Sanitatis*, Antwerp, 1521, showing flies taking their share of a meal.

FIG. 41. "Hence came the invention of that which some make of Leather, rushes, or bristles which we call a Fly-flap, . . . of which Propertius of old, makes a mention . . . 'That which forbids the nasty fly thy dish to lick.'" (T. Muffet, 1658). Drawing by Francesco del Tuppo, Aesop, *Vita et Fabulae*, Naples, 1485. (Pierpont Morgan Library.)

moved into the area. Previously there had been few flies in the area, but the improper operation of straddle trenches quickly changed that. The epidemic had these fly-borne features: it developed gradually, was widespread, reached a peak in two weeks, and declined as flies gradually disappeared. Of 292 pools of trapped flies, 9, or 3.08%, yielded the same *Shigella* type found in patients. Flies in a number of the positive pools were from field mess areas and from active latrines (Kuhns and Anderson, 1944).

FLIES IN INSTITUTIONS

Moving from the military to institutional life, we note that both have a communality of living and eating, and thus share some epidemiologic patterns.

In 1910, Orton and Dodd studied a dysentery epidemic in a mental hospital in Massachusetts. The outbreak did not appear food-borne because of its gradual development and scattering of cases. Water and milk were conclusively excluded, and fruits and vegetables reasonably so. House flies were present in great numbers, and it was impossible to prevent their access to the feces of some of the disturbed inmates. To demonstrate that contaminated flies could actually spread through the hospital, broth cultures of *Serratia marcescens* were exposed in the laundry room (where contaminated bedding was handled), and flies carrying the same organism were trapped for the next six days in various parts of the hospital, including the kitchen and five dining rooms. In a similar analysis of institutions in Cook County, Illinois, in 1910, Dick (1911) noted that a resident physician "by using extreme care as to screening and other means of avoiding flies, . . . was able to greatly reduce the number of cases of dysentery in one ward as compared with control wards where ordinary precautions were used." He also noted that most outbreaks occurred during the fly season.

The Minnesota Board of Health (Anon., 1911) cites examples of typhoid outbreaks in state institutions, which ended when cold weather annihilated the flies. In one institution men and women dined separately. "The men's dining rooms were alive with flies, the women's comparatively free. The flies were not abolished and typhoid fever continued until cold weather eliminated them. At no time were the women affected." In another institution the authorities were so impressed with the fly danger that sixty inmates were set to work to destroy flies in the men's dining room. Not a single case developed later than the end of the incubation period of typhoid fever following elimination of the flies. This was long before cold weather stopped similar outbreaks in other institutions.

As Medical Officer of an Indian jail, Ross (1916) instituted a number of antifly measures, including fly-proofing the kitchen and pouring crude oil into the trenching pits. These measures resulted in substantial reduction in the number of dysentery cases from a previous five-year average of 27.6 to 5.

During an outbreak of shigellosis among lumbermen in a camp in British Columbia, *S. flexneri* and *sonnei* were isolated from flies caught in food storage rooms; the same organisms were recovered from stool specimens of the men. Bacteriologic examination of the water showed no evidence of fecal pollution, and the outbreak was attributed to contamination of food by flies. The flies apparently transported the pathogens from an area adjacent to the kitchen that had been polluted by sewage through a break in the sewage system (Gibbons, 1937). Similar outbreaks traced to flies through poor sewage and feces disposal have been reported by Brock (1957) and Dale (1922).

During a severe outbreak of diarrhea of the newborn in a hospital in the United States, the same pathogenic *E. coli* serotypes were recovered from infants and from *Drosophila*. The flies were trapped in a disposal can in which soiled diapers were collected in a ward (Ewing, 1962). In a mental hospital *S. dysenteriae* was carried by flies near wards where there were active cases of bacillary dysentery (Hardy et al., 1942).

FLIES IN THE COMMUNITY

During the first decade of this century, the public image of the fly declined to its lowest level in history due to the dedicated efforts of L. O. Howard in the United States, C. Gordon Hewitt in Canada, G. S. Graham-Smith in England, and many others. The house fly was rechristened the "typhoid fly," and health departments in many countries promoted spirited and imaginative campaigns to drive home to people the danger of "the filthy feet of fecal feeding flies" (Fig. 35). In 1912, Italy passed a law requiring that various foods be "protected from contact with flies in stores, public streets, etc., by means of glass, metal or the like." Philadelphia had enacted a similar law in 1905.

Under the auspices of the Merchants' Association of New York, Daniel D. Jackson (1909) published an interesting document entitled "The house-fly at the bar. Indictment. Guilty or not guilty?" It presented the views and experiences of responsible health officials throughout the country in a powerful indictment of the fly. In terms of water supply and general sanitation, Jackson characterizes three classes of cities toward the end of the 19th century: Dresden, Vienna, and Munich had good water supplies, good sewage disposal and general sanitation, and had a uniformly low typhoid rate, regardless of season; Baltimore, Boston, and New York had good water, but poor sewage disposal and/or general sanitation, with a high summer typhoid rate that followed the fly curve; Cincinnati, Chicago, and Philadelphia had poor water and a uniformly high typhoid rate throughout the year.

The previous year, Jackson called attention to a curious distribution of fatal cases of typhoid and diarrhea that hugged New York's harbor areas. A survey of the entire waterfront revealed many points where sewer outflows were above the low water mark, and exposed raw sewage was found covered with flies. Because proper toilets were not available to dock workers, a large amount of human excreta was found along the docks, and these, too, invariably swarmed with flies. The map of Manhattan that accompanies Jackson's report shows that practically all cases were within a few blocks of the waterfront and were most numerous where fly density was outstanding and sanitation was poorest. Where fatal cases clustered away from the waterfront, as in the Bowery, it was found that the area had numerous outdoor privies clogged with excreta and covered with flies. Jackson's plea for better sanitation extended beyond the city to dairy farms (see also Smillie, 1916).

Other studies in Providence (Anon., 1915) and in Cincinnati (Osmond, 1909) pinpointed the hotspots of gastroenteritis and typhoid as those areas that were unsewered, "privied," and had high fly densities. For a graphic account of comparable rural squalor the reader is referred to Washburn (1911), who studied immigrant workers and their families living on the Iron Range of Minnesota.

What is the relative impact of unsanitary homes versus flies on diarrheal disease? To find out, Hudson (1915) made a survey of 1000 infants in New York City families. The families were divided evenly between a fly-protected and a fly-exposed group. Each group had the same number of large or small families, twins, diseased members, and clean or dirty homes. It was found that diarrheal incidence was about the same in clean homes whether or not they were fly-protected. This is consistent with the presence of fewer flies in clean homes and the smaller burden of microbes that they probably carry. In dirty homes, however, twice as many cases of diarrhea existed among fly-exposed infants as among fly-protected infants. Furthermore, there was almost twice as much diarrhea in all dirty homes as in all clean homes, and two and a half times more cases among infants in fly-exposed, dirty homes as among infants in fly-protected, clean homes. What or how the babies were fed did not seem to influence the statistics.

In 1903, Hamilton published her study of a typhoid epidemic in Chicago. The epidemic was most severe in the 19th ward, which had 1/36 of the city's population but more than 1/7 of the deaths. The water supply, although not beyond reproach, was not considered the source of infection; the milk, though often diluted, was bacteriologically good; and the constant rains ruled out dust as a factor. The distribution of cases showed a preponderance in those streets where removal of sewage and disposal of excreta were most deficient. Only 48% of the houses had sanitary plumbing, 23% had privies, and the rest had defective plumbing facilities. Hamilton caught house flies in privies, in a typhoid patient's room, and elsewhere in the neighborhood that were carrying *S. typhi*.

A typhoid outbreak during autumn in a small town near Turin, Italy, was the setting for another study of possible vehicles. In one home where almost everyone was ill, water, milk, and vegetables were studied and excluded. This house was a milk-collecting station, and attracted huge numbers of flies, from whose legs and heads the pathogen was readily recovered. In one case, the legs on direct streaking yielded more than 100 typhoid colonies (Bertarelli, 1909, 1910).

"The practices and circumstances favoring the spread of enteric diseases are often linked with time-honoured beliefs and customs that may have deep roots and purposes in the society concerned . . ." (Anon., 1964). Practices among immigrants of the Minnesota Iron Range provide an example of just such a crucial ethnic pattern. The Italians and Austrians were filthier in their habits than the Swedes and Finns, yet they suffered much less from typhoid. The Italians and Austrians ate three hot meals a day, while the others frequently lunched during the day on cold food that had been left exposed on the table. Although mosquito netting was often thrown over the food, it served rather to confine large numbers of flies under the netting to do their worst (Washburn, 1911).

Another example of how custom may influence epidemiology is an outbreak of dysentery in a shepherds' hamlet in Hungary, in which 30 of its 150 inhabitants became ill. Living in the village at the time were 80 construction workers employed on a drainage project hardly ten meters from the village houses. While 29 of the villagers were stricken, only one construction worker was affected, and he was quartered in a house where there were already four cases. Both villagers and workers obtained their food and water from the same source and lived in the same manner, with one significant difference. The construction workers kept their food in fly-proof kitbags, but the food of the villagers stood on the table all day and was covered with flies (Johan, 1933).

Fly involvement in an endemic focus of enteric disease in a Ukrainian mining community is described by Boikov (1932). He found *S. typhi* on *Limosina* sp. taken in a mine shaft and in the gut of *M. domestica* caught in a toilet that served a barracks where typhoid patients were housed. *S. paratyphi* A was recovered from four house flies in a kitchen of a children's home.

The relative ease with which flies pick up enteric pathogens from the contaminated environment of institutions is convincingly shown by Gandel'sman et al. (1947) and Zaidenov (1961). The latter examined synanthropic flies taken mainly in lavatories of hospitals and nurseries for children with chronic dysentery in the town of Chita. Seventeen isolations of *Shigella* spp. were made predominantly from *M. domestica* (8♀, 3♂), and to a lesser extent from *M. stabulans, F. scalaris, P. terraenovae*, and *C. vomitoria*. The contamination rate for *Shigella* in all flies examined was 6.39% and in the house fly, 5.4%. In this setting the house fly was probably far more dangerous than exophilic, coprobiontic flies. However, this does not preclude the latter as possible vectors. In the middle zone of the U.S.S.R., *C. uralensis* breeds rather exclusively on human feces, shown by the fact that the intestines of as many as 40% of both sexes may contain fresh feces. The fly swarms on food in outdoor markets and is one of the most numerous exophilic coprobionts. Sukhova (1957) reports the isolation of a possible *S. flexneri* from this fly taken where there were dysentery patients. A fuller account of the pathogens carried by various flies is given in the following sections.

Differences in the behavior of the northern and southern subspecies of *M. domestica* in the Palearctic region are of great epidemiologic importance. In the Leningrad region, house flies generally do not breed in latrines, preferring garbage and animal manure. Because they infrequently come in contact with human feces, they are less likely to carry human pathogens. Thus, in a study by Shura-Bura (1951) in a northern village where each pair of houses shared a privy and dysentery was known to be present, 0 out of 328 house flies were positive for *Shigella*. This, despite the fact that the privies often contained feces with typical mucous and blood. Only 9% of flies collected in privies were house flies, and about 30% of these were lightly contaminated with *Escherichia* types. This does not completely exonerate the northern house fly, for some were found to harbor pathogens when caught at a children's hospital for chronic dysentery in the same area. Although flies were not numerous indoors they did occasionally enter through breaks in the screens. Among 99 such flies, 10 carried *S. sonnei* and 1 carried *S. flexneri*. In 9 flies the number of pathogens ranged from 30 to 500; the other two contained 1250 and 10,000 *Shigella*. It is a pity that Shura-Bura fails to describe the technique that enabled him to enumerate 35 to 500 *Shigella* organisms on plates that were undoubtedly overgrown with larger numbers of gram-, non-lactose fermenters. He merely mentions that he was able to distinguish by colony type and by a "widely used method of agglutination screening," neither of which is illuminating. Returning to the hospital study, he found that female flies were four times more frequent carriers than males. In this hospital, even though sources of contamination were difficult of access, chance exposure to human feces was enough to bring the northern house fly into the dysentery cycle.

A comparison study of the southern house fly was made in a village on the Crimean coast. Here, the major type visiting privies was *M. domestica vicina*, which crawled over the feces and fed with voracity upon blood and mucous. The same species was also numerous in nearby dwellings. Of 125 such flies taken in toilets of the sick, in dining rooms and other quarters, 22 carried serologically confirmed *S. flexneri*. Twenty of the positive flies were females, and all positive cultures came from the gut, none from the exterior.

Hungarian investigators consider *M. domestica* to have no epidemiologic significance in their country because it is not common in open-air markets and infrequently visits human feces (Aradi and Mihályi, 1971). Morhardt (1936) observed that *M. domestica* could hardly be trapped by using human feces as bait, and it constituted only 17% of all flies in privies. Lörincz and Makara (1936) could not recover *S. typhi* from 50 house flies trapped in typhoid fever wards nor from 2500 flies trapped in rooms of typhoid patients.

Nevertheless, Hungarian health authorities agree that water- and food-borne typhoid are of little importance compared to fly-borne typhoid in their country. Warmer climates may, in fact, favor the spread of typhoid fever by extending the fly breeding season. In southern Europe, according to Petrilla (1933), people make less use of enclosed toilets, and feces are therefore more accessible to flies. Scholtz (1933) is also deeply convinced of the importance of the fly in the epidemiology of typhoid in Hungary. He writes, "Flyproofed latrines are scarcely to be found in our villages—indeed there are not even the simplest pits. On manure piles and next to the walls of buildings human feces are freely deposited. To be sure, typhoid germs can reach the springs by way of ground-water or vegetables by fertilization with liquid manure; but there can be no doubt that the germs can be massively transported by the innumerable flies swarming around the feces to the foods lying uncovered in kitchens and dining rooms." The Royal Hungarian State Institute of Public Health reported over 22,000 cases of typhoid fever in 1932 and only *one* definitely water-borne epidemic; "in the other cases the most probable carrier of infection was the fly" (Johan, 1933).

Jones (1941) has described an epidemic of typhoid fever in a concentrated Rhodesian community of 10,000 persons, where the water supply was good and feces disposal consisted of open-bucket latrines. Cases were confined mainly to natives although a few Europeans were afflicted.

All circumstances accused the fly as broker, in support of which it was found that flies trapped near the native hospital and in kitchens and latrines carried *S. typhi*.

In the Japanese Amami (Oshima) Islands, the open straddle trench is widely used, and flies have ready access to human feces and food. Under these conditions, the endemicity of bacillary dysentery among the islanders seems to be less a consequence of transmission by food handlers than by flies (Shimizu et al., 1965). *Salmonella* spp. were isolated from 2.27% of fly pools and *Shigella* from 0.85%. Sarcophagids, calliphorids, and muscids were tested separately, and the most frequent carrier of pathogens proved to be the latrine fly, *C. megacephala*. In Manchuria, 8% of latrine flies examined by How carried dysentery microbes (Sukhova, 1950).

On Fiji Island, several decades before, Bahr (1912, 1914) found a number of conditions favoring the fly hypothesis: a high fly density concurred with a high dysentery morbidity; the water supply was good, but sanitation was poor. Supporting evidence was the isolation of *S. dysenteriae* from the hindgut and from the legs and wings of a small number of flies near dysentery patients.

OTHER SIGNIFICANT RECOVERIES OF ENTERIC
PATHOGENS FROM WILD FLIES

In addition to the isolations of salmonellae and shigellae, a large number of isolations have been made from flies under circumstances that strongly reinforce their epidemiologic importance. The list of places is a long one and includes farms, slaughterhouses, outdoor markets, bakeries, animal feed plants, food stores, and sick rooms, and their environs. The best we may hope to do, given limitations of space, is to highlight certain studies, particularly those that are less accessible for language or other reasons, and to cite other equally relevant ones for the reader who wishes more information.

Food destined for human consumption runs a dangerous gauntlet from the farm to the table, with the slaughterhouse, food processing plant, and market intervening to compound the risk of contamination. Numerous studies have shown how extensively contaminated with enteric pathogens each of these environments can be. In most studies, no attempt has been made to measure the degree of participation of flies, rodents, or other commensals as contaminators of these environments because investigators have been loathe to undertake a systematic study of such a complex environment. Yet the involvement of flies must be significant if one can gauge from the relative ease with which pathogenic organisms are isolated from them. The list of isolations would be many times greater if more attention were given to the fly-phase of the cycle. One wonders whether the widespread occurrence of salmonellae in lizards, turtles, and other reptiles is not associated with a diet that includes flies.

Farms

During an investigation of an outbreak of *S. typhimurium* at two turkey ranches, the same organism was found in a gopher snake, a garter snake, two cats, and house flies. There had been a yearly recurrence of the disease on one of the ranches, which was not directly traceable to the poults. The evidence suggests instead that *S. typhimurium* was possibly being maintained in the other animals. One of the infected snakes was caught in a brooder yard at about the same time that several poults turned up sick. Six weeks later, 320 house flies were caught on fly paper and divided into 8 lots for testing. *S. typhimurium* was isolated from 4 of the lots (McNeil and Hinshaw, 1944).

Ostashev (1956) studied the microflora of flies caught in piggeries in the Molotovsk region of the Soviet Union during a period of four years (see section on Swine erysipelas, Chapter 5). In August 1951, on a farm considered satisfactory with respect to swine paratyphoid, *S. choleraesuis* was found in the crops of 3 out of 68 flies caught in the piggeries. A month later, illness broke out among the suckling pigs on this farm, and the same organism was isolated from piglet corpses. Cases of swine paratyphoid had occurred that spring and summer among piglets on a nearby farm 1½ km away. Since there had been no economic or operational contact between the two farms, the suspicion was strengthened that dispersing flies had introduced the infection.

In Uruguay, Hormaeche and co-workers (1944) found a new type, *Salmonella carrau*, in a pool of 20 unidentified flies and in normal pigs and man. *S. typhimurium* was also recovered from the flies.

In southwestern Finland, *S. typhimurium* was detected in *Lucilia illustris* captured near a mink farm (Ojala and Nuorteva, 1966). The organism was also recovered from *P. sericata* and *P. terraenovae* caught on the grounds of the Zoological Museum in the heart of Helsinki.

Slaughterhouses

In an attempt to unravel the cycling of salmonellae in a slaughterhouse we (Greenberg et al., 1963) studied the flies, rats, and livestock in a Mexican slaughterhouse. Twelve *Salmonella* types were detected in 10 species of flies, while 4 types were recovered from slaughtered animals and from the indigenous rats, showing that flies can serve as superior sensors. Flies captured exclusively on offal had a higher frequency of contamination and more types than did flies taken on manure. It appeared that the prime source of salmonellae for flies was carrion derived from the livestock and discarded in an adjacent dump. The secondary screwworm fly, *C. macellaria*, yielded 8 types, *M. domestica* and *P. regina*, 5 each, and *O. aenescens*, 4 types; sphaerocerids and sepsids were also contaminated. The population density and contamination rate of *C. macellaria* would qualify it as a dangerous vector were it not for its symbovid, exophilic habits, which tend to confine its activity to outdoor markets, etc.

In a follow-up study, about 100,000 flies on the grounds of the slaughterhouse were sprayed with a fluorescein dye (see Dispersal, Chapter 2). During the following week, 543 tagged flies of the above species, in addition to *S. calcitrans* and *P. sericata*, were captured in residential neighborhoods, a market place, a dairy, and in a village 3 miles away. The flies traveled across a city of about 80,000 inhabitants in less than a day. Seven *Salmonella* and two *Shigella* isolations were made from 28 pools of these flies.

Because there is mixing and spread of infection among livestock as they are moved from widely separated farms to the slaughterhouse, the slaughterhouse becomes, in effect, a *Salmonella* re-distribution point. Contamination rate is enhanced by close packing in transit and in the holding pens. Dispersal of contaminated flies from slaughterhouses thus places an entire community at risk (Greenberg and Bornstein, 1964).

An investigation by Brygoo et al. (1962) in Tananarive, Madagascar, demonstrated the role of flies in the transmission of fecal organisms. A total of 2800 *Chrysomya chloropyga* was collected from diverse places, including a slaughterhouse, markets, kitchens, houses, and hospital grounds. Pools of the flies were triturated in saline solution and centrifuged at 3000 RPM, the supernate being used for virus isolations and the centrifugate for bacteriology. Besides the impressive isolation of 35 virus strains (for details see sections on Coxsackie, Polio, and Enteroviruses in this chapter), 10 pathogenic serotypes of *E. coli* (types 9 III B4 and I 26 B6), 6 *Salmonella* strains (3 *typhimurium* and 1 each of *choleraesuis*, *typhi*, and *give*), and 1 *S. flexneri* were isolated. *S. typhimurium* was also detected in 2 pools of *M. domestica* (Brygoo and LeNoc, 1962; also Coulanges and Mayoux, 1970).

The following studies provide additional support for the need for fly control programs in the slaughterhouse. In 1911, Horn and Huber captured 25 flies (probably *M. domestica*) on the grounds of a Veterinary Institute in Leipzig and found that 2 contained *S. paratyphi* B. The same organism was found in other flies taken in a slaughterhouse, cattle stalls, and nearby homes. Of 660 pools of 10 flies each, caught in a slaughterhouse and market in Pernambuco, Brazil, 12 pools contained *Salmon-*

ella, 1 pool contained *Shigella,* and 30 pools contained two pathogenic serotypes of *E. coli* known to be associated with human diarrheal disease (Maroja et al., 1956). A comparable study of house flies in a slaughterhouse, animal room, and market in San José, Costa Rica, yielded similar results and a higher fly carrier rate (Bolaños, 1959). In Quezon City, Philippines, *S. derby* was found in a sample of blue bottles, *Calliphora* sp., in a poorly sanitated slaughterhouse (Tacal and Meñez, 1967).

Food processing plants and warehouses

A survey of warehouses in a number of German cities revealed the presence of the typhoid organism and more than a half-dozen other *Salmonella* types in fly feces and in undetermined flies. The warehouses handled fish meal, cadaver meal, bone meal, etc., destined for animal feeds (Steiniger, 1957). In West Berlin packing plants, muscid and calliphorid species were found carrying *S. typhimurium, cottbus,* and *chester* (Bulling et al., 1959). In Peru, the role of flies in food contamination has been studied by Quevedo and Carranza (1966). Flies were taken in fish meal plants, canneries, and other places of epidemiologic interest. *Proteus* and *Escherichia* were the most common isolates, but *salmonellae* were recovered from flies in the fish meal plant, as well as in a cow stable and a food store.

Markets

Flies are serious contaminators of meats, fruits, and vegetables sold in markets. During a 5-month investigation of the problem in New Orleans' markets, Beyer (1925) isolated *S. typhi* from the secondary screwworm, *C. macellaria,* and two species of *Shigella* from various blow flies, a flesh fly and a *Drosophila.* In all, 36 out of 100 flies belonging to 13 species harbored 6 or more potential pathogens. Beyer urged cleanliness, screening, and other measures to protect food from flies.

In Santiago, Chile, analysis of pooled house flies collected in markets, dairies, restaurants, and other locales revealed that a large proportion carried one or another member of the Enterobacteriaceae and that a smaller but significant percentage (10.6) carried *S. flexneri* and the agents of typhoid and paratyphoid (Prado and Jimenez, 1955).

Shura-Bura (1952) has written about the habits of flies in the Crimea, particularly the dangerous predilection for fruits of *P. sericata, M. stabulans,* and other common synanthropes. Many of these flies contained fecal bacteria, which they freely deposited on grapes. Anastas'ev (1952) also emphasizes the danger from exophilic, coprobiontic flies. Aradi and Mihályi (1971) have made a detailed survey of the composition of the fly population frequenting open-air markets in Budapest, which is treated in detail in Volume I, Chapter 2.

In Bukhara, various synanthropic flies carrying *S. flexneri* and *dysenteriae* were captured in markets, restaurants, latrines, and hospitals. The flies included the muscids, *M. domestica vicina*, *F. canicularis* and *scalaris*, and *M. stabulans* (Sychevskaĭa et al., 1955, 1959a). And in Fergana, Uzbekistan, analysis of single flies caught in the market disclosed a dysentery carrier rate of 5.8% in *M. stabulans* and 5.4% in *M. d. vicina* (Sychevskaĭa, 1959b; see also Lobanov, 1960). House flies captured on food in the market of Tashkent were found to be highly contaminated with *S. dysenteriae* externally (18 of 170), and slightly less so internally (18 of 200) (R.M., 1937). In 1949, Shterngol'd published the results of his fly studies in Tashkent in 1935 and 1936. In those two seasons of observation, 32 agglutinable strains of *Shigella*, including *flexneri* Y and *dysenteriae* were isolated from *M. domestica vicina*. Shterngol'd noted a much higher internal than external carriage of the pathogens.

In Manila, de la Paz (1938a) describes the threat from *Musca*, *Chrysomya*, and *Sarcophaga* species, which carried 6 *Shigella* types in markets, groceries, restaurants, and food stores. He also found 9 *Salmonella* types in the first two fly genera.

H. M. Jettmar of the League of Nations Epidemic Commission (1940) observed a serious outbreak of enteric disease in southern Shensi, China, which occurred in the absence of *M. domestica* and in the presence of swarms of latrine flies. This fly was present in homes and markets "sitting in great numbers on vegetables, fruits, meats and other victuals and provisions." In Peiping, China, all four batches of the latrine fly captured in public latrines carried *S. dysenteriae* (Chow, 1940).

Parry emphasizes the hazard of flies in bakeries, especially in summer, even when bakery personnel are careful to reduce splash and soiling to a minimum. Five flies contaminated with salmonellae were caught in one bakery (Hobbs, 1961).

Sick rooms and environs

Exposure of the excreta of enteric patients in the presence of flies is going into partnership with contagion. We have learned enough to condemn this as an irresponsible and antisocial act. But for the reader who is yet to be convinced there is the following additional information.

During the summer of 1911, flies obtained from diarrhea infected houses in Birmingham and Cambridge, England, harbored *Proteus morganii* at least nine times as often as flies from noninfected houses. In 1912, the summer was cold and wet, flies were scarce, and epidemic diarrhea was uncommon; *P. morganii* was not isolated from flies in dwellings of either the sick or the well (Graham-Smith, 1912, 1913). Galli-Vallerio (1905) recovered dysentery organisms from flies he caught over human excrement; the excrement had been transported from sick rooms and deposited near peasant dwellings (see also Lauber, 1920). In a central Italian town, Ara (1933) reported that 8 out of 19 house fly pools, totaling 102 flies, from the rooms of typhoid patients were positive for *S. typhi*; flies from homes without typhoid were all negative (see also Ara and Marengo, 1932).

In Cuba, Curbelo and Arango (1945) demonstrated *S. typhi in* house flies they caught in the vicinity of typhoid patients in a hospital. The flies were held for 72 hours before they were tested (also Joós, 1936). Sukhova (1954) in western Turkmenia studied flies in foci of acute intestinal illness where the patients had not yet been hospitalized. A strain of *S. flexneri* W was detected in one lot of *M. domestica vicina*, and a strain of *S. typhi* was isolated from the external washings of 3 specimens of *P. sericata*.

The community

The microbial cargo of flies provides a good index of neighborhood sanitation. House flies from unsanitary sections of Liverpool contained 800 thousand to 500 million aerobic bacteria, while those from cleaner sections or from the suburbs contained 21,000 to 100,000 bacteria. The number of enteric bacteria ranged from 10,000 to 333 million in the former and from 100 to 10,000 in the latter (Cox et al., 1912; also Parisot and Fernier, 1934). There did not appear to be much dispersal of flies from unsanitary to sanitary quarters. In the Bronx, New York, surface washings of flies caught in homes with diarrheas averaged about 1,100,000 organisms compared with 14,000 from healthy homes, a nearly 100-fold difference (Armstrong, 1914). In Peiping, China, Chow (1940) noted that all specimens of *C. megacephala* taken from public latrines carried *E. coli*, but about 80% of the same species found on fruits, vegetables, and garbage were similarly contaminated. In Leningrad, Tarasov and Chaïkin (1941) observed that 7.5 to 16.2% of house flies from poorly sanitated locales contained *E. coli*, while only 2% of the flies from a clean, well-kept apartment building were contaminated with *E. coli*. This has been further borne out by Lysenko (1958) in Czechoslovakia and by Lysenko and Povolný (1961), who have shown that different species of flies from the same biotope harbor a similar composition of microorganisms.

A few attempts have been made to learn the source of flies from the bacteria they carry. In Michigan it was not possible to distinguish farm-bred from beach-bred stable flies on the basis of their coliform and enterococcal counts (Love and Gill, 1965). Hoffmann (1950) made a bacteriologic analysis of the surface of almost 3000 flies caught during a four-year period in the Swiss town of St. Gall and obtained interesting data on the probable sources of contamination. The average number of bacteria on flies from stables was 12,540 (*E. coli* 29.7%); from garbage pails, 1,423,000 (*E. coli* 80.7%); from a confiscatory room, 43,680 (*M. domestica*) and 102,890 (*Calliphora*) (*E. coli* 45.1 and 81.8%, respectively); and from human excrement, 153,500 (*E. coli* 54.9%). Others have reported a much lower proportion of fecal types from flies on garbage (also Post and Foster, 1965).

One of the chief difficulties in re-tracing the pathway of dissemination during scattered outbreaks of enteric infection in a community is that the contaminated food or other vehicle has disappeared by the time illness becomes apparent. We are therefore left without the source. But flies may be the link to that source, and they may also be useful as indicators of pathogen prevalence in a human population. We have discussed their superiority as *Salmonella* indicators in the slaughterhouse zoocoenosis. Are they equally useful elsewhere?

In gauging pathogen prevalence in a human population, esthetic considerations and convenience favor surveys based on flies rather than on rectal swabs or privy samples. To test the validity of this approach, Richards et al. (1961) selected two Indian villages in southcentral Arizona as the study site for *Shigella* prevalence. A total of 65,273 flies collected outdoors (pooled in lots of 20 or fewer specimens) yielded 69 shigellae, and 5,664 flies taken indoors yielded 12 shigellae. In all, 11 different serotypes were found. The house fly was the most numerous and medically important fly present; second in importance were *Ophyra* spp. There was a concurrence in the frequencies of most *Shigella* types obtained from flies and from the rectal swabs from children. For example, *S. flexneri* 6 was the most common type isolated from flies and from human sources; other common types were *flexneri* 3, *flexneri* 4a, and *sonnei* 1. However, *flexneri* 2a, the second most common human type, was found only twice in flies. The fact that this type produced the severest illness may have worked to reduce fly-fecal contact. Despite the general parallels, the conclusion reached was that the fly index was too low to afford a sufficiently sensitive and effective means of measuring *Shigella* prevalence among children. The same flies yielded only 13 isolations of *Salmonella*, showing that *Salmonella* was circulating at a much lower frequency than *Shigella* in these communities (DeCapito, 1963).

Flies proved better indicators of *Shigella* in another study in Arizona.

During the latter part of November, 13 persons living in a migrant farm laborers' camp were hospitalized with gastroenteritis. *S. flexneri* 3 was recovered from 8 patients and from the operator of a chuck wagon who served meals to the workers in the field. The same organism showed up in 7 out of 37 fly pools and in 9 out of 36 samples of fresh human feces on the ground (Coleman and Maier, 1956).

There are obvious advantages to using flies as indicators, since their mobility and habits give them wide access to infectious matter and their size permits undiminished recovery of the pathogen from a pool of up to about 30 specimens. However, there is probably an attenuation of pathogenic organisms in the fly that must be greater than in the bowel, which increases the difficulty of isolation. As shown in our study of the Mexican slaughterhouse, a truer fly-pathogen index will be obtained in areas of high fly density and high pathogen prevalence. This is supported by studies we shall now consider.

In Cairo, the frequency with which *Salmonella* and *Shigella* were recovered from *M. d. vicina* reflected the incidence of cases of gastroenteritis due to these two groups (Floyd and Cook, 1953). Of 156 fly pools, 12 pools yielded *Shigella*, 6 pools yielded *Salmonella*, and all but one were positive for *E. coli*. *S. typhi* was a frequent isolate. Flies from urban areas and from rural areas were equally dangerous and the fact that all were randomly trapped around the city brings into sharper focus the public health significance of these findings. Lackany (1963) obtained fewer positives in Alexandria, but he may have used too many specimens (from 100 to several thousand) per pool.

An enterobacteriologic analysis of flies in Alchevsk in the Ukraine found *E. coli* in 57% of the flies (Vasil'ev, 1935); but the reported recovery of *S. schottmuelleri* and a number of *Shigella* types is questionable because adequate serological confirmation appears to have been lacking.

In Montevideo, Uruguay, Hormaeche and his group, in collaboration with R. V. Talice (1943), found *Salmonella* in 40% of about 150 fly pools, a most impressive statistic; the flies were taken in various parts of the city. In Guayaquil, Ecuador, a number of common flies, including the house fly, were trapped on the grounds of the Institute, and 17.5% were positive for *Salmonella* (46 times) and/or *Shigella* (60 times). A minimum of 19 *Salmonella* types and 6 *Shigella* types were reported. It is interesting that the house fly yielded first place as most frequent carrier to *C. macellaria*, the secondary screwworm. It is also noteworthy that *Shigella* was isolated 15 times from this fly, which prefers carrion to human feces for feeding and breeding (Alcivar and Campos, 1946).

The incidence of enteric diseases is generally characterized by a single or bimodal summer peak, although in some parts of the United States and England, the incidence of diarrheal disease has been reversed, with the peak occurring during winter months. In the latter instances, the fly factor can be definitely excluded. But in typical situations, warm weather brings on both flies and diarrheal diseases. The species composition of the synanthropic fly population varies with time and place, some species reaching maximum abundance in spring and fall, while others are most numerous in midsummer (see Chapter 2). The overlapping or concurrence of enteric disease peaks with fly peaks is a folk observation of considerable venerability. For centuries it was and perhaps remains the basis for our dim view of flies, and, indeed, the amount of work on this aspect alone has been impressive, if not always illuminating. If there is weakness in this work it is the tacit acceptance of the erroneous premise that seasonal concurrence proves causality.

Bacteria, like flies, have certain temperature constraints. Pathogens outside their host survive best in cold; but if they are to multiply it will have to be under summer temperatures attended by the risk of overgrowth and suppression by other more competitive types. As for the host itself, we know little about physiological changes in the human or animal gut that may predispose to summer infection. If seasonal changes do occur, are they solely host physiological alterations, or do modifications of diet and gut flora enter into the picture? Our eating habits are certainly different in the summer, when there is a greater consumption of raw fruits and vegetables that are often eaten unwashed. In any event, a summer increase in host susceptibility, hypothetical at this time, could parallel an increase in fly density, both being temperature dependent, yet the two phenomena could be totally unrelated.

We are not suggesting that this is so—on the contrary, the evidence is overwhelmingly against it—but we wish to emphasize the futility of phenological studies that merely show two curves that happen to coincide. Confidence that such concurrence is significant can be increased if the fly curve can be manipulated, as in fly suppression, and the incidence of disease responds appropriately; or when two "roller coaster" curves continue to be parallel despite weather changes and other vicissitudes. Experiments on the effect of fly suppression on disease incidence have yielded information important enough to deserve separate treatment. The rest is discussed here.

The bacteriological burden of house flies follows a Gaussian seasonal distribution with the greatest number of organisms and largest percentage of coliforms being carried by house flies during summer months, as shown in Washington, D.C. (Scott, 1917), New York City (Torrey, 1912), Seijo, Japan (An, 1933), and Valencia, Venezuela (Cova Garcia, 1956).

Insomuch as air temperature influences the breeding and activity of flies, it is a factor of considerable importance. Nash (1903, 1905) explained the absence of infant diarrheal mortality in Southend, England, in the summer of 1902, and 23 deaths the year before on the basis of weather-suppression of flies. The summer of 1902 was cold and wet, and house flies were almost completely absent. In September, when flies appeared, 13 infant deaths were recorded in 3 weeks; with the disappearance of flies at the end of September, infant mortality again returned to zero. Nash was one of the first to consider malnutrition a key factor in diarrheas, "accentuated 100-fold in the diarrhea season by flies" (see also Niven, 1910; Dudfield, 1912; Adelheim, 1919; and above sections on Military and Institutional Dysentery).

A close correspondence between fly density and bacillary dysentery, which followed a bimodal course, in Egypt was observed by Aly (1940; Perry and Bensted, 1928, 1929). Ledingham (1920) observed a sharp fall in fly density during the hottest part of summer in the Near East, while disease incidence declined much less. He explained this by assuming that a large number of chronic cases and human carriers continue to spread enteric infections during the summer hiatus until the fall pulse of flies causes another upsurge.

According to the observations of Sukhova (1954) and others, the seasonal incidence of dysentery in western Turkmenia exhibits a greater spring-summer peak and a lesser summer-fall peak. The upsurge of dysentery always follows upon the high tide of flies, and this is especially clear when fly populations in nearby areas reach their maxima at different times. Thus, in one settlement 45 km from the city, flies did not become numerous until the middle of June, and only after that was the seasonal dysentery upswing observed. Again in a southern suburb, the upsurge of dysentery coincided with the second fly peak in August. Finally, in a western suburb where flies were almost absent there was no second seasonal peak of dysentery, although, Sukhova writes, the climate, conditions of life, and medical service were identical with the other areas. Since flies were almost absent in this suburb, one wonders whether the conditions of life were, in fact, identical.

Investigators have generally observed that the fly peak precedes the dysentery and diarrheal disease peak by a few days at most. Niven (1905, 1906) tabulated fly counts in homes with the numbers of fatal cases of diarrhea on a weekly basis and found that the former led by 3 or 4 days. Vanskaĭa (1947, 1957) made a five-year study in a large Russian city and found that the upturn and peak of the fly population slightly preceded or coincided with the upturn and peak of dysentery incidence. The closeness of the two curves is explained by the short incubation period. Similar observations have been made in Gor'kiĭ (Semenov, 1945), in a town in western Siberia (Zhovtyi, 1954), and in British Guiana (Bodkin, 1917). The graph in Figure 42 illustrates very well what we have been discussing.

The peak incidence of typhoid and paratyphoid lags behind the fly peak by one to four weeks, presumably because of the longer incubation period (Vanskaĭa, 1943; Yavrumov, 1947; and Semenov, 1945). Tesch (1937-1941) has made similar observations in Batavia, Java.

Zinchenko and Nestervodskaya (1956) (also Zmeev, 1943) suggest that dysentery incidence follows more closely the increasing numbers of *M. domestica vicina* indoors, while Arskiĭ and co-workers (1961) convincingly link the seasonal increase in dysentery morbidity to fly infection rate. In central Tadzhikistan, morbidity was more closely correlated with fly infection rate (0.92 ± 0.09) than with fly density (0.62 ± 0.25). Thus, a high morbidity rate in July coincided with a high fly density and a high infection rate. Later, despite a high fly density, the morbidity was lower as was the fly infection rate. In October, the fly density was low, but the infection rate increased and with it there was a second increase in dysentery morbidity. Fly density was determined by catches on fly paper at 24 village sites. Fly infection was determined in a novel way. Ten specimens of a species taken from one locality were triturated in broth. The broth was then incubated for 12-16 hours with phage, which lysed most local strains of *S. flexneri* and some *sonnei* strains. In the presence of these types of pathogens, phage would have multiplied and could then be detected by titrating against a known *flexneri* or *sonnei* on agar plates. Arskiĭ's imaginative treatment of an old problem leaves one point unclear. Since shigellae in human feces are the sole source for flies, isn't the cart before the horse if we base the cause of an increase in dysentery morbidity on an increase in the fly infection rate?

Not all investigators are convinced that enteric disease peaks have a real relation to fly peaks. Some go further in discounting the fly factor altogether. In England, Hamer (1908) observed that, while fly density remained stationary during two consecutive summers, diarrheal incidence increased one summer and actually declined the next.

If Morison's (1915, 1916) observations in Poona, India, had been less complete, coincidence could have easily been mistaken for causality. In tracing the relationship between flies and dysentery, he noted that fly density from May 29 to June 10 coincided with a rise of dysentery. Both curves rose and fell together until July 31, and then rose again. Fly density reached its peak the third week of August, and two weeks later the flies had disappeared. Yet, three distinct epidemics of diarrhea occurred after their disappearance. The next year, the water supply was treated with calcium hypochlorite, and, though flies were as much a nuisance as ever, the epidemic in one cantonment ceased entirely, while, in other parts of Poona, the usual July-August outbreaks failed to materialize. Water supply, not flies, proved to be the crucial factor in Poona.

In Tashkent the peaks of typhoid incidence and house fly activity are widely disparate—the former occurs in October and the latter occurs in July. Even when allowance is made for the incubation period, it is still impossible to establish a connection between the two phenomena (Shterngol'd, 1949). The difficulty of assuming an invariate one-to-one relationship between fly density and dysentery incidence is illustrated in Figure 43.

Sykes (1910), too, felt that the importance of flies was exaggerated. Although fly density in various parts of Providence, Rhode Island, correlated well with the level of sanitation, there was no correlation with the seasonal or spatial distribution of typhoid or dysentery cases. He felt that the spread of these diseases in a city equipped with modern sanitary facilities had a more fundamental vehicle than the fly, but he did not identify the vehicle.

In Latin America, the epidemiology of diarrheal disease in Guatemalen highland villages shows that the high incidence at weaning results from massive exposure of malnourished babies to enteric pathogens. Yet water and milk are minor factors as shown by the scattering of cases and absence of explosive outbreaks, and flies reach their peak during the rainy season, when the incidence of diarrhea is lowest. Later, in September and October, flies are still numerous when the dysentery curve rises (Bruch et al., 1963). In Guayaquil, Ecuador, there was no correlation between fly density or fly infectivity with diarrheal mortality over a two-year period despite the high infectivity of flies with *Salmonella* and *Shigella*.

In summary, the arguments against the fly factor are: 1. fly densities are sometimes significantly out of phase with disease incidence in spring, summer, and fall; winter outbreaks occur in the absence of flies; 2. in hot climates, e.g. South and East Africa, flies are present the year around, yet seasonal cycles of enteric disease may still occur (Pach, 1935).

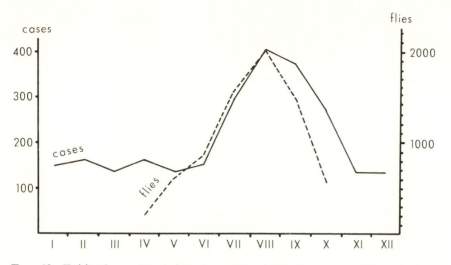

cases

400

300

200 — cases

100

flies

flies

2000

1000

I   II   III   IV   V   VI   VII   VIII   IX   X   XI   XII

FIG. 42. Epidemic curve of dysentery cases in Budapest in 1967 and the curve of the monthly total of all flies collected in open-air markets. (After Aradi and Mihályi, 1971.)

% of total

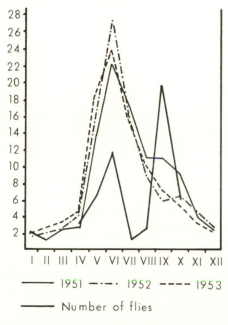

28
26
24
22
20
18
16
14
12
10
8
6
4
2

I   II   III   IV   V   VI   VII   VIII   IX   X   XI   XII

———— 1951   —·—· 1952   ———— 1953

———— Number of flies

FIG. 43. The seasonal incidence of dysentery and numbers of flies in the Turkmen S.S.R. show a close correspondence from May until September. In September and October flies reach their second seasonal peak, however the incidence of dysentery continues to decline. (From Elkin, 1961.)

The problem is more complex than graphs purport to show because high fly densities are obviously of little consequence if access of flies to sources of infection or food is blocked.

EXPERIMENTAL TRANSMISSION

In this chapter and the next, our discussions of flies and specific diseases usually include laboratory data on persistence, multiplication, etc., of various agents in flies. In the case of enteric infections, the accumulated information is so extensive that we have included it in the previous chapter on The Fly as Host.

EFFECT OF FLY SUPPRESSION ON DISEASE INCIDENCE

Thus far, we have reviewed various aspects of fly transmission of enteric infections in military, institutional, and civilian environments. Perhaps the most impressive evidence of fly involvement is derived from tests conducted at the community level, often involving many communities in a large-scale experiment. The availability of synthetic insecticides since the late 1940s has provided us with a number of grandly conceived experiments, but, long before the advent of synthetics, investigators achieved fly control by other means and were able to measure its effect on the incidence of disease.

We previously alluded to the British Army experience in India. Faichnie (1909b) refers to an article by Jones in 1907, in which he points out that, in Nasirabad and Mbow, control of flies since 1904 had markedly reduced the incidence of typhoid fever and this had been maintained up until 1909, the time of writing. Control had been achieved for the most part by effective systems of trenching the night soil. In both places, the water supply had been tested and was good. At Jubbulpore, in the same area, the water was also good, but flies swarmed everywhere, including at the trenching grounds. This was the only station in the area that suffered from typhoid outbreaks. Odlum (1908) describes his medical experience in an army camp, also at Nasirabad, and points out that typhoid cases disappeared with eradication of flies and reappeared with their return. He not only recovered *S. typhi* from flies but found *E. coli* in each of several hundred specimens (also Luxmoore, 1907).

The victory of the British forces at El Alamein in 1942 brought with it serious sanitation problems and flies. The seasonal reduction in fly density and the clean-up of fouled areas reduced the incidence of dysentery and diarrhea (Gear, 1944). In a letter to the British Medical Journal, Logan (1944) describes a cavalry camp plagued by flies and 30 dysentery cases a day. This number was rapidly reduced to zero after a Medical Officer launched a campaign of fly extermination.

As Medical Officer with the Dutch army in Java, van Voorst Vader (1951) was given an opportunity to test the effect of fly suppression on the number of cases of bacillary dysentery and infectious enteritis in a garrison. In a part of the garrison, extermination of fly larvae in latrines during February and March was carried out with 10% cresol in kerosene. The month before treatment started, the test group had 74 cases per day (1.8%) and the control group had 73 cases (1.1%). During February and March, the number of cases fell to 26 (0.6%) and 17 (0.4%), respectively, in the test group and 94 (1.4%) and 47 (0.7%), respectively, in the control group. The reduction in the test group was considered to be appreciable despite the fact that fly control was initiated before the epidemic reached its height, and the implementation of control measures left something to be desired.

In the United States, the typhoid picture in Jacksonville, Florida, in 1910 could be summed up as follows: fly and typhoid peaks were parallel; water and milk were beyond suspicion. In 1911, a law was passed that required flyproofed privies and fly paper in rooms of typhoid victims. With the implementation of this law, the number of cases of typhoid fell from 329 in 1910 to 158 in 1911; 55% of the latter were cases originating in other places. Although this does not constitute proof that the cases were fly-borne, the suggestion is strengthened by the fact that the only antityphoid measures taken during this time were those directed against flies (Terry, 1913).

Armstrong's (1914a, b) study in New York City is a model pioneering effort to gauge the effect of fly suppression on diarrheal diseases in a community. He selected two areas in the Bronx that were comparable in population size, level of sanitation, similarity of ethnic groups, etc. For 8 weeks, from July 21 to September 13, 1913, the "protected area" was the focus of an antifly campaign that included the following tactics: a well-attended film on the life history and dangers of flies; extensive fly trapping in the neighborhood; fly swatting and fly paper in the home; screening of every window and door in the block, 1700 in all (this part of the program was nullified because, during warm weather, it was impossible to keep the screens in or closed, and often they were destroyed); destruction of maggots in horse manure and privies with iron sulfate; the Department of Street Cleaning made a special effort to maintain clean streets; the Sanitary Division of the Health Department cooperated to bring about a general clean-up of accumulated refuse in the area, including installation of many new garbage cans with lids; and clean courtways, yards, and stables were maintained. Although the 8-week observation period was admittedly brief and the total population studied was small (3469 persons in both areas), some striking results were obtained. With respect to diarrheal morbidity, which data were obtained by weekly survey of each family, the "protected" area had a total of 20 cases compared with 60 in the control area. In contrast, non-communicable diseases remained about equal in the two areas. We should be cautious in fully crediting fly control for these results, since no effort was made to assess a "sanitation psychology," which may have developed among residents in the "protected" area. And, as Armstrong admits, a longer baseline should have been established before fly control was instituted. Nevertheless, he is to be applauded for his total approach to a community problem.

Decades later, Barinskiĭ (1952) applied this gestalt to a Russian community, with a view to dysentery prophyllaxis. In the process, more than 30,000 laboratory tests were made to obtain background data on human carriers, soil and water contamination, and bacteria in flies. The climate of the study area is characterized by long, very hot summers, scanty precipitation, and frequent hot winds that raise clouds of sand and dirt. The sanitary level was poor with respect to feces disposal and the presence of flies. The past incidence of enteric disease was high. From 287 flies, with *M. d. vicina* and *P. sericata* predominating, 6 strains of *S. flexneri* were isolated. Against this background was launched a campaign of fly suppression with the use of DDT and the widespread application of petroleum to feces, garbage, and liquid wastes. Oiling was carried out regularly and proved to be effective in the destruction of fly larvae. Conducted jointly with fly control were programs aimed at systematic waste removal (including the emptying of the contents of privies) and general clean-up, public education, and extensive efforts to sever the human link in the epidemic chain. This was done by treating foodhandler carriers, convalescents, and chronic cases, and by hospitalizing those with gastro-intestinal complaint. Under this program, the incidence of dysentery was significantly less than in nearby communities without such a program.

In a study that extended from 1959 to 1962, cities in the Dnepope-trovsk region were subjected to one of three methods of sanitary clean-up: systematic dwelling unit-based clean-up; systematic courtyard-based clean-up followed by insecticide treatment of garbage bins; and courtyard clean-up without insecticide treatment. The mean fly density index in zones employing each of the above methods was 1.8 to 2.1, 4.3 to 7.1, and 4.9 to 10, respectively. Only in cities that used the first method was there a decided lowering of the dysentery rate (Petrov, 1964).

Pavlovskiĭ and Bychkov-Oreshnikow in 1934 noted that successful fly control at Kushka led to reduction of acute intestinal illness in summer by more than 50% (Sukhova, 1954).

In southern United States, installation of bored-hole privies sharply reduced house fly breeding and the number of house flies frequenting the privies but had no measurable effect on house fly density in the community. Nevertheless, the exclusion of flies from human feces resulted in a 52% reduction in the shigellosis rate among children under ten, and in the diarrheal morbidity rates (McCabe and Haines, 1957).

The preceding investigations share emphasis on the attainment of fly control through sanitation. Following the second World War, DDT became available for civilian use and public health workers in many countries enlisted the chemical in a re-declared war on disease vectors. In just a few years large-scale, fly control-enteric disease programs were conceived, executed, and reported in the literature. Some of these have achieved the status of text book models, others, equally good, are less well known. Most are positive reports of disease reduction through fly suppression, but some are negative. Both kinds expose pitfalls in our thinking and provide a sobering and sometimes surprising appraisal of how complex the human ecosystem actually is, even from manipulation of the simplest communities.

The byproduct of an extensive malaria control project in Kharwar and Kanara districts in Bombay Province, India, was a reduction by 600 in the number of deaths due to diarrhea and dysentery in 1947. This reversed the previous mortality figures, as well as the general trend elsewhere in the region, and represented a reduction of 40 to 60% in mortality (Viswanathan and Rao, 1948). Recent fly control efforts in a rural Indian community were attended by a reduction in diarrheal disease morbidity in children under 5 years of age (Kumar et al., 1970).

The effect of an antimalarial program on the incidence of dysentery in Greece was analyzed by Vine (1947). The number of officially reported cases for the summer of 1945 (before DDT spraying) as compared with 1946 (after spraying) were, respectively: July, 268:151; August, 585:85; September, 510:2; and October, 152:6. In commenting on this, Simmons (1959) says: "This spectacular decline of dysentery following the use of DDT is certainly indicative of considerable fly transmission and of the efficacy of DDT in controlling the disease."

Fly control in refugee camps in Lebanon, Syria, Jordan, and Gaza reduced the fly index from 5.8 during May to September, 1951, to 3.6 the following year. The destruction of flies was attended by a notable decline in the incidence of summer diarrhea and dysentery, but the level of typhoid and eye diseases remained the same (West, 1953).

DDT spraying took place in the Lebanese town of Marje in spring 1946 in conjunction with an antimalaria program. From September 1 to October 15, following this treatment, there were about 60% fewer gastrointestinal illnesses in Marje than in a control town nearby (Berberian, 1948). These results, though suggestive, are difficult to evaluate because of the brief observation period and lack of comparability within the village itself (before and after treatment).

During 1950, the use of chlordane and DDT in the Egyptian village of Sindbis effectively controlled flies. At the same time, infant mortality (due primarily to diarrheal diseases) fell from 227 per 1000 live births in 1949 to 115 per 1000 in 1950. By 1951, the fly population had become resistant and returned to its former density, and infant mortality resumed its previous level (Higgins and Floyd, 1955; Weir et al., 1952).

The province of Latina in central Italy was the locale for studies by Corbo (1949, 1951a, 1953) of gastrointestinal illness in children less than 2 years of age. Zone A, the treated area, contained some 210,000 inhabitants and 15 towns; zone B, untreated, contained 68,000 inhabitants and 13 towns. With the application of insecticides, the number of cases in zone A fell from 45.67 per 1000 live births in 1945, to 9.09 in 1946. With the rise of DDT resistance in 1947, cases increased to 21.06 but fell again in 1948 to 12.10 and to 7.90 in 1949, when chlordane was used with DDT. Confirmation of the importance of the fly factor was obtained by extending fly control into 1952 in zone A by means of a combination of four insecticides, and by suppressing flies in zone B solely during 1950, both of which resulted in a tangible reduction in relevant mortality figures (Fig. 44). Although the suppression of gastrointestinal illness through fly control is clearly indicated by Corbo's data, a corresponding and rather large reduction in the number of cases in the untreated area from 1945 to 1946 indicates that unidentified factors were also involved.

Prior to the initiation of an antimalarial campaign, the incidence of enteric illness in a large number of Sardinian villages had changed little over a period of 8 to 10 years. Then, from November 1946, until June 1947, DDT was sprayed in 272 out of 318 villages. With fly suppression, the following reductions in disease incidence were noted from 1946 to 1947, respectively: typhoid, from 3486 to 820; paratyphoid, from 729 to 349; bacillary dysentery, from 75 to 10; and infant gastroenteritis, from 559 to 129. Additional positive evidence was the fact that relaxation of insecticidization in August 1947 coincided with an increase in these diseases (Spanneda and Marchinu, 1948). However, Spanneda pointed out in 1950 that the decline in typhoid and paratyphoid morbidity could not have been strictly due to fly suppression since there had also been a decline in some 46 untreated villages. In Sardinia, as in Latina, unidentified but powerful factors had contributed to the declining incidence.

Ylppö and co-workers (1950) undertook a program of fly control to determine to what extent flies are involved with infant diarrhea in Finland. The observation period was only two months, too brief for any firm conclusions, but the evidence suggested that the small difference in the diarrheal rates between infants in the test and control towns was due more to diet than flies. This small difference disappeared among infants who were fed human or powdered milk instead of dairy milk, regardless of fly treatment. Gorbatow (1951) describes a similar experience in Helsinki, where effective fly control did not influence the diarrheal rate. Instead, the cases seemed to be grouped around milk stores whose product had very high bacterial counts; the same neighborhood also had the most questionable wells.

An investigation in northwestern Russia revealed that, at most, fly control may have played a role in depressing the dysentery peak but did not prevent it. In a 4-year study by Sorin and Rundkvist (1956), dysentery incidence in treated towns was sometimes greater than in untreated towns, despite successful suppression of flies! The most afflicted group was the peat-dryers, who suffered a dysentery attack rate 2 to 2.5 times higher than that of other workers. By providing the peat-workers with good water and hot meals, and easing their arduous work, the incidence was cut to one half.

After a thorough study of dysentery in the Moscow district, Uvarova (1958) also concluded that flies contributed in only a minor way to summer incidence. We may recall that *M. domestica* in these northern regions is not attracted to human feces, and other synanthropic flies may only be peripherally involved. In a southern city the usual seasonal upswing of dysentery cases did not occur after flies were suppressed (Derbeneva-Ukhova, 1952).

The above studies should caution us against an exclusively entomological approach to the epidemiology of enteric diseases. There are many potential vehicles and their importance depends in part on the habits and life style of a people.

Turning to the Western Hemisphere, the inception of antimalaria programs in Venezuela provided additional opportunities to evaluate the impact of insect control on enteric disease trends. Gabaldon and his colleagues (1955) selected communities that had the same sewage systems and sanitation, and had suffered about the same enteritis and dysentery mortality rate from 1941-1945. The one clear difference between the two groups of towns was that one group was malarious while the other was not. With the application of DDT between 1946 and 1950 to the malarious towns there was a much greater reduction in enteric and diarrheal disease rates than in the untreated areas. The results are shown in Figure 45A and B. The *coli*metric index of *M. domestica* (percentage

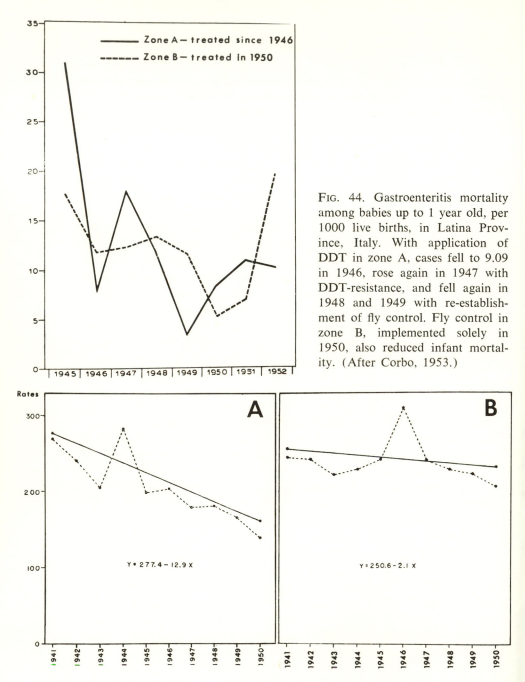

Fig. 44. Gastroenteritis mortality among babies up to 1 year old, per 1000 live births, in Latina Province, Italy. With application of DDT in zone A, cases fell to 9.09 in 1946, rose again in 1947 with DDT-resistance, and fell again in 1948 and 1949 with re-establishment of fly control. Fly control in zone B, implemented solely in 1950, also reduced infant mortality. (After Corbo, 1953.)

Fig. 45. Rate of reduction of diarrhea and enteritis in Venezuelan towns treated with insecticides (A) and untreated (B). (From Gabaldon et al., 1955.)

of flies bearing *E. coli*) was lower in the treated areas, apparently because DDT repelled the flies. Another Venezuelan study reported by Cova Garcia (1956) claims corresponding reductions in these diseases with the suppression of flies, but this is not clearly demonstrated by the results, nor is there a convincing parallelism in the fluctuations of the fly and diarrheal disease curves. He does note, however, that with the resurgence of DDT-resistant flies during 1952 to 1954, diarrheal rates were much higher than in 1951 when fly control was effective.

A WHO team concludes that high fly counts indicate a low level of sanitation and that it is the latter that is more directly related to the spread of diarrhea. Within a zone (in Venezuela) of high fly density, the percentage of dysentery cases depended on availability of water, toilets, etc. In support of this view, comparison with another zone where the fly index was half, and which presumably had comparable sanitation, showed no significant difference in percentage of diarrheal cases among children (van Zijl, 1966). This study overlooks what may be an important feature of fly transmission—the transmission of subclinical, undetected and unreported infections.

Flies were unequivocally guilty in Hidalgo County, Texas, an area of high diarrheal morbidity. In a study by the U.S. Public Health Service, DDT-treated towns had fewer flies and a significant reduction in diarrheal infection, disease, and death in children, compared with children in control towns (Table 5). This difference was maintained from 1946 to 1947, when a reversal of the treatment plan resulted in a corresponding reversal of fly density and disease incidence. Although the effect on *Shigella* prevalence was marked, salmonellosis incidence was not influenced by fly control (Watt and Lindsay, 1948; Watt, 1949). There were no significant differences in *Salmonella* recovery rates from human and animal populations in treated and untreated towns (Watt and DeCapito, 1950).

The proven association of flies with shigellosis in an area of high morbidity has been shown to apply to an area of moderate morbidity, as well. Three towns in southern Georgia were selected for treatment with insecticides, and comparable towns were chosen as controls. Fly control became increasingly difficult, requiring combinations of insecticides, but was maintained for about one year. *Shigella* and *Salmonella* prevalence was measured, as in Texas, by taking rectal swabs of children under 10, before, during, and after fly control. The results again demonstrated a significant lowering of *Shigella* prevalence and cases of diarrhea in the treated towns. With release from fly control, diarrheal incidence returned to the levels of untreated towns. As in the previous study, pediatric *Sal-*

TABLE 5

EFFECT OF FLY CONTROL ON SHIGELLOSIS AND DIARRHEAL MORBIDITY IN
INSECTICIDE-TREATED AND UNTREATED TOWNS IN HIDALGO COUNTY, TEXAS*

Illness	Fly control	No fly control
Shigella recovery per 100	1.6	4.1
Cases of diarrhea per 1000	10.0	18.4
Deaths (among children under 2) per 1000 (under 2 years) from diarrhea and enteritis	16.1	31.4

*From a study by Watt and Lindsay, 1948.

monella infections were too few to permit firm conclusions, but did not appear to respond to fly control (Lindsay et al., 1953).

Hemphill (1948) has gathered additional evidence in support of the fly factor. He has analyzed downward trends in diarrheal disease mortality in rural areas in southern United States and finds the decrease was twice as great in counties where antimalarial spray programs were in operation, as compared with areas without such programs.

CONCLUSION

We conclude as we began, by emphasizing the epidemiologic complexity and multivehicular spread of enteric diseases generally. In this or that outbreak, it may be possible to trace dissemination through a specific vehicle, e.g. water, milk, food, or flies. But it is more difficult to rate these vehicles as to their general importance because this will change from one outbreak to another, from one season to the next, from region to region, and from one socio-economic or ethnic group to another within the same community. Thus, although the evidence is overwhelming that flies are important vectors of enteric organisms, outbreaks can be identified in which they have taken no part at all. The aim of these final remarks, therefore, is to leave the reader with a fair and balanced perspective.

Poverty contributes to the prevalence of enteric disease in several ways. In Chile, the diarrheal mortality is two times greater among infants whose fathers are laborers than among infants whose fathers are nonlaborers. And in a moderate-sized city in the United States, the diarrheal death rate among toddlers in poverty areas was six times higher than among toddlers in well-sanitated environments (Schliessmann, 1958). Malnutrition plays an important role in lowering the resistance of the individual, and other debilitating factors may also be involved. Sir Malcolm Watson has said, ". . . when we cleared malaria out of the Malay States, we closed our dysentery wards," meaning that malaria weakens the individual and makes him more susceptible to enteric disease.

Backing up individuals' defense mechanisms is the environmental perimeter of defense. Here, sanitation, expressed through availability of water for personal hygiene and sanitary disposal of feces, is of the utmost importance. Drinking water is infrequently a vehicle for dysentery and diarrheal disease, but the availability of wash-water has been identified as a crucial factor in studies of children in California migratory labor camps, in Georgia towns, and in Costa Rica. Those who lived in units with an outside water supply had up to twice as many *Shigella* infections as others who had water inside their dwellings (Beck et al., 1957; Stewart et al., 1955; Moore et al., 1965). The report of a WHO team that carried out extensive surveys on diarrheal diseases in seven countries firmly supports the importance of the wash-water factor in reducing intrafamilial transmission through finger contamination.

Data collected by a WHO team in Mauritius (1966) offer proof that the type of excreta disposal is important in the spread of diarrhea. The diarrheal disease rate was significantly greater among those who used no toilets (sugar cane fields) than among those who used outdoor toilets (pail system, water closet, pit privy, etc.). The report goes on to say that "sanitation applied in the wrong way poses a danger. . . ." In a non-water-supplied area in the Sudan 594 persons without toilet facilities reported 19 cases of diarrhea during a 1-month period. At the same time 185 persons using an unsanitary privy reported 30 cases of diarrhea. The difference was highly significant. Another report, this one by the International Cooperation Administration (Anon., 1956), links a 50% reduction in mortality due to enteritis and diarrhea in Costa Rica (between 1942 and 1954) to the installation of 10,455 privies.

Privies must be properly constructed to exclude all synanthropic flies, and this is not always possible. As we have seen, nonflyproof privies, or inadequate trenching, may be worse than none. Bored-hole privies are an improvement, for the deeper the latrines, the fewer house flies and sarcophagids will be found breeding. However, they produce more *Drosophila*, and Berti (1958) cautions that in Venezuela pit privies will not eliminate *D. ananassae*, from which *Shigella* has been recovered and which is often found in kitchens, on fruits, etc.

Elimination of the privy and installation in each home of water-borne methods for excreta disposal is, obviously, the best solution. In Colombo, Ceylon, typhoid fever diminished significantly only when sewage became water-borne (Spaar, 1925). This is the most effective means of preventing fecal contamination of water supplies, flies, soil, and the environs of human dwellings (Fig. 46).

In the case of salmonellae, we are dealing with zoonoses that require a broader approach and even greater vigilance. Here, the contaminative food chain (e.g. contaminated fish meal to livestock to man) is likely to be international in scale, and food processing may involve many geographically separated feeder operations. Adequate controls are required over all phases of food production, extending to importations of food and feed, and extermination of flies, cockroaches, stored-grain insects, and rodents.

EYE DISEASES

Certain flies have long been regarded with suspicion as possible disseminators of human eye diseases, especially trachoma and the various forms of conjunctivitis. Much of the research on this problem has been complicated by difficulties in determining the specific causal organism and by problems of cultivation and human transmission experiments. We will attempt to give the general epidemiology of the diseases under consideration and review the significant fly research.

Trachoma

Trachoma (granular conjunctivitis) is a chronic disease of the eye limited to man and characteristically involves inflammation and then granulation of the conjunctivae, leading to gross deformity of the eyelids, visual disability, and possibly blindness.

The virus-like agent of trachoma, one of the psittacosis-lymphogranuloma group, is a large particle. It is visible by light microscope in the conjunctival and corneal epithelium and in the exudate in the form of elementary bodies. The sources of infection are secretions from the eyes, mucoid or purulent discharges of nasal mucous membranes, and also the tears of infected persons. Susceptibility is general, but children are more frequently affected than adults. The incubation period is 5 to 12 days as determined by human volunteer experiments. The disease occurs irregularly throughout the world and high prevalence is generally associated with poor hygiene, poor nutrition, and crowded living conditions, particularly in dry dusty areas. The highest prevalence has been reported in Egypt and the Middle East, where more than 90% of the population may be affected.

The mode of transmission is direct contact with secretions of infected persons or by fomites; carriers have not been demonstrated. Flies have been shown to be probable vectors in eastern countries.

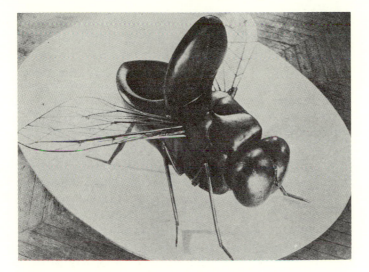

FIG. 46. Whimsy in the water closet. From amulets to ultra-modern commodes—incorporation of the evil object.

ACUTE BACTERIAL CONJUNCTIVITIS

Conjunctivitis of multiple bacterial etiology presents a more complex picture. Acute bacterial conjunctivitis (sore eyes, pink eye) begins with lacrimation, irritation, and vascular injection of the conjunctivae. This is followed by edema of the lids, photophobia, pain, and a mucopurulent exudate that runs its course in 2 to 3 weeks, but many patients have no more than vascular injection of the conjunctivae and slight exudate for a few days. Confirmation is by bacteriological culture or microscopic examination of smears of exudate.

Haemophilus aegyptius, the Koch-Weeks bacillus, appears to be the most important agent. *H. influenzae, Moraxella lacunata*, staphylococci, streptococci, pneumococci, *Corynebacterium diphtheriae*, and others may produce the disease. *H. aegyptius*, a small gram-negative bacillus first found by Koch in 1883 in smears from infected eyes, has been shown conclusively to be a different species from ordinary *H. influenzae*. *M. lacunata*, the Morax-Axenfeld bacillus, is also gram-negative, occurring characteristically as a diplobacillus in the pus from conjunctival and corneal infections in man.

The period of communicability is during the course of active infection. The incubation period is usually 24 to 72 hours. Children under 5 years of age are most affected, and the incidence decreases with age. The debilitated and aged are particularly susceptible to staphylococcal infections. The disease is widespread throughout the world, particularly in warmer climates; it is frequently epidemic.

The sources of infection and modes of dissemination are the same as for trachoma. The importance of flies probably varies in different regions and seasons, and, in some places, eye flies, e.g. *Musca sorbens*, are replaced by the eye gnats, *Hippelates* or *Siphunculina*. At least a dozen species of eye-frequenting moths from Africa and the East are now under suspicion.

Other etiologic agents of acute conjunctivitis are the gonococcus, which is important but not necessarily the most frequent among the infecting agents, meningococci, hemophilic bacilli and inclusion blenorrhea virus. The gonococcal infection is transmitted during passage of the newborn through the birth canal. Gonococci remain viable for a short time on inapparently soiled clothes, towels, and hands. Flies could be involved in spreading the infection by this means or by contact with infected and then healthy eyes of defenseless infants.

FLIES IN GENERAL

The Book of the Ten Treatises on the Eye, ascribed to Hunain Ibn Is-Haq (809-877 A.D.), is considered the earliest existing systematic textbook of ophthalmology. Concerning ophthalmia and the swelling that follows, it has this to say (p. 56): "The first kind occurs suddenly and usually as the result of a predisposing condition in the corner of the eye brought about by the bite of a fly or a bug; and it occurs mostly during the summer and in old men." Larrey, Napoleon's military surgeon, mentioned earlier in connection with maggot therapy, considered small flies one means for the spread of ophthalmia in Egypt.

In more recent times, Howe (1888) observed that the number of cases of conjunctivitis in Egypt increases in proportion to the number of flies prevailing. Demetriades (1894) in discussing purulent ophthalmia in relation to trachoma in Egypt reported that it was particularly flies that transmit the gonococcal ophthalmia, carrying it from infected to uninfected persons. Germann (1896), from personal observations, remarked that flies spread eye diseases (trachoma) in Syria and Palestine. Meyerhof (1914) stated that in Egypt during an epidemic of conjunctivitis in May 1909 there was also a plague of flies. Butler (1915) was of the opinion that in Palestine flies are the important agents in the dissemination of ophthalmia; they were seen to be abundant and showed a preference for the eye and were commonly in numbers on children's eyes.

Nicolle and Cuenod (1921) stated that the fly season in Tunisia in

September and October was marked by peak incidence of acute conjunctivitis. Petit (1923) described his researches on trachoma in Tunisia, reporting that the incidence of eye infections was highest in October with the fly population reaching its peak during September-October. The natives attributed the disease to figs, dates, Barbary figs, and pomegranates, all of which have a sweet juice rich in tannin. The natives transferred the fruit juice with their fingers to the eye, thus increasing the attraction for periocular flies; Barbary fig even has fine hairs forming a down that is irritating to the conjunctiva. Petit also relates the account of a Tunisian doctor who observed a small epidemic among hospitalized children in September and who, considering the hygienic precautions taken, ruled out contact infection and attributed the cases to flies.

Later investigators made more exacting phenological correlations between disease and fly prevalence. Rao (1931) reported the results of a study of 1000 cases of acute conjunctivitis in Bangalore, India. The seasonal epidemic started about the middle of May, reached its height in June and July, and then receded to the pre-epidemic level about the first of November. The Koch-Weeks bacillus was found in almost 60% of cases; a few cases showed *Pneumococcus* or *Streptococcus*. Because of the seasonal incidence, he implicated mango gnats (*Siphunculina funicola*) and house flies, but believed that the majority of the cases were due to direct intrafamilial transmission.

The regularly recurring epidemics of acute ophthalmia in the spring and autumn of each year in a number of provinces in Egypt are considered to depend on the reservoir of carriers of Koch-Weeks bacillus and gonococcus in each community, on the presence of climatic conditions favorable to these organisms, and on the sharp seasonal increase in the number of flies (Wilson, 1936).

Lyons and Amies (1949) investigated the epidemiology of conjunctival infections caused by the Koch-Weeks bacillus and the gonococcus in Egypt. The climate was found to be remarkably constant, and there was little variation year to year in the timing and severity of ophthalmia epidemics. Seasonal clinical findings indicated that epidemics did not arise in the usual way from a few individuals in the acute stage to a large number of susceptible individuals. It appeared, rather, that, with onset of warm weather and the concomitant increase in the number of flies, repeated passage of organisms between the numerous carriers and others led to progressive virulence of the strains, and, thus, a large proportion of children acquired the disease in its more acute form. Clinical acute ophthalmia was confined mostly to younger children, the highest incidence occurring between the ages of 3 months and 2 years; there were many more or less symptomless carriers among school-age children, and few acute cases among them. Acute ophthalmia was rare among adults, with a higher attack rate occurring among mothers of affected infants. It is significant that the eyes of young children were the most prone to attack by flies, and that as children grow old enough to be able to move about and brush away flies, the incidence of the disease decreases.

Antitrachoma campaigns in Turkey were geared to epidemic cycles and were most intensive during July to September, when heat, dust, and flies reached their maxima. The campaign required that shops have fly traps and that manure be deposited at a 15-minute walk from dwellings and covered with dirt (Kural and Ayberk, 1936).

One of the most convincing kinds of evidence of flies as vectors derives from the effect on disease incidence of community-wide reduction of flies. Gaud and Faure (1951) reduced the flies in a small village in southern Morocco to 1/100 of their normal number and this was followed by a reduction of trachoma and acute conjunctivitis in children from 88% in April to 36% in September–October, while incidence in an untreated village remained the same. Extensive fly control efforts in two Egyptian villages likewise caused a threefc'd decrease in the incidence of acute gonococcal and Koch-Weeks ophthalmia as compared to villages with no fly control (Lyons and Abbine, 1952). Reinhards et al. (1968), in a well-planned and thorough study, suppressed fly populations in selected villages in southern Morocco from June to November. The results showed that village units under fly-suppression averaged less conjunctivitis than corresponding ones without fly-suppression. Fly-suppression alone failed to show any direct effect on trachoma during a short observation period, but did decrease the risk of infection where it was used in conjunction with mass antibiotic and chemo-prophyllactic treatments. Additional examples of the positive effects of fly control are given below.

FLY SPECIES INVOLVED: MUSCA

In Eastern countries *M. sorbens* and *M. domestica* have been implicated as disseminators of eye diseases. Zimin (1944) studied the habits and seasonal population dynamics of *M. sorbens* in relation to the incidence of eye diseases in Tadzhikistan and concluded that it was a probable vector. In western Turkmenia, the seasonal distribution of *M. sorbens* and the incidence of acute epidemic conjunctivitis in inhabited areas coincided completely (Sukhova, 1953). Extensive fly control conducted in one area sharply depressed the *M. sorbens* population, and the incidence of eye disease was reduced to 1/5 that of previous years. Smears of gut contents of flies caught at foci of infection showed gram-negative

218 FLIES AND HUMAN DISEASES

bacteria morphologically similar to Koch-Weeks bacilli; the same organism was found in smears from patients' eyes. The author considered *M. sorbens* the prime vector of this disease. The house fly, *M. domestica*, also closely associated with man in this area, predominated in spring and autumn, when disease incidence was low. West (1953) also believed that *M. sorbens* rather than *M. domestica vicina* was probably the species largely concerned in conjunctivitis and trachoma. Gaud et al. (1950) considered *M. sorbens* the likely vector of ocular infections in Morocco and reported that when antifly campaigns were extended to yards and other open areas there was a marked reduction of acute conjunctivitis. Bakry (1955) and Decoursey et al. (1956) collected *M. sorbens* from diseased eyes. The latter authors found *M. domestica vicina* predominating on faces from December through March, and *M. sorbens* from April to July in Egypt.

McGuire and Durant (1957) studied the gross numbers and kinds of bacterial flora externally and internally in the two common Egyptian house flies, *M. sorbens* and *M. d. vicina*, captured from the faces of children, but in only a few instances did the flies examined contain *N. gonorrhoeae* or the Koch-Weeks bacillus. They concluded that, although these two bacteria were present in infected eyes and in flies, the greater number of bacterial ophthalmias associated with fly infestation of the faces and eyes of Egyptian children appears to be caused by streptococci and staphylococci and possibly in some instances by enteric bacteria. The organisms most commonly isolated from either eyes or flies were the hemolytic micrococci, several of which produced ophthalmia in experimentally infected rabbits.

Ponghis (1957) conducted studies in 1954 in 24 Moroccan villages to provide data for evaluation of three fundamental measures for the control of epidemic eye infections: antibiotics, chemotherapy, and fly control. Fly control measures were carried out in 8 of the 24 villages. Reduction of the fly population was best achieved by residual insecticides (chlordane), but larviciding of manure heaps was disappointing and was abandoned. *M. domestica* was always the most prevalent fly seasonally, while *M. sorbens* ranged from 1% of the total captures in June to 30% in October. A statistically significant reduction in the *M. domestica* population was obtained in the experiment, but a significant reduction of *M. sorbens* was not obtained until October, when human defecation areas were treated with insecticide. A correlation was established between the number of flies (mainly *M. domestica*) and the frequency of acute and subacute conjunctivitis. It was reaffirmed that children 2 to 8 years old were the most vulnerable to fly attack because they made little effort to rid themselves of flies that settled around their eyes. They were, therefore, the last age group to benefit by a reduction in fly population. The 8- to 15-year-old group showed the greatest drop in infections by the end of the study, presumably a measure of increased intolerance for flies.

The results obtained by Hafez and Attia (1958) from fly catches on diseased eyes, the seasonal prevalence of both *M. sorbens* and ophthalmias, and the incidence of eye disease in residential districts of low and high living standards also suggested the role played by this fly in the transmission of ophthalmia in Egypt. *M. sorbens* predominated in spring, summer (August), and autumn (October). In March, April, and June, over 80% of the flies caught on the eyes were *M. sorbens*; in February, May, and August, over 70% were this fly; while in October it was about 55%. It was found that the numbers of *M. sorbens* caught around the eyes generally reflected the natural population level during the year. The seasonal incidence of eye diseases studied at a hospital in Cairo over several years clearly demonstrated that the occurrence of the disease reached a peak in the spring, summer, and autumn, markedly correlating with the abundance of *M. sorbens* in nature, or attracted to the eyes. In winter, the disease incidence was lower, and the numbers of *M. sorbens* were markedly decreased. Of interest is the increase in abundance of *M. d. vicina* attracted to the eyes during winter, which may indicate that this species in this and other seasons does not play an important role. Eye diseases were found to be most prevalent in districts with no sewage disposal system, where defecation occurred in the open, and where *M. sorbens* as a consequence bred extensively.

Laboratory experiments suggest generally limited survival of the bacterial pathogens and probably a longer survival of trachoma in flies. Wollman (1927), working in Tunis, placed house flies in tubes containing cultures of *Bacillus aegyptius*? (= *Bacterium conjunctivitidis*) and the Morax-Axenfeld bacillus and observed that the flies lost their infectivity after three hours. Welander (1896) had found that *N. gonorrhoeae* was carried on the feet of a fly (probably *M. domestica*) for three hours after being soiled with human secretions. Flies retained trachoma virus on their legs and proboscis at least 24 hours after exposure to fresh or 6-hour old ocular secretions (Nicolle and Cuenod, 1921). However, the Koch-Weeks bacillus was not transferable by flies 24 hours after exposure due to the more delicate nature of the organism.

Gear et al. (1962) studied the ability of *M. sorbens* to transmit trachoma under experimental conditions. These flies are common in South Africa, where the disease is prevalent, and are often seen feeding on eye secretions of infants and even older children; they are less common in the mountains, where trachoma is also rare. The experiments failed to demonstrate trachoma in flies by injection of homogenates of legs, gut, etc., into embryonated eggs and by other means after flies had fed on infected yolk sac suspensions. Zardi (1964) reported successful transmission of the trachoma agent by *M. domestica* that had been sterilized externally with corrosive sublimate and fed penicillin and streptomycin in sterile food. Under these axenic or near axenic conditions, trachoma survived in the gut up to 15 days, but apparently for only four hours on the legs and proboscis. Larvae, similarly sterilized, were put through the same tests with positive results. No virus ingested by normal maggots has been shown to survive to the adult stage. Therefore, Zardi's work should be repeated with naturally contaminated maggots and adults.

While our primary interest is in the role of flies in eye diseases, we must guard against a presentation that may distort their importance. Siniscal (1955) regards trachoma in the Near East as a family disease resulting from poor sanitary practices under crowded conditions, and Werner et al. (1964) consider that trachoma can be transmitted from person to person by flies, but more often by direct contact between family members. For Lyons (1953) the weight of circumstantial evidence is so great as to leave little doubt that, in the majority of cases, trachoma and acute conjunctivitis are transmitted together and that the common fly is the principal agent in spreading infection from eye to eye.

Kupka et al. (1968), speaking of eye infections in Morocco, have stated: "The eyes of young children, particularly if they are full of purulent or mucopurulent discharge, are constantly covered by a great number of flies and the number of flies seems to increase the more the children are grouped.

"All this leads to the natural conclusion that the main source of bacterial and trachomatous eye infections in these areas is in the eyes of the younger children (between the ages of 6 months and 5 years) and that transmission by flies or otherwise occurs more easily to persons who are in a continuous and close contact with them."

Reinhards (personal communication, 1970) sums up some of his views based on long experience with the problem and indicates areas that need further study. He states that a better knowledge of the respective roles of *M. domestica* and *M. sorbens* is needed to control trachoma and the bacterial infections, which usually occur together and cause so much misery. The circumstantial evidence is based only on a statistical evaluation of clinically and bacteriologically measurable fluctuation in the incidence and prevalence. He feels that difficulty in cultivating the trachoma virus and in adapting it to tissue cultures is probably the main reason why no direct experiments have attempted to prove its presence somewhere in the transmitting fly or to find the mechanism of transmission. Similarly, little is understood about transmission by the fly of *Haemophilus, Neisseria,* and other organisms important in eye pathology. He thinks that only under certain conditions are flies the main vector in transmission. Moreover, as there is more and more indication that frequent superinfection plays a most important role in the development of case-severity and sequelae of trachoma, it is becoming increasingly important to know more about the mechanism of the transmission, its possible frequency, and the volume of inoculum. Unfortunately, there is a lack of research in this direction, especially in those countries in which the need is greatest and where it is difficult to undertake serious laboratory research.

HIPPELATES AND SIPHUNCULINA

Considerable circumstantial evidence involves chloropid flies and conjunctivitis in India, Ceylon, Java, and in the United States. *Hippelates collusor* and *H. pusio* are probably the most important species in North America, and *Siphunculina funicola* fills this role in the Orient.

In the United States, *H. pusio* swarms in great numbers in many southern states and is particularly abundant in Florida, settling on the eyes, sores, and natural body openings of man and animals. The first published accounts were from Florida and strongly incriminated *Hippelates* flies in the epidemic spread of sore eyes there (Schwarz, 1895; Schwarz et al., 1895; Neal, 1897). Moreover, it was considered that the fly produces an irritation that aggravates the disease, resulting in serious cases and possibly permanent eye weakening. There are unsubstantiated reports that in Cuba a species of fly frequents the flowers of *Euphorbia ferox* and may carry the poisonous pollen of the plant to the eyes of sleeping people, causing severe vesicular eruptions.

A seasonal acute conjunctivitis occurs in certain parts of the United States, notably in southern Georgia, Florida, other parts of the South, and in the Coachella Valley of California. Bengston (1933) found organisms resembling Koch-Weeks bacilli principally concerned in cases of eye disease in Georgia; the Morax-Axenfeld bacillus, pneumococci, and pleomorphic streptococci were present in some of the cases. Attempts were made to isolate the Koch-Weeks bacillus from gnats collected from the vicinity of the eyes of children, among whom the disease had its greatest incidence, but numerous saprophytes growing on the plate made the demonstration impossible.

Herms (1926) described the problem with *H. pusio* (= *collusor*) in the Coachella Valley of California, its activity from 9 to 10 months of the year, and its persistent attraction to the eyes and secretions of man and animals. He noted that these flies had become increasingly numerous over the previous ten years or so (resulting from increased tillage of the land) and this was related to the numerous cases of pink-eye seriously affecting the people of the valley. An example is given of a great increase in the number of *Hippelates* related to an outbreak of pink-eye in 15,000 school children at Thermal, California.

The incidence of acute conjunctivitis in various sections of the United States seems closely related to the seasonal and geographic abundance of *Hippelates* (Dow and Hines, 1957). Correlations between outbreaks of pink-eye and adult *Hippelates* abundance in southern California have been found by several investigators.

Ayyar (1917) considered *S. funicola* of India, Ceylon, and Java as responsible for the spread of the severe forms of ophthalmia found in these countries. From a study of its habits and the presence of *Staphylococcus aureus* and streptococci among other types, Syddiq (1938) felt that this fly was involved in transmission of eye diseases and yaws. Rao (1931) also considered that these flies, referred to as mango gnats, were likely agents in transmission in Bangalore, India. This fly is also a serious pest in Assam, India, during May-July and is generally held responsible for epidemic conjunctivitis. Roy (1928) gives a chart showing the incidence of this disease peaking in June, when *S. funicola* was at peak abundance.

It is probable that the flies mentioned in this section are important vectors of eye diseases purely on the basis of their habits and phenology. But the critical experiments have yet to be made, and they should include studies of the disease agents in flies as well as the effect of fly suppression on the incidence of eye infections. From what we know of the breeding habits of *Hippelates* species, experimental control may not be easily accomplished. It should be emphasized that persistence of the agent in or on the fly is not crucial to the mechanical mode of transmission and this is the mode that fits the habits of these flies quite well. It is strange that, since 1895, when "sore eyes" was first linked with *Hippelates*, not a single isolation of an eye pathogen has been reported from these flies.

STAPHYLOCOCCAL INFECTIONS

In 1914, Beresoff noted the presence of staphylococci and streptococci in the digestive tracts of flies hibernating in a Leningrad hospital, and house flies in the surgical wards of a British hospital were found harboring coagulase positive staphylococci (Shooter and Waterworth, 1944). In the United States, Scott (1917) first called attention to the fly as a possible agent in the transmission of suppurative organisms in the human population, through the isolation of *S. aureus* from house flies in the Washington, D.C., area. And in Egypt there was a high frequency of recovery of pyogenic staphylococci and streptococci from house flies, where these microorganisms appear to be associated with nonspecific conjunctivitis and keratitis (McGuire and Durant, 1957). Again, in Peru, 8 out of 10 fly pools yielded coagulase positive *S. aureus.* The flies were collected at slaughterhouses, canneries, food stores, homes, and other epidemiologically important places (Quevedo and Carranza, 1966).

Under experimental conditions, house flies harbored a food-poisoning strain of *S. aureus* for 8 days in their gut, and there is little doubt that, under suitable circumstances, they may deliver an inoculum that can produce an outbreak of food poisoning (Moorehead and Weiser, 1946).

Taplin et al. (1967) call attention to the potential importance of *Hippelates* flies as transmitters of pyodermas among rural populations in the tropics. They studied the skin flora and skin infections of 150 men before, during, and after a 3-month military exercise in a tropical jungle environment in Panama. It was noteworthy that any denuded area of skin, particularly infected lesions, attracted large numbers of *H. flavipes*. Flies captured while feeding on lesions were kept in sterile containers for 3 days and then cultured. Phage type 8, coagulase positive staphylococci were recovered from several flies. This finding assumes special significance because the organism was the predominant type in both the military and native populations, and yet it was absent from everyone in the group before they came to Panama. The frequency of insect bites, abrasions, and superficial lacerations of the skin, and the avidity with which myriad *H. flavipes* feed upon these sores, combine to make it highly probable that the fly is an important vehicle of staphylococci under field conditions in tropical areas.

STREPTOCOCCAL INFECTIONS

There is a growing suspicion that *Hippelates* flies are involved in the transmission of post-streptococcal glomerulonephritis. In cases where flies are suspected, the streptococcal infections are limited to impetiguous and other types of skin lesions. Bassett (1970; personal communication, 1967) has studied the epidemiology of the infection in Trinidad and finds that such lesions are common in children in rural areas. Nephritogenicity has usually been considered an attribute of certain types of *Streptococcus* within Lancefield's Group A, but Bassett has found other types and, quite commonly, more than one type in a patient. It should be remembered that only a small percentage of such infections lead to kidney involvement.

Fifteen percent of the children in Bassett's study had skin infections with *Streptococcus pyogenes*, generally on exposed parts of the body. Three species of *Hippelates*—*flavipes, currani,* and *peruanus*—were caught flying around the children or feeding on their skin lesions. *S. pyogenes* belonging to three different T-typing patterns (25/Imp. 19, 3/13B3264, and 9) was recovered from 5 of 32 *H. peruanus* and *H. currani.* The same types were recovered from skin lesions on the children. Streptococci of Lancefield Groups C and G were also isolated from flies. These isolations were made by allowing flies to walk on crystal violet blood agar plates, usually overnight (Fig. 47). Among five flies captured in a home, one *H. flavipes* yielded *S. pyogenes.* Bassett compared *Hippelates* density with the percentage of children with *S. pyogenes* skin lesions during October to June. Figure 48 shows that there was a general correspondence between flies and streptococcal lesions, particularly in fall and winter. In spring, the fly density began to increase about one month before the human cases.

Within the family, contact is undoubtedly the most common means of spreading the streptococci. In the larger environment, *Hippelates* flies may be important vectors, considering their mobility, numbers, and passion for sores. Some additional evidence of their importance in the spread of pyodermas is given under Staphylococcal Infections. Taplin et al. (1967) have also recovered β-hemolytic streptococci from *H. flavipes* in Panama. Any exposed break in the skin, even minor scratches and insect bites, may be contaminated by these gnats and progress to a skin infection with potentially serious sequelae.

We have found one study with biting flies, in which it was shown that *S. calcitrans* can transmit infection for up to 24 hours after feeding upon the spleen of a rabbit fatally infected with streptococci. Flies that fed upon an infected live rabbit were capable of transmitting the infection only by interrupted feeding (Schuberg and Boing, 1913).

A brief but very interesting paper on possible house fly transmission of streptococci in a British hospital was published by Shooter and Waterworth in 1944. They examined 27 flies caught in surgical wards and found that 9 carried hemolytic streptococci, 3 of which proved to be Group A. A control group of flies caught in the hospital laboratory did not carry hemolytic streptococci. Two of the three Group A strains from flies were Type 4. The authors point out the significance of this finding:

"Since a close check was being kept on all wound and throat infections in these wards, all Group A streptococci being typed . . . , it is possible to trace a hypothetical relationship between a carrier condition in these flies and infections in the human population of the wards. Type 4 has been identified only once previously—in a sore throat in a nurse ten

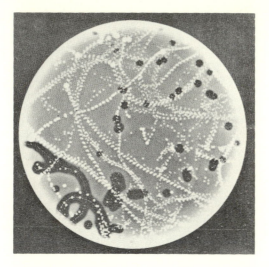

Fig. 47. Blood agar plate showing colonies of *Streptococcus pyogenes* (surrounded by darkened zones of hemolysis) deposited by *Hippelates flavipes* caught on a child without skin lesions. (From Bassett, 1970.)

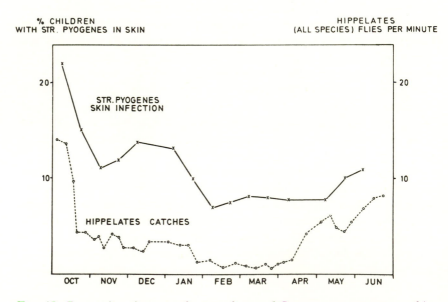

Fig. 48. Comparison between the prevalence of *Streptococcus pyogenes* skin infections among children in Maracas St. Joseph, Trinidad, and catches of *Hippelates* flies, using a mongrel bitch as bait. (From Bassett, 1970.)

226   FLIES AND HUMAN DISEASES

days before the infected flies were caught; it is possible that this nurse conveyed the infection to a wound which escaped recognition. During the following two months Type 4 was recovered from eight further persons, six having wound and two throat infections. All these cases occurred in one of the two wards studied, and it was in this ward that both of the Type 4 carrying flies had been caught." Aside from calling attention to the danger from flies (and possibly from nurses and other personnel) in critical areas of a hospital, this study reveals the value of working with highly defined types of microorganisms in tracing avenues of dissemination.

DIPHTHERIA

Man is the only natural host of this primarily respiratory disease. Historically, it has been considered that flies may be involved in spreading this disease. This is a disease of autumn and winter months, and, of course, flies are absent during winter outbreaks. Interestingly, an outbreak is occurring in Texas as this is being written, August 24, 1970.

Graham-Smith (1910) conducted a series of experiments that seemed to indicate that *C. diphtheriae* does not remain alive for more than a few hours on the legs and wings of house flies, but may live for 24 hours or longer in the intestinal tract. He recovered the organism from the feces 51 hours after ingestion. He stated (1913) that there was no evidence that, under natural conditions, flies are concerned in the spread of this disease, but that, under suitable conditions, it is possible that the disease may be occasionally conveyed by them. As far as we are aware, no further work with flies has been reported, or needs to be.

LEPROSY

This is a somewhat infectious disease, which is characterized by lesions of the skin and internal organs. The agent is *Mycobacterium leprae*, and demonstration of acid-fast bacilli in suspected lesions is usually confirmatory in lepromatous cases. The source and reservoir of infection are the discharges from lesions of infected persons.

Little is certain about the method of transmission. The question is complicated by the long incubation period of the disease. It seems to vary from a minimum of a few months to a maximum of 30 years, with an average of 2 to 7 years. Another problem has been that there is no animal in which progressive leprosy can be produced experimentally (though the organism can be cultivated in the foot pad of the mouse), and human experiment has generally failed, except in rare instances. Long and close contact with a leper is usually the history of the affected person. It must be emphasized, however, that man actually is very resistant to the development of leprosy even when exposed over a period of years.

Most observers believe that the common mode of entry is probably through accidental abrasions or lesions of the skin. Leprosy bacilli are continually discharged from ulcerated nodules, as well as from nasal lesions, in at least 80% of lepromatous cases.

It has been claimed that leprosy may be transmitted by flies, mosquitoes, bed bugs, fleas, cockroaches, ticks, lice, itch mites, and chiggers. There is bacteremia, particularly during the febrile periods of leprosy, and any blood-sucking insect might ingest the organisms. Nonbiting insects may become contaminated from the lesions. The research on flies has a long history, and evaluation is complicated by the epidemiological problems of the disease.

The Leprosy Commission in India (1893) tested the possible spread of leprosy through flies by examining flies that had settled on ulcerated sores of lepers; but no bacilli were found within them. Joly (1901) in Madagascar found bacilli in the intestinal tract and on the feet of flies that fed on leprous ulcers, and considered fly contamination of food as one means of spreading the disease. Tucker (1903) emphasized leprosy as one human disease that could surely be carried by flies. The leprous lesions are large and insensitive and exude fluid; flies therefore can settle and remain undisturbed for long periods on these parts. The lesions give off a peculiar goat-like odor that is extremely likely to attract flies from a distance. He observed in the Leper Asylum in Pondicherry the frequent presence of small black flies on leprous ulcers and counted as many as twenty on an ulcerated foot.

Wherry (1908) investigated common flies in the transmission of leprosy and used rat leprosy (*Mycobacterium lepraemurium*) as a model. The presence of enormous numbers of lepra bacilli in the exudations of cutaneous ulcers suggested the possibility of their being taken up by flies. Adult *Lucilia caesar* (= *illustris*), *Calliphora vomitoria*, and *Musca domestica* were placed on the carcass of a leper rat and then removed to a clean jar and fed on the liver of a normal rat. Examinations of the specks deposited during the second 24 hours showed that the lepra bacilli were almost completely voided during the first portion of this period; only the older dried specks contained acid-fast bacilli. Apparently the bacilli do not multiply in the flies. The larvae of *Calliphora vomitoria*, which hatched out in the carcass of a leper rat, became heavily infected with lepra bacilli. When these larvae were removed and fed on uninfected meat, they soon eliminated most of the bacilli; flies from these larvae were generally uninfected, but an occasional fly deposited some lepra bacilli. When the larvae were fed almost continually on a leprous rat carcass, they remained heavily infected with bacilli, but were unable to develop normally during pupation. He also reported that a house fly caught on the face of a human leper was found infected with lepra bacilli; at first few in numbers, but on the third day more than 1115 bacilli were present in each speck deposited. However, only one bacillus was found between the third and sixth days. The bacilli in this fly were not infective when injected into the subcutaneous tissue of a guinea pig nor did human antileprosy serum agglutinate rat lepra bacilli except at low titre.

Currie (1910) conducted an extensive series of experiments on flies in relation to leprosy. He worked with five species of flies found in Honolulu and Kalawao, including *Musca domestica*, *Sarcophaga barbata*, and *Lucilia* sp. Control examinations of 20 *M. domestica* and 12 of each of the other species not in contact with lepers or tuberculosis showed they did not contain any acid-fast bacilli. A series of seven experiments with *M. domestica*, fed broth suspensions of lepra bacilli and scrapings from leprous ulcers, demonstrated generally that a majority of fly specks contained lepra bacilli and that the intestinal contents of such flies showed the bacilli; specks were positive 72 to 96 hours after feeding. In experiments with *S. barbata*, which was fed on an experimental leprous ulcer of a rabbit, 2 of 5 flies showed a large number of bacilli in their guts. One of three of the *Lucilia* sp. that fed on an experimental ulcer had 98 bacilli in its intestinal tract and two specks examined contained fairly large numbers of bacilli (168, 1800 +); to Currie, it appeared that this species freed itself of the organism more rapidly than the other species. He concluded that these flies when given the opportunity to feed on leprous fluids will harbor the bacilli in their intestinal tracts and feces for several days, and may frequently convey immense numbers of leprosy bacilli, directly or indirectly, to the skin, nasal mucosa, and digestive tracts of healthy persons. He, of course, could not determine whether such fly-borne bacilli are or are not capable of infecting persons who might be thus contaminated.

Sanders (1911), on the other hand, conducted much simpler experiments with *M. domestica*, *Stomoxys calcitrans*, *Haematobia irritans*, as well as with mosquitoes, fleas, and bedbugs. He enclosed 70 flies over ulcerated leprotic surfaces and found only two acid-fast bacilli in the gut of one fly and one in the gut of another (species not given), although 20 of 75 bedbugs contained acid-fast bacilli. Direct contact and transmission by flies, fleas, mosquitoes, or other insects are possible modes of spread that are accidental and exceptional. However, Minett's (1912) work with *M. domestica* and *Lucilia* sp. confirmed Currie's findings, but he did not succeed in proving that the lepra bacilli present in the flies' feces could cause infection in guinea pigs. Leboeuf (1912, 1913) observed the frequency with which house flies settled on leprous ulcers and recovered bacilli from 19 of 23 flies that had settled on such ulcers known to contain bacilli. He also found that the feces deposited by these flies contained bacilli that showed no signs of degeneration 2 days after feeding, and reached conclusions similar to Currie's.

Honeij and Parker (1914) continued along these and other lines of research; they examined *M. domestica* (59), *Muscina stabulans* (3), *S. calcitrans* (30), *Lucilia* sp. (2), and an undetermined sciomyzid in their studies. They examined fecal spots from flies caught in patients' rooms and on lesions of patients and from flies fed on the contents of human ulcerous material that contained large numbers of bacilli. Of 12 flies of different species caught in patients' rooms, only 2 *M. domestica* showed acid-fast bacilli in their excreta; of 6 house flies allowed to feed on patient's lesions, none showed bacilli in their excreta. Of 21 house flies fed on the contents of ulcers, 15 were negative and 6 positive; of 20 *S. calcitrans*, 4 were negative and 16 positive; of another series of 6 flies (3 of each of these two species) two were positive, two negative, and two gave questionable acid-fast bacilli. No acid-fast bacilli were found in the excreta of the other species of flies. From these limited experiments, they concluded that *S. calcitrans* probably plays a more important role than the other species as a carrier of leprosy bacilli.

Two Asian investigators studied leprosy and flies in sanitariums. Arizumi's (1934) observations at the Leprosy Sanitarium near Taihoku in Formosa led him to conclude that adult flies are the most likely spreaders of leprosy bacilli. He examined *Musca*, *Lucilia*, and closely related genera. Of 689 flies caught in serious-case wards, 28% carried the bacilli in their guts, and, of 723 caught in slight-case wards, 24% were positive; infected flies were also found in buildings 200 meters from the institution. The flies caught in the wards had a higher rate of bacilli on the outside of their bodies than flies caught elsewhere. Confirming Currie, Arizumi found that, when the flies were fed on an emulsion of leprous nodules, the bacilli in their guts were most numerous 24 hours after feeding; after 39-48 hours their number was greatly reduced, but some flies still contained bacilli as late as 72 hours. From stools in the toilet of serious-case wards, 307 fly larvae were taken, 17.5% of which harbored acid-fast bacilli. Examinations of 300 pupae that in the larval stage had been fed on material containing lepra bacilli gave negative results. In Japan, Asami (1934) found that 26% of the house flies in serious-case wards of a leprosarium contained the bacilli; 4.6% in mild-case wards, and 2.3% of flies in homes of patients living there. Examinations of the external surfaces of flies caught in wards showed that fewer bacilli occurred on the surface than inside the flies. Several species of flies were washed and fed on leproma pulp rich in bacilli and examined every 24 hours thereafter with these positive results: *Calliphora lata*, 6 out of 10 infected; *Lucilia argyrocephala*, 44 out of 98; *M. domestica*, 16 out of 53; and *Fannia canicularis*, 8 out of 78. Bacilli on the surface of the flies disappeared completely in 3 days; they were seen in the feces up to the end of the second day.

Species other than those mentioned above have been implicated. De Mello and Cabral (1926) found that 40% (number not given) of the *Musca bezzii* captured at a leper asylum or fed on lepromata contained lepra bacilli in their intestinal contents. In Nyasaland, Lamborn (1937) obtained positive results with the feces of *Musca sorbens* 24 hours after they were fed on leprous material. In Brazil, Souza-Araujo (1944) caught many wild flies, which he reported as Tachinidae, on the lesions of the left leg of a lepromatous case. Examination of abdominal material from these flies showed rather great numbers of lepra bacilli. Among 186 specimens of *M. domestica* collected in a Chilean leprosarium, one fly was found which had numerous acid-fast bacilli presumed to be Hansen's by several tinctorial tests.

Vedder (1928) presented arguments for the possibility of the transmission of leprosy by insects. He cites the following facts as relevant. In perhaps the majority of cases a sharply localized cutaneous infection is the first demonstrable lesion of leprosy and, in about 75% of early cases, the first lesion is found not only upon the skin but upon an exposed part of the body. The skin, then, is the common portal of entry. Epidemiological studies show that, if leprosy is transmitted by contact, it is only with great difficulty under very special circumstances, and usually after long and intimate association with a leper. This and other data indicated to Vedder that leprosy may be more commonly transmitted by some intermediate agency such as the bites of insects. He did not favor the idea that nonbiters such as house flies and others carry the infection and contaminate persons. He believed that if a blood-sucking insect transmits leprosy, it can only do so by first biting a leper through a leproma and then transmitting the disease mechanically while its proboscis is still contaminated. That biting insect transmission can happen only rarely under natural conditions agrees with the epidemiological facts, partially explaining the rarity of the disease and the necessity for a long and close association. Finally, presumed lepra bacilli have been found with a considerable degree of frequency in a number of biting insects (see Vedder, 1928; Steinhaus, 1967). Vedder (1928) found acid-fast bacilli in 41% of the *Aedes aegypti* he fed on lesions of leprosy, and thought mosquitoes were likely transmitters. St. John et al. (1930) found that the bacilli survived in the gut of this mosquito for at least 24 hours, but not after 7 days.

It is conceivable that in some instances the bacilli of leprosy may be transmitted by biting insects from one person to another, but most of the evidence regarding nonbiting species is unconvincing. Even if all the reported isolations of the organism from flies were valid, and this is questionable because all of Koch's postulates could not be satisfied and acid-fast saprophytes have often been mistaken for the pathogen, we are left with only one certainty—lepers may contaminate flies. That flies transmit leprosy remains unproved. It is an interesting fact that Mexico and Guatemala are among the countries in the Western Hemisphere with low leprosy rates, despite the large fly populations that accompany poverty in these places.

TUBERCULOSIS

This is one of the most common and important communicable diseases of man, endemic in practically all populations. It is caused by *Mycobacterium tuberculosis*, which is present in all tuberculous lesions, in the pus of abscesses, in affected glands, in the sputum and expectorations, in diseased pleural sacs and joints, and in the skin lesions of man. Human, bovine, and avian types occur in warm-blooded animals.

The sources and reservoirs of infection are the respiratory secretions of persons with pulmonary tuberculosis and milk from tuberculous cattle; patients with the extrapulmonary form are usually not sources of infection. Transmission occurs via coughing and sneezing of patients with the pulmonary form, setting up a cloud of infectious material; minute particles may be inhaled directly or after settling and resuspension as dust.

In 1720, Dr. Benjamin Marten advanced the theory that tuberculosis is transmitted by flying insects. Spillman and Haushalter (1887) were the earliest investigators into the possible role of flies as disseminators of the tubercle bacillus. They examined the excrement and intestines of *Musca domestica* that had fed on the contents of spit-cups used by consumptive patients and found the bacilli in abundance; the bacilli were also found in the dried excrement of flies scraped from the windows and walls of patients' rooms. Hofmann (1888) demonstrated TB bacilli in the gut contents of 4 of 6 flies caught in the room of a recently fatal case; fly specks in the room also showed the bacilli. Flies were fed on TB sputum, and the bacilli were found in the excreta within 24 hours. Only one of 3 guinea pigs inoculated with the intestines of these flies developed TB: inoculation of specks that had dried for 6 to 8 weeks gave a negative result. Celli (1888) reported experiments made under his direction by Alessi, who fed flies on tubercular sputum and subsequently inoculated the flies' excretions into two rabbits which developed the disease.

Aylett (1896) smeared a cover-glass with sputum from a tuberculosis patient and placed several clean covers around it. These were made accessible to flies and quite a number fed on the sputum. Numerous specks on the clean cover-glasses were examined and found to contain one to three thousand bacilli each. Similarly, Buchanan (1907) allowed *M. domestica* to walk across a surface soiled with TB sputum, and afterwards across a surface of clean agar. The agar was washed, and the washings were inoculated into a guinea pig that died of TB 36 days later. Similar results were reported by Hayward (1904) for *M. domestica* and *Lucilia caesar* (= *illustris*), and by Tison (1950) for the house fly.

Knopf (1899) theorized that, since the tuberculosis expectoration has been exposed to repeated freezing and thawing and to a temperature of $-8°C$ without loss of virulence, infected insects may die and crumble to an infective dust that may thus enter the system through the respiratory tract.

In a limited test by Lord (1904), spontaneous liberation of TB bacilli from fly specks (even when agitated by a current of air) did not occur, since a guinea pig (only one animal was used) in proximity did not become infected by inhalations. He also found that flies (presumably house flies) may ingest TB sputum and excrete the virulent bacilli for at least 15 days (checked by inoculation of fly specks into guinea pigs). Lord cautioned greater attention during the fly season to the screening of TB wards, rooms, and laboratories against flies. Cobb (1905) believed that flies may play a part of some importance by infecting human food after feeding on tubercular sputum. He observed that flies were everywhere on sputum on the streets of Los Angeles and in restaurants and on exposed food.

André (1908) found no acid-fast bacilli in wild-caught house flies before he fed them on infected sputum. The flies began passing large numbers of tubercle bacilli six hours later and for the next five days, "plenty of time," as the author points out, "to carry these bacilli to a great distance." Flies caught at random in a TB ward produced tuberculosis when injected into guinea pigs.

Jacob and Klopstock (1910) visited 19 peasant houses in Germany and caught many flies, which they ground and inoculated into guinea pigs in the knee region. Five to seven weeks later, the animals were examined microscopically and histologically; the flies obtained from 6 of the 19 houses caused the disease in test animals; in three of these houses TB patients resided; in two others, phthisics only slightly attacked were living; and in one, there had been no TB case for many years. Considering the circumstances, this is a rather high proportion of positive flies.

Lamborn (1938) studied tubercle bacilli in *Musca sorbens* in Nyasaland. Flies fed fresh tubercular sputum passed the bacilli in the feces up to 15 days, in numbers greater than in the original medium.

Alexsandrov (1938) collected fly specks as smears from electric light bulbs in the kitchen, toilet, bath, and serious-case rooms of the tuberculosis section of a hospital and cultured them for TB bacilli. The ratio of bacilli to fly specks ranged from 1:13 to 1:1.23; the highest contamination was 23 bacilli in two fly specks from a toilet lamp. Examination of sputum flora in fly specks showed that the flies light on sputum that misses spittoons and carry away microscopic organisms. Morellini (1952) extended the previous positive findings of his predecessor, Omodei-Zorini (1944). He collected 1781 adult house flies, 888 from a tuberculosis sanitarium in Rome and the rest from different parts of the city. The homogenates of 705 flies gave a positive result for *Mycobacterium* without an evident percentage difference in groups of different origin. In tests on guinea pigs carried out for 664 isolates of which 250 had been obtained from flies caught inside the sanitarium and 406 from flies caught outdoors, 25 cases of tuberculosis occurred: 21 infections from flies inside (2.36%) and 4 from flies outside (0.44%).

Morellini and Saccà (1953) and Morellini (1956) have summarized a more moderate view of the involvement of flies in tuberculosis. They object to three recurring statements in the literature on the house fly as a vector: 1) Flies are particularly attracted to sputum, especially tubercular sputum; 2) flies that feed on TB sputum suffer a form of diarrhea; and 3) fly vomit is more important than diarrhea as a spreading factor. The experimental studies of these authors showed that flies are not particularly attracted to human TB expectoration. It was found that the amount of excreta varies with different diets, but it is not influenced by the presence of TB bacilli. It was also concluded that the fly cannot spread pathogenic microorganisms through its vomit or through regurgitation from the gut, since this regurgitation was never observed to be deposited upon the surfaces where the insect was setting.

We agree with their view on the second of the above statements. The output of excreta, measured by fecals spots deposited in a given time, certainly is highly variable. It depends on diet, ambient temperature, and possibly other physical factors. We have some data that suggest it may even vary with age. As to the degree of avidity with which flies feed on infected sputum, this seems academic considering the frequency of isolation of tubercle bacilli from wild-caught flies, which numerous authors, including Morellini, have reported. The flies certainly are picking up the organism, and sputum seems to be the likeliest source. As regards deposition of vomit by flies, we have seen it often enough in caged populations (see Fig. 23), and it has been recorded as a normal occurrence in photographs by Hewitt (1914, figure 35) and Graham-Smith (1914, plate 18, figure 1).

James and Harwood (1969) state that, though flies have the mechanism and habits for the transmission of the tubercle bacillus, no conclusive work has established the relationship of flies to such transmission. Several workers have also reported studies indicating that the bacillus passes through cockroach intestines unharmed. The role of the house fly in TB is probably that of a standby vector called to active duty when people are careless or sloppy. With improving nutrition, sanitation, and chemotherapy in tuberculosis management, it is increasingly unlikely that the relationship of flies to TB will be conclusively established.

RECURRENT FEVER

The organism *Borellia recurrentis* declined in numbers during 26 hours in the tract of the house fly and was no longer demonstrable after 28 hours by microscopic examination or by mouse inoculation (Nattan-Larrier, 1911). *S. calcitrans* failed to transmit relapsing fever by biting a sick, then a well woman, although the guts of 2 flies contained the organisms after biting the donor (Frankel, 1912).

PINTA (El Carate, Mal del Pinto)

Pinta is considered a contagious, inoculable skin disease caused by *Treponema carateum*, a morphologically indistinguishable but immunologically distinct relative of the agent of syphilis, *T. pallidum*. The disease usually runs a prolonged course characterized by the appearance of dry, scaly lesions of the skin, which becomes depigmented or variously tinted brown, pink, or blue, hence the name "pinta." There may also be cardiac and central nervous system involvement. The disease attacks various age groups and was originally thought to be restricted almost entirely to the warm, moist areas of Mexico, Central America, Cuba, and parts of South America. It is now known to occur in India, Africa, the Philippines and the Pacific islands.

Pinta's resemblance to yaws, which is caused by the related *T. pertenue*, has naturally inclined investigators to assume similar means of dissemination. Since the organism is present in primary and later lesions, most transmissions are probably accomplished by direct contact. The involvement of biting and wound-frequenting arthropods has been seriously considered, however. As early as 1860, León related the transmission of the disease to a biting dipteran, "jejen," tentatively assigned by some to the genus *Empis*. This hypothesis received moral support from Gomez (1879) and Ruiz Sandoval (1881). The possibility that pinta may be mechanically transmitted by such hematophagous types as *Cimex*, *Ornithodoros*, and simuliids was suggested by Peña Chavarria and Shipley (1925) but was not confirmed experimentally by León y Blanco (1940) in *Cimex lectularius*, *Ornithodoros talaje*, or *Triatoma palidipennis*, nor by Gonzalez Herrejón and Ortiz Lombardini (1938) in *Simulium haematopotum*.

Hippelates flies continued under suspicion. An experiment with a human volunteer was performed in Iguala, Mexico, by León y Blanco and Soberón y Parra in 1941. A pinta lesion, which was suspected of containing a large number of treponemes, was excoriated until the serum flowed. The subject was taken outdoors and immediately *Hippelates* flies landed on the wound and began to feed. When their abdomens were distended, 10 or 12 flies were aspirated into a tube that was then inverted on the excoriated skin of a volunteer's forearm. The flies fed on the wound during the 15 to 30 minutes that they were confined. Two series of six experiments each were performed on the same volunteer. The first series was negative, but, in the second series, pinta developed in one of the scored sites. In a follow-up study (Soberón y Parra and León y Blanco, 1944), *M. domestica* was also found capable of transmitting the disease. In the gut of *Hippelates* flies, the organisms begin to lose their motility after 70 to 80 minutes and are completely immobilized at the end of two hours. Dark-field observations of the vomit drops of recently fed flies revealed treponemes in 3 out of 11 tubes of flies. It is indicative of the delicate nature of the organism that it could not be found in the gut of 31 *Hippelates* flies captured while feeding on a pinta lesion. One may tentatively conclude from these results that only the vomit of the fly is potentially dangerous.

FRAMBOESIA TROPICA (Yaws, Pian, Boubas)

This widely and unevenly distributed disease predominantly occurs in children and sometimes in adults and is associated with granulomatous or ulcerative lesions of the skin. Two to 8 weeks after exposure, a primary lesion appears at the site of inoculation as an ulcer. In several weeks to months, mild constitutional symptoms appear, with generalized skin eruptions often in successive crops lasting from a few months to several years. Yaws is primarily a disease of rural peoples of the tropics and subtropics; the lowest social and economic groups have the highest incidences. It is particularly common in equatorial Africa, the Caribbean area, parts of India, the Philippines, and countries in between, and throughout the South Pacific Islands; there are endemic foci in parts of several countries in Central and South America; sporadic cases occur in North America and Europe from infection contracted elsewhere.

The etiologic agent is *Treponema pertenue*. The source and reservoir of infection is an infected person, particularly the surface and exudates of his early skin lesion. The period of communicability is variable and may extend intermittently over several years while relapsing moist lesions are present; treponemata are not usually found in late ulcerative lesions. The mode of transmission is thought to be principally by direct household or other contact with infectious lesions and perhaps also by fomites. A considerable body of evidence indicates vector transmission by *Hippelates* and possibly other flies.

Some authorities consider that, in some instances, what is described as leprosy in the Old Testament (Leviticus, 14) is probably yaws. The first unequivocal reference to yaws and its association with flies is that of de Sousa (1587), who wrote of his travels in Brazil:

"We must now refer to those mosquitoes which are called nhitinga, and which are very small and of the form of flies; these do not bite, but are very troublesome, because they settle on the eyes and in the nostrils, and will not let one sleep by day in the open unless there be a wind. They are very fond of sores, and suck the poison which is in them; and if they then go and settle upon any abrasion on a healthy person, they leave the poison in it, and then many people are seen covered with boubas."

About ten years before, Mercurialis had described fly transmission of plague in somewhat similar terms. After de Sousa, we have no record of yaws and flies for more than a century and a half until 1768, when Sauvages published the following account sent to him by Virgile, a surgeon who practiced in St. Domingo: "Another method of contagion, and this the more frequent one, is the mediation of flies; for if a fly that has settled on a yawy ulcer passes on to a healthy person and by chance alights on some slight wound of his, this wound, previously simple and clean, is changed into a mother of yaws; thereafter growths sprout up not in the ulcer but on the face, arms, trunk or elsewhere."

In the following year, Edward Bancroft published his "Essay on the Natural History of Guiana in South America," which contains the following epidemiologic observations: ". . . a small quantity of yellowish pus is usually seen adhering to their surface, which is commonly covered with flies, through the indolence of the Negroes. Almost all the Negroes, once only in their lives, are infected with it, and sometimes the Whites also, on whom its effects are much more violent. It is usually believed that this disorder is communicated by the flies, who have been feasting on a diseased object, to those persons who have sores, or scratches, which are uncovered; and from many observations, I think this is not improbable, as none ever receive this disorder whose skins are whole; for which reason the whites are rarely infected; but the backs of the Negroes being often raw by whipping, and suffered to remain naked, they scarce ever escape it."

The next account, by Schilling (1770), refers to a neighboring locale in South America. The excerpt appears in an excellent history of yaws and flies by Barnard (1952), who renders the original Latin text as follows: "Now you will marvel at the method by which the disease is propagated. The region of Surinam produces an abundance of small flies, to which the name of Yaws Flies is given (because they propagate the cause of the disease). Natural instinct impels these flies to suck open places of the animal body, such as ulcers or wounds. Hence they invade the naked bodies of negroes which afford them nourishment, and, having drawn the poison from the pustules known as Mother Yaws, they transfer it into the ulcers and wounds of healthy bodies."

Bertrand Bajon (1777) made the same observation following his travels in Cayenne and French Guiana: ". . . the matter that is continually running from the yawy pustules attracts the flies, whose feet become clogged with particles of virus, which they then go and deposit on some part of healthy negroes; in that case the disease develops in them fairly quickly, especially if the latter have some ulcer or even a slight scratch; for the flies settle for preference on effected parts, which are those where the virus acts quickly and where it begins to work its havoc."

Variations on this theme can be found subsequently in the works of Nielen (1780), Dazille (1792), and Stedman (1796), all on the Guianas and French Antilles. Possibly the first description of the epidemiology of yaws in Africa was that of Winterbottom (1803), who wrote of Sierra Leone: "The complaint is sometimes inoculated by means of a large [?] fly, called in the West Indies the yaw fly. When this insect alights upon a running yaw, which the Africans never keep covered, and afterwards settles upon the body of an uninfected person, it introduces the poison, if there happens to be a wound or scratch there, as effectually as the most dexterous surgeon."

In 1806, Aibert wrote: "The contagion of yaws, it is claimed, is significantly facilitated by a kind of fly called 'Framboesia Flies,' and which

238   FLIES AND HUMAN DISEASES

are very abundant in the tropics. These flies constantly alight on the horrible pustules that arise during the disease, and they inoculate the virus into healthy persons, whom they bite until they draw blood. Is it also by this means, that it has been transmitted to domestic animals, as some claim to have observed?" He mistakenly implicates biting flies and adds nothing new except the following interesting reference: "Loeffler states that there are certain places in America, where the law forbids individuals with yaws from going out, or even having access to hospitals. One finds, in effect, that this precaution has considerably diminished the spread of the disease."

Following this, we have the direct observations of Koster (1817), Walsh (1830), and Sigaud (1844) in Brazil, Williamson (1817) in the West Indies, and Maxwell (1839) in Jamaica, which support, always without experiment, a popular belief in the importance of flies in the transmission of yaws. As late as 1873, this uncritical acceptance was manifest in a "Report on Leprosy and Yaws in the West Indies," which contained papers presented at a scientific meeting (Milroy).

"That yaws cannot be disseminated through the medium of the atmosphere is the firm belief of the people generally. . . . They also believe that, if flies from the body of an infected person alight on any part of the naked skin where there is the smallest abrasion, yaws may be then readily communicated. The sick-nurse, who has charge in that capacity of the Yaws Hospital near Roseau, an intelligent middle-aged black woman, positively affirms that she contracted the disease in this way."

Von dem Borne (1906), was apparently the first to search for the agent of yaws in an insect. Unfortunately, he chose mosquitoes, an unlikely vehicle for yaws. He writes: "In the gastric juice of a couple of mosquitoes of the species occurring here (Dutch East Indies), the *Stegomyia fasciata*, which I examined, I was able to observe living spirochaetes of the *pallida* type. Both mosquitoes were caught in the immediate neighborhood of framboesia patients." Though rather little and late, this may be considered the turning point in the entomology of yaws.

The following year Castellani published his landmark study, and a number of scientific investigations followed. Working in Ceylon, he took scrapings from a patient's yaws lesions, which were observed to contain the spirochaetes, placed them with ten *Musca domestica* into a sterile petri dish for one-half hour; the flies fed greedily, and an examination of their legs and mouth parts showed 9 flies with other spirochaetes and two also with the specific pathogen; examinations of control flies were negative. A similar experiment demonstrated 14 of 15 house flies contaminated with spirochaetes. He also fed 30 flies on scrapings from non-ulcerated papules containing only *T. pertenue*, and applied five de-winged flies to numerous scarifications made over the left eyebrow of each of five monkeys, only one of which showed a lesion containing the pathogen after 45 days. Also, 23 flies were deprived of their wings and set to feed on two small lesions of a patient; they readily sucked the fluids and one hour later were removed; similarly, several flies were placed on scarifications of seven monkeys for two hours. Two of the monkeys gave positive results, thus demonstrating the possibility of the house fly acting as a vector; but Castellani made no attempt to reproduce natural conditions. Robertson (1908), in the Gilbert and Ellice Islands, also demonstrated that house flies could become contaminated by feeding on patients' lesions. Yasuyama (1928) observed the treponeme in the house fly gut 8 hours after a meal, but he felt the extreme sensitivity of the pathogen possibly limited its transmission by flies to a shorter period. Oho (1921) concluded from his studies in Formosa that *M. domestica* and *M. xanthomera* (?) were involved in transmission.

Hunt and Johnson (1923) state that the principal transmitter of yaws in Samoa is the fly, but do not identify it. They found that the flies mostly breed in bread fruit on the ground during November or December, April or August. Swarms of flies appear during bread fruit seasons and yaws becomes epidemic during or shortly after these seasons, though it is endemic at all times. Sores on native children were always covered with flies, and white children, who were close to but did not come into contact with natives, had the disease.

In St. Lucia, Nicholls (1911) made a detailed study of another possible vector of yaws, *Hippelates flavipes*, which, no doubt, is the "small fly," "yaws fly," and "fly" of previous observers in the New World. He distinguishes two ways it can convey germs, via the gut and by external contamination, and he conducted experiments to test the first method of transmission. He removed scabs full of spirochaetes from a patient's lesion and allowed the flies to feed on them. In only two of numerous fly gut preparations could the pathogen be demonstrated and then from the upper tract within 15 minutes after the flies fed. He also gave a full description of the habits of this fly and epidemiological data on the greater incidence of the disease in wet weather and in unsanitary situations, the same conditions under which this fly is prevalent, and concludes that *H. flavipes* may play the principal role in the dissemination of the disease but that the infection may be conveyed by other mechanical means.

Wilson and Mathis (1930) confirmed this opinion with observations in Haiti of swarms of *H. flavipes* around lesions; they noted also that yaws became less prevalent at altitudes about 2,500 feet, which could be due to the greater numbers of fly carriers at the lower level. Araujo

(1934) in Brazil found numerous *H. brasiliensis* and *H. currani* feeding on a large number of yaws lesions.

The careful and extensive observations and experiments by Kumm and his associates on the Jamaica Yaws Commission have established that, at least in this country, yaws is carried by *H. flavipes* (corrected from *pallipes* by Sabrosky, 1941). Kumm (1935) reported that this was the most common fly found feeding on yaws lesions and ulcers in Jamaica; it was suspected as a vector because of its feeding habits and particularly because of the mechanism of regurgitation of a vomit drop. Kumm et al. (1935) found that the majority of the *T. pertenue* ingested by this fly after an infective feeding remained motile for about 7 hours in the esophageal diverticula, but those that entered the mid- or hindgut lost their motility very quickly; no motile treponemes could be demonstrated at intervals of 18 or 24 hours after ingestion. Kumm's (1935) further experiments did not demonstrate the pathogen in the fly later than 48 hours after the infective meal, nor was there any evidence of the invasion of the salivary glands or proboscis, nor of cyclical development of the spirochaetes in the fly, up to 28 days after the initial meal on a yaws lesion. Under natural conditions *H. flavipes* becomes infected with *T. pertenue* from human lesions on any part of the body, though the fly feeds by preference on the perineum or on the lower extremities. Of 500 flies dissected, 71% were found naturally infected. Kumm and Turner (1936) found that rabbits were relatively insusceptible to infection when inoculated directly with material from granulating wounds of patients; of 63 animals inoculated, only three developed yaws, but 7 of 68 animals fed on by infected *Hippelates* flies developed positive lesions. At times the infected esophageal diverticulum of a single fly was sufficient to infect a rabbit with yaws. The investigators concluded that man to man transmission of yaws through the mediation of flies is a definite possibility.

Saunders et al. (1936) determined that yaws has a widespread, patchy distribution in rural Jamaica, which is correlated with heavy rainfall, a fertile, moisture-holding soil supporting abundant vegetation, a peasant population living under unsanitary conditions, and many *H. flavipes* flies. But they could not decide whether the distribution of this fly determined the distribution of yaws, or whether both were affected by the same environmental factors, causing their distribution to coincide.

Chambers (1938a), from his experience in Central America, Cuba, and Jamaica, thought that a warm, humid climate was the main factor necessary for the dissemination of yaws and that the increased number of *Hippelates* flies in these areas was merely incidental. However, in a second report (1938b), he states that, out of 62 new cases, he found only four in which no direct contact with diseased persons could be traced, and suggested that this may represent the proportion of cases in which the disease is spread through intermediaries such as flies rather than direct contact.

Syddiq (1938) states that he has seen the chloropid, *Siphunculina funicola* (eye fly) sucking serum from the wounds and ulcers of persons in Hyderabad (Deccan), where the sylvan population suffers much from yaws and an abundance of these flies, and he thought that they may play some part in dissemination. Fox (1920-21) mentions a Dr. Hall Wright, who actually infected himself through the agency of these flies. Hamilton (1939) investigated the breeding places of this fly in Assam because it was considered of possible importance in the spread of yaws. Field (1951) in Malay also reported that *Siphunculina* was suspected in the spread of yaws, settling on lesions and carrying the pathogen mechanically to any moist break in the skin of uninfected persons.

Vargas Cuella (1941) stated that *Ornithodoros*, *Aedes aegypti*, the house fly, and *Hippelates* have been incriminated as carriers of yaws in Colombia, and added that *H. flavipes* is very abundant in the Pacific coast and abounds on yaws ulcers to a great extent; it was his opinion that it was almost the only transmitter on the coast.

Lamborn (1936a, b; 1937) described experiments on yaws transmission by *Musca sorbens* in Nyasaland. He fed these flies on exudate from a yaws lesion and later allowed them to feed on scabies lesions and scratches made on a human volunteer. On the 30th day, a primary lesion appeared, and spirochaetes were found in a secondary lesion on the 76th day. Examinations of fly guts showed that cyclic development in the fly does not take place. The observed feeding habits of this fly led Lamborn to conclude that it must play a very definite part in the transmission of yaws; after an infecting meal the fly regurgitates a vomit drop that it then draws up again thus contaminating the tip of its proboscis.

Barnard (1952a) has made an extensive literature survey on yaws and flies. He tabulated the content of 86 of what he calls primary statements, including those based on observation and experiment, those that record a popular belief held by local inhabitants, and even personal opinions unbacked by any experience if they appear to be based on a consideration of extant evidence; 31 come from the eastern hemisphere, 46 from the western hemisphere, and 9 make no mention of location. Five statements from the western hemisphere refer to small flies; 26 from this same area specifically mention *Hippelates* or "*Chloropidas*," or give some name that can be identified as such. Fourteen (with one exception in the eastern hemisphere) mention house fly or another species of muscid. There are 3 mentions of *S. funicola*, all in India, 2 of *A. aegypti*

(Dutch East Indies and Colombia), 3 of *Culicoides* (Panama, Cuba and Brazil), and one each, all in the western hemisphere, of *Sarcophaga*, *Simulium*, and *Phlebotomus*. If small flies can be included with *Hippelates*, the total score is 31 for these and 18 for muscid flies. In a second paper (1952b) he reports the results of questionnaires sent to medical officers in all countries where yaws is prevalent, asking what were the most important factors in yaws transmission and how important are flies in such transmission in their experience. Of 37 answers to question two, transmission by flies is regarded as important by 10. His general conclusion based on the literature survey and the questionnaire was that various species of flies can and do act as mechanical transmitters of yaws in most parts of the tropics, but the extent to which this is an important or even common mode of dissemination is uncertain and probably varies in different localities; the principal vector in the western hemisphere is *H. flavipes*, that in India and Malaya, *S. funicola*, and that in Africa, Asia, and Australia generally the house fly or other Muscidae; other species, such as *Sarcophaga*, *Simulium*, *Culicoides*, and even *A. aegypti* may be concerned but apparently less commonly.

Later, Satchell and Harrison (1953) experimented with yaws transmission by flies in western Samoa. They found that *M. sorbens*, *M. domestica*, and *Atherigma excisa* may pick up live spirochaetes on the proboscides while feeding on a yaws lesion and retain motile spirochaetes in their crops for up to two hours. *M. domestica* and *M. sorbens*, if interrupted during a feed on a yaws lesion, may fly off and subsequently feed on some other type of lesion. They thus showed that the prerequisites for yaws transmission in this area are present and suggested that spraying selected villages with a residual insecticide and comparing the incidence of yaws there with that of villages in which the domestic muscids were left to flourish would be a promising experimental approach to answering the perennial question of the insect transmission of yaws.

Very recently, Gourlay and Marsh (1965) reported on an outbreak of yaws in a suburban community in Jamaica in 1964, where the disease does not commonly occur. Circumstantial evidence indicated that yaws was imported to this community by infected patients from other areas, and the authors found it interesting to note, in light of the work of Kumm et al. (1935), that there was a considerable increase of *H. flavipes* toward the end of 1963, due to favorable breeding situations (moist grass roots) resulting from an exceptionally heavy rainfall in October; they thought this may have been the reason for the delayed spread of the disease from the original case.

The evidence presented here leaves little doubt that *Hippelates* and other flies have the appropriate habits to disseminate yaws. Now needed are definitive community-wide studies such as proposed by Satchell and Harrison to obtain a possible measure of the degree of fly participation. The unresolved question of fly involvement is an old one, almost 400 years old in fact.

AMEBIC DYSENTERY AND RELATED DISEASES

Most of the Protozoa discussed here, with the exception of *Trichomonas foetus*, have an intestinal residence and a fecal dissemination. The trypanosomes are handled separately since their life style is quite different. Although diarrheas and gastroenteric diseases have various etiologies, the avenues of dissemination are similar, and there is an advantage to a unified treatment of their epidemiology. The reader will therefore find references to the prevalence of amebic dysentery in relation to seasonality, effects of fly suppression and sanitation, etc., in the section on Enteric Infections. The remainder of the material is presented here because the Protozoa, like other denizens of the gut—enteroviruses, bacteria, and helminths—have natural histories, adaptations, and responses in flies that are peculiar to themselves.

Four genera of Protozoa—*Entamoeba, Giardia, Trichomonas*, and *Chilomastix*—have been frequently isolated from flies caught around food destined for human consumption, in markets, kitchens, etc. Species within a genus range from innocuous to pathogenic; some form cysts, others do not. It is therefore desirable to highlight the characteristics of the pathogens before we discuss their fly associations.

Amebic dysentery is almost entirely referable to *Entamoeba histolytica* except for infrequent cases caused by *Dientamoeba fragilis*. The disease has a worldwide distribution. It is more common in the tropics, but this is probably influenced less by climate than by lower standards of sanitation. *E. histolytica* is primarily a human parasite, with carriers also serving as reservoirs, although it turns up in a surprising range of animals including dogs, rats, monkeys, and chimpanzees. A prevalence of 8.4% in dogs in a Tennessee survey suggests that canine infection, which is usually symptomless, is of potential public health importance.

Infants are rarely infected, and the incidence of amebic dysentery increases throughout childhood to a maximum in young adults. However, reliable data on incidence are difficult to obtain for several reasons. Cases may go unreported since carriers are so common; but even when feces are examined, variations in laboratory skills and diagnostic procedures frequently result in the amoebae being missed, mistaken for benign species, or vice versa. An intensive study of amebiasis in a rural community in Tennessee comprising 374 persons and 75 families revealed that 38% of the inhabitants harbored both large and small cysts of *E. histolytica*. There was no correlation, however, between the occurrence of dysenteric symptoms and the harboring of the pathogen. In fact, many cases resembled bacillary rather than amebic dysentery (Milam and Meleney, 1931). A statewide survey, the following year, revealed that more than 11% of the rural population was infected (Meleney et al., 1932). An incidence approaching 100% of the population over one year of age has been found among the inhabitants of Egyptian villages (Chandler and Read, 1961).

It is now accepted that the large-cyst race (mean size 12 μ) is generally pathogenic and the small-cyst race (7-8 μ) is generally benign. Infection is acquired by ingestion of the cysts in contaminated food or water, and excystment takes place in the small intestine probably as a result of contact with the alkaline digestive juices. The trophozoites penetrate the epithelium of the large intestine, principally of the caecum and ascending colon, and reach the submucosa, where they produce lesions. There may be even deeper penetration to the gut wall, followed by bacterial invasion and inflammation, and if the gut wall is breached, metastatic infection of the liver and lungs usually follows. Cysts are produced in the lumen of the bowel, and not in the tissues. Epidemiologically, it is noteworthy that carriers rather than cases usually produce cysts.

Giardia lamblia is considered the most common flagellate of man. It occurs more frequently in children, and exposure to it seems to build resistance. Chandler (Chandler and Read, 1961) found the parasite in 16% of Egyptian villagers below the age of puberty and in only 3% of those above. Soulsby (1968) speaks of a general prevalence of 2 to 60% or more. Here also, it is the cyst that passes between hosts. The trophozoites settle on the epithelium of the small intestine, particularly the duodenum. They do not lyse the cells but mechanically interfere with absorption of fats, which may lead to fat-soluble avitaminoses. Other species of *Giardia* are found in dogs, cats, goats, rabbits, mice, rats, and the ox. Evidence of pathogenicity in these hosts is doubtful or inconclusive, except *G. chinchillae* has been reported to cause a severe blood diarrhea with a 38% mortality in chinchilla (Shelton, 1954). It is interesting that nematodes sometimes become infected with *Giardia* when both parasites coexist in the intestine.

Among the other Protozoa frequently carried by flies, two flagellates deserve brief mention. *Chilomastix mesnili* inhabits the large intestine of man, monkeys, and pigs and is transmitted as a cyst. Globally, human infection rates vary from 2 to 25%. Though it is considered a benign commensal, there is evidence that it may be associated with occasional diarrheas.

Trichomonas hominis makes its home in the lower bowel of man but can be established in monkeys, young cats, and rats. Pediatric infection rates of 10% or more are reported for the tropics, less in temperate zones, and the question of the organism's pathogenicity is unresolved. There is no cyst and dissemination is by means of a relatively hardy trophozoite.

Survival of Protozoa in flies—Entamoeba histolytica

Werner (1909) discounted any role for flies in the spread of amebic dysentery because of the poor survival of trophozoites in the fly's gut. Subsequent workers have confirmed the delicate nature of the active ameba. Roubaud (1918) found that trophozoites become inert an hour after they are ingested by adult house flies, and die within a few hours, never encysting. Root (1921) obtained similar results with the house fly and other synanthropic flies. When trophozoites in mucous material taken from a patient were fed to flies, the amebae died in the fly's gut in less than 30 minutes. Pipkin (1949) finds that survival time in the crop ranges from 15 minutes in the house fly to 40 minutes in *Sarcophaga misera*, and in the midgut from 5 minutes to 30 minutes. Larger flies have a longer carrying time presumably because their initial intake is greater. Only a few motile but no cultivable trophozoites were found in the rectum of any fly (see also Jausion and Dekester, 1923).

Though Werner's conclusion concerning *histolytica* trophozoites in flies has been amply confirmed, he failed to reckon with the more durable cyst. It is interesting that the first study of cysts in flies produced only negative results. Kuenen and Swellengrebel (1913) found that none of the *histolytica* cysts in the house fly's gut was viable and those on the outside died from desiccation in less than an hour. They introduced the eosin test as a means for determining the viability of cysts, the rationale being that only dead cysts take the stain. The eosin test has been widely used, but Sieyro (1942) considers that it is not clear-cut, and Root (1921) found neutral red superior.

Thomson and Thomson (1916) and all subsequent investigators have found that viable *histolytica* cysts are passed in the feces of flies. Working with *Musca, Fannia, Calliphora,* and *Lucilia,* Wenyon and O'Connor (1917a) discovered that cysts appear in the feces 20 to 30 minutes after feeding and continue to be voided for at least 16 hours. Roubaud (1918), Aleksander and Dansker (1935), and Sieyro (1942) extended the infective period to 24 hours, and Roberts (1947) to 31 hours in the house fly. Root's (1921) careful study was made by first determining the proportion of viable cysts in the material fed to the flies. He found that in the guts of 29 flies, no deaths occurred during the first two hours; half the cysts were dead after 15 hours; and the last living cysts were observed in two flies after 49 hours. He obtained similar results when he used the commensals, *Entamoeba coli* and *Endolimax nana,* except that the latter species was totally killed in 39 hours.

Viable cysts may turn up in the vomit for about nine hours. There is evidence, however, that the majority of cysts that enter the crop are held back by the pseudotracheae when the fly vomits. Thus, of 312 vomit spots deposited by infected flies, 52% were negative and the rest contained 1 to 8 cysts. Larger cysts, such as those of *E. coli,* are even more effectively excluded from the vomit. The same flies averaged 4 cysts per fecal spot (Roberts, 1947).

The duration of the fly's carrier state can be increased by concentrating the cyst material in the infective meal and lowering the ambient temperature. Perhaps, anything that affects gut motility—the fly's state of satiety, the diet, and frequency of feeding—will influence its retention of cysts. Sieyro claims to have found cysts in the feces of normally fed house flies one minute after the cysts were swallowed. Yet Roberts (1947) found cysts, five minutes later, in the feces of house flies that had been starved for 24 hours. The influence of weather (particularly temperature and relative humidity) and availability of food on the persistence of the carrier state in flies are worth studying.

Moisture also affects the survival of cysts. They are quickly destroyed in vomit and fecal spots deposited in dry places. Under these circumstances, flies perform the novel function of purifying agents, as Roubaud pointed out. Successful transmission requires that cysts be deposited on moist or liquid foods. Cysts survive in drowned flies for at least a week (Roubaud, 1918; Root, 1921), but are no longer viable when the flies begin to disintegrate about a month later. In relatively unpolluted water, cysts may survive for about 30 days (Wenyon and O'Connor, 1917b; Kuenen and Swellengrebel, 1913).

OTHER PROTOZOA

Stiles and Keister (1913) first proved that flies can carry viable intestinal Protozoa, using cysts of *Giardia lamblia.* They emphasized the movement of flies from privies to houses as an important avenue of infection. Root (1921) found that *G. lamblia* succumbs twice as rapidly as *E. histolytica* in the gut of various flies—half the cysts were dead after 8 hours and the last living cyst was found at 16 hours. Sieyro (1942), however, recovered viable cysts of *G. lamblia* in house fly feces 30 hours later. In drowned flies, the cysts remained viable for at least 4 days, about half the survival time of *E. histolytica*.

The trophozoites of *Chilomastix mesnili* are also extremely delicate. Motile forms may be excreted less than seven minutes after ingestion, but those which remain in the tract are killed within an hour (Root, 1921). The cysts, however, are tougher than those of *E. histolytica* and *G. lamblia*. Half the cysts survived 36 hours in the fly's gut, and some were still alive after 80 hours. Resistance seems to be correlated with thickness of the cyst wall in these Protozoa.

Trichomonads do not form cysts, and face the risk of transferring from one host to another as trophozoites. They have met this challenge to a limited extent, for although they are by no means as tough as cysts they are tougher than other trophozoites. Hegner (1928) fed various flies on feces heavily contaminated with *T. hominis*. After the meal he placed each fly in a separate test tube until it vomited or defecated, when it was again transferred and the spots were immediately examined. He found that undulating trichomonads could be passed for 5.5 hours by *Cynomyopsis cadaverina*, for 2.75 hours by *P. sericata*, and for 2 hours by the house fly. Four hours was the maximum time that the organism could remain in the gut of any fly and still be infective. According to Simitch and Kostich (1937), *T. hominis* survives 8 hours in the house fly gut, but death occurs in ten minutes on the legs and proboscis.

In his treatise on the Bobwhite quail, Stoddard (1931) mentions a severe epizootic of trichomoniasis among artificially reared quail and considers the possibility that flies may have been involved. Flies caught near the brooders contained numerous *Trichomonas* in their gut, and birds may have become infected by eating such flies.

Trichomonas foetus is normally transferred from cow to bull and vice versa during coitus. Nonvenereal transmission also occurs, but the means by which this is accomplished are not known. Morgan (1942) has looked into the possibility that flies are involved. He recovered infective organisms in the gut for 17 hours and in the feces for 8 hours after house flies fed upon a culture with a titer of about 2×10^6 organisms per ml (see also Akatov, 1955). The flies began vomiting live trichomonads almost at once and may continue to do so for five hours (Holz, 1953). It would seem, therefore, that both the vomit and excreta of the fly can convey infection if deposited on the moist vulva of a cow in heat or on the sheath of a bull, but the possibility of such transmission has not been tested. Bartlett (personal communication) considers that nonvenereal transmission by flies is rare if it ever occurs under the conditions of barns and corrals. He points out that it is not considered essential to isolate infected from noninfected cows during the course of a "clean-up" program for bovine venereal trichomoniasis in either dairy or beef herds.

OCCURRENCE OF PROTOZOA IN WILD FLIES

Isolations of Protozoa and helminth eggs from the same flies are not infrequent. Therefore, much of the work discussed here has its counterpart in the section on Helminth Diseases, presented elsewhere in this chapter. The first recorded isolations of Protozoa are attributed to Wenyon and O'Connor (1917a), who captured house flies near a hospital kitchen adjacent to an Egyptian village. No details are given except that cysts of *E. histolytica*, *G. lamblia*, and *E. coli* were found in the feces of these flies.

In Amara, Iraq, Buxton (1920) dissected no less than 1027 house flies and found that only three contained cysts of *E. histolytica*. The figure is low, but he argues that one or two cysts in a fly's gut could easily escape detection and that the actual frequency is probably higher. However, one could argue that a few cysts are of little epidemiologic importance since they would rarely initiate a human infection because of the intervening hurdles of the external environment and the acid stomach. In the same collection, 4.09% of the flies harbored helminth eggs; the figure is higher partly because eggs are easier to detect and partly because they are more hardy than cysts. Despite the low infection rates in flies, several circumstances favored Buxton's view that the fly was a major carrier of enteric disorders in Amara: fly density was high; 63% of the flies had recently visited human excreta, as shown by examination of the material in their tracts; and the flies were caught near kitchens and mess halls. In a smaller collection of 198 flies captured in kitchens, five flies were contaminated with human intestinal Protozoa and one carried *E. histolytica*.

Metelkin (1935) mentions an investigation of the gut contents of 339 flies by Burova and Kassirskiĭ in 1931 in Tashkent. The flies (probably *M. domestica* for the most part) were caught in dumps, latrines, and living quarters and were found to be carrying, in addition to helminth eggs, cysts of *E. histolytica* (2.3%), *E. coli* (5.8%), and *Giardia* (3.2%). Another study of 157 specimens of *M. domestica vicina* in Ashkhabad also showed considerable previous contact of the flies with human feces and an infection rate of 5%. The species of Protozoa and their frequency were: *E. histolytica*, 4; *E. coli*, 1; and *G. lamblia*, 3 (Pletneva, 1937). Sukhova (1951) mentions but does not cite Korean authors who demonstrated the carriage of viable cysts of *E. histolytica* and helminth eggs by *Chrysomya*, *Sarcophaga*, and *Lucilia*. In Brazil, Coutinho et al. (1957) found only a few cysts of nonpathogenic amebae in over 1200 flies caught in a public market. Working in Omsk, Fedorov (1962) called attention to the possible danger of transmission by flies that are driven indoors by cold weather. He recovered 32 protozoan cysts from 347 house flies taken during November and December. Twenty-three cysts belonged to *G. lamblia*, and the rest were those of nonpathogenic amebae. House flies collected in slum areas in Peiping were examined in batches of 100, and, in 12 out of 70 batches, intestinal Protozoa including some *E. histolytica* were recovered from the surface and the gut of the flies (Yao et al., 1929).

The most thorough field studies on the role of flies as vectors of entero-pathogenic Protozoa are those of Frye and Meleney (1932), Chang (1940), and Harris and Down (1946).

Frye and Meleney investigated house flies and other animals as possible reservoirs and transmitters of *E. histolytica* in a rural community in Tennessee where 38% of the inhabitants were known to be carriers. Water was eliminated as a probable mode of transmission because most of the families had their own wells. Flies were numerous, screens were scarce, indoor toilets nonexistent, and privies were few, so that conditions were ideal for fly transmission. Fly trapping was concentrated in the kitchens and dining rooms of those homes where persons were known to be positive for *E. histolytica*. Three methods were used for examining flies: 1. They were confined to bottles for 24 hours and then homogenized and filtered, the filtrate being examined for cysts in iodine-eosin solution; the fly spots in the bottles were also studied. 2. Groups of fly intestines were homogenized and treated as above. 3. The digestive tracts of individual flies were so treated. In the first method, a total of 7420 flies were collected in 36 bottles, and 15 of these bottles contained fly specks with cysts of one or more intestinal Protozoa, including *E. histolytica* and *Giardia*. Use of the fly homogenate was not successful. With the second method, cysts of *E. histolytica* were found in one group of 425 flies. Individual dissections of 103 flies gave negative results. In exploring the possibility that other animals serve as carriers, the authors discovered a very low frequency of cysts of *E. histolytica* and *Giardia* in pigs, but rats, mice, chickens, and dogs were negative.

During the course of two summers, Chang assessed the importance of synanthropic flies as carriers of human intestinal parasites in Chengtu, China. He collected 423 flies, and in the digestive tracts of 6 he found cysts of *E. histolytica* and in another 10, *G. lamblia* was present. The frequency of helminth eggs was considerably greater, about 50 flies carried hookworm, the whipworm, *Trichuris trichiura*, and the human round worm, *Ascaris lumbricoides*. In Chang's experience, examining flies by first homogenizing them was less effective than dissecting individual flies.

The majority of flies harbored one or two species, but some flies carried as many as six different parasites! The most important fly in his study was the latrine fly, *Chrysomya megacephala*, with a few positives turning up among *P. sericata* and *Sarcophaga*. Not a single one of 153 *M. domestica vicina* was positive, which is explained by the fly's preference for cooked food rather than garbage and human feces. It is called the "rice fly" by the people of Szechwan because it has a predilection for cooked rice. The fly also prefers to breed in pig manure, whereas *C. megacephala* prefers liquid human feces.

The latrine fly also plays the star role in the Pacific islands. On Guam, cultivable cysts of both races of *E. histolytica* were repeatedly obtained from the feces of these flies. Positive flies were caught in various quarters and once on a kitchen table. *G. lamblia* also turned up with considerable frequency; *E. coli*, *E. nana*, *T. hominis*, and *C. mesnili* were less prevalent. The technique used for the detection of cysts is worth mentioning. Flies were lured from a darkened trap into a lamp chimney in which they were kept for up to four hours. In the interim, they deposited fly specks on a piece of gauze moistened with a physiological salt solution. The flies were then anaesthetized, counted, and identified. The gauze was thoroughly washed in salt solution, the solution was centrifuged at 2000 R.P.M., and the supernate was discarded. A portion of the sediment was examined microscopically and another portion was mixed with a little serum on a glass slide, fixed in Schaudinn's fluid, and stained with Heidenhaim's hematoxylin (Harris and Down, 1946). An obvious improvement would be to separate the flies according to species before placing them in the lamp chimneys. Heinze (1949) and Attimonelli (1940) describe other methods for staining Protozoa of flies.

Flies may harbor a veritable zoo of Protozoa, according to the experience of Attimonelli in Bari, Italy. He found a variety of flagellates, rhizopods, and ciliates in the sterile water washings of a large number of wild flies. Part of his success was that he allowed the washings to stand several days to allow decystment.

EXPERIMENTAL TRANSMISSION OF PROTOZOA BY FLIES

Surprisingly little work has been done in this area. We know of only two studies, and the first, by Wenyon and O'Connor (1917b), is too preliminary to justify any conclusion. Two kittens were fed for almost a week on a daily ration of milk and bread that was first exposed to flies that had fed on dysenteric stools. Although kittens are generally quite susceptible to *E. histolytica*, these two remained healthy.

The other study, by Rendtorff and Holt (1954), was well designed and executed under carefully controlled conditions, as one might expect since they were working with human volunteers. Volunteers were selected who were free of E. coli and G. lamblia, the test organisms, and they were housed and fed under conditions intended to keep them free. One-week-old house flies were fed cysts in such a way as to reduce external contamination and to ensure transmission primarily by means of fly feces and vomit. Six lots totaling 127 flies were released during an eight-day period into a room where the volunteers took their meals. None of the 18 men became infected. Fly-food contact was considered to have been too brief, and therefore a second test was devised. This time, caged flies were given a concentrated suspension of cysts mixed with an equal volume of sterilized human feces. After the flies were given the infective meal, the volunteers' food was placed for about an hour in their cages. Even under these conditions, only one of the sixteen volunteers became infected. He developed an unusually heavy infection with E. coli, which persisted for about 80 days.

The results of these experiments cast doubt on the fly's effectiveness in transmitting protozoan infections, but there is far too little evidence to justify a verdict of "not guilty." Although the second attempt by Rendtorff and Holt was meant to assure considerable fly-food contact, there is nothing in their report to assure us that this actually happened. We also do not know whether the fly-exposed food actually contained cysts.

The evidence cited earlier establishes without question that flies are capable of carrying pathogenic Protozoa and frequently do so under natural conditions. There is an obvious need for additional carefully executed transmission studies that take into account the number of cysts at each stage in the chain of infection: feces $\longrightarrow$ flies $\longrightarrow$ food $\longrightarrow$ host. At the present time, the concensus among those who have studied its epidemiology is that amebic dysentery is spread through polluted water, contamination of food by food handlers, and by flies. The usual setting is a poorly sanitated environment, where feces are exposed, flies are plentiful, and personal hygiene is neglected.

LEISHMANIASIS (Oriental Sore, Espundia, Kala Azar)

The causative agents of cutaneous leishmaniasis are, in the old world, *Leishmania tropica* and, in the new world, *L. braziliensis*. This disease is a localized superficial infection of reticuloendothelial macrophages producing nodular and ulcerating lesions of any exposed part of the body. In kala azar, caused by *L. donovani*, the viscera are also attacked. Sand flies are the immediate source of infection (see Strong, 1944, and James and Harwood, 1969, for good accounts of the research on *Phlebotomus* as biological vectors). Each lesion is the result of a bite by an individual infected insect vector, except for secondary lesions produced by scratching. When the parasites are ingested by these flies in the leishmania form, they develop into flagellated forms and multiply in the stomach, gradually moving forward to the pharynx and tip of the proboscis.

Persons with exposed lesions containing the parasites are reservoirs of infection, as well as dogs, cats, and Turkestan gerbils; the disease is communicable as long as the parasites remain in the lesions, a year or longer in untreated cases. The incubation period is from a few weeks to many months. It is also thought that transmission may occur by direct contact of abraded skin with the lesion of another person, but this is doubtful. Adler and Theodore (1957) find no evidence of endemic leishmaniasis in the absence of sand flies, though other flies may transmit the organisms mechanically.

Wenyon (1911a) in Bagdad reported observations on oriental sore as it occurs there. House flies collected from the faces of children with open sores nearly always showed the parasites in the gut, but the parasites quickly degenerated and failed to develop into flagellate forms. He believed that the house fly could only mechanically transmit the parasite from a sore to a wound within a short time, but the disease was limited in distribution in Bagdad although the house fly occurs everywhere. *Stomoxys* was very limited in its distribution and failed to show any parasite development in the gut after as many as ten feedings. Wenyon (1911b) also found that house flies fed on blood saline rich in *Leishmania* passed the unaltered parasites in the feces three hours later and that unaltered parasites could be detected in the gut upon dissection; the parasites degenerated beyond six hours. The gut contents of flies that had taken up the parasite were infective and produced the disease in monkeys up to three hours after feeding but not after. Cardamatis and Melissidis (1911) fed house flies on sores and found by dissections what they believed to be *Leishmania* retaining viability for up to 9 days, but Wenyon (1911b) showed that these forms were really *Herpetomonas*, a parasite of the fly itself.

Patton (1912) in Cambay allowed house flies (*Musca domestica nebulo* and *Musca* sp.) to crawl over sore discharges and then let them feed on scratches on his face and hands every day for about a month, without his developing any sign of infection. He concluded that the house fly does not carry viable parasites on its legs and proboscis and deposit them on a scratch or abrasion of a susceptible person. He did, however, find that the parasites could be recovered only in the fly midgut, unchanged up to 6 hours after ingestion, and that after that they degenerated and disappeared. He tested other insects (including *Phlebotomus*) and concluded that bed bugs (*Cimex rotundatus*) were the only probable insect transmitters of the disease because they harbored viable parasites.

Thomson and Lamborn (1934) reported that leptomonad forms of *L. donovani*, *L. infantum* (= *donovani*, dog strain), and *L. tropica* were passed viable in the excreta of *Musca sorbens* for several hours after ingestion of cultures. They concluded that both kala azar and oriental sore could be actively transmitted through the agency of these nonbiting hematophagous muscids, which were abundant in Nyasaland and favored man. Wollman (1927) could not recover *L. tropica* one day after feeding a culture of house flies. Later, Lamborn (1935) reported unsuccessful transmission experiments on dogs using this fly. In 1955, he reviewed the evidence that muscid flies with blood-sucking habits may be vectors of leishmaniasis, but added nothing substantial in favor of these flies as vectors (Lamborn, 1955).

Berberian (1938) was not entirely satisfied with the sand fly theory of transmission primarily because critical experiments designed to demonstrate the transmission by the bite of the sand fly had regularly failed at this time. Attracted by the possibility that sucking or biting flies, which were commonly attracted to skin lesions, mechanically transmit oriental sore and kala azar, he experimented with *Stomoxys calcitrans* and obtained successful transmissions of cutaneous leishmaniasis by its bites. He allowed stable flies to feed on his own lesions and immediately transferred them to a volunteer, who was bitten a total of 11 times by 7 flies and developed at least 2 positive lesions 5 and 6 months later. Because of the ease of transmission, he believed that this method frequently occurred under natural conditions. He reported two more successful transmissions by three stable flies to volunteers the following year.

In the same year, it was discovered that sand flies produce a bumper crop of parasites if they are fed raisins after the infective meal. And such flies readily transmit leishmaniasis to human volunteers.

Very recently, Lainson and Southgate (1965), although agreeing that sand flies are the natural vectors of leishmaniasis and that cyclical development of the parasite is necessary before transmission can occur, confirmed Berberian's work with *S. calcitrans*. They placed 4 flies individually on a large dermal lesion of a hamster infected with *L. mexicana*. Twenty seconds after they began to feed, the flies were removed and allowed to feed on an uninfected hamster; six weeks later the second animal developed papules containing large numbers of Leishman-Donovan bodies at the sites of three fly-bites. The investigators believe that, since in cutaneous leishmaniasis, much of the body may be covered with nodular lesions heavily infected with parasites, transmissions by interrupted feeding of this fly may occur.

254 FLIES AND HUMAN DISEASES

The helminth diseases of man and animals in which flies figure include a number of cestodes and nematodes whose complex life cycles are beyond the scope of this book to describe. The interested reader should consult a standard parasitology text for the biology and importance of the various parasites discussed. We have combined the discussion of human and animal diseases in a single treatment because worms often have a range of hosts, which includes man, and flies make even fewer distinctions. We shall deal in the main with parasites that are lodged in the intestine and therefore rely on fecal dissemination. Diseases caused by *Thelazia* and *Habronema* are discussed separately in Chapter 5.

HELMINTHS IN ADULT FLIES

Grassi (1883) may have been the first modern investigator to point an accusing finger at flies as disseminators of helminths of higher animals. After exposing *Trichuris* eggs in his laboratory, he discovered them later in fly spots in a nearby kitchen and in the intestines of flies. He also found that flies transported oxyurid eggs. A few years later, Grassi and Rovelli (1889) demonstrated that the house fly can disseminate eggs of *Choanotaenia infundibulum*, the fowl tapeworm, by swallowing them.

Chapter 3 described the structure of the fly's proboscis and the constraints it imposes on the fly's vector capacity. Ingestion of helminth eggs by flies is limited by the dimensions of the egg and the anatomy of the proboscis. Eggs above a certain size are excluded by the interbifid spaces on the pseudotracheae, though they can still be swallowed if the fly spreads its labellar lobes and applies its prestomum directly. The size of the prestomum, then, determines the largest egg that the proboscis will accommodate. How eggs are carried is important because they are usually retained longer in the fly's gut than on its surface, thus increasing both the duration and the range of the fly danger. From an evolutionary point of view, admission to the gut of the fly opened a way to its exploitation as a true intermediate host. Many of the helminth eggs we shall be discussing are illustrated in Figure 49.

Early investigators thus bent their efforts toward understanding the fly's swallow. Galli-Valerio (1905a) exposed flies to feces that contained eggs and larvae of the hookworm, *Ancylostoma duodenale*, and found them only on the exterior, not in the gut. However, larger eggs of *Hymenolepis nana* (Calandruccio, 1906) and *Diphyllobothrium latum* (Léon, 1908) were not only ingested but appeared undamaged in the excreta of the fly. Unfortunately, the fly species were not given in these accounts so we cannot be sure that the results are comparable.

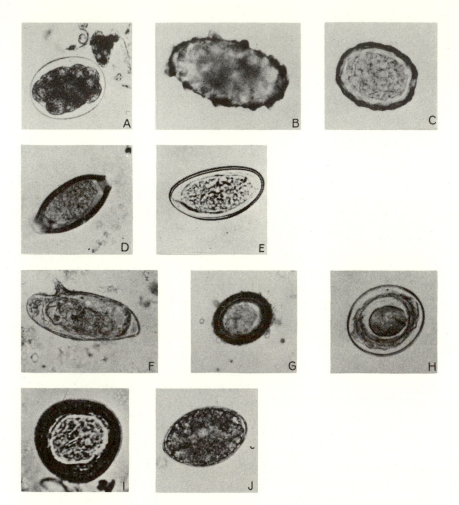

FIG. 49. Photomicrographs of helminth eggs. A. Hookworm, *Necator ameri-canus* (× 400). B. Roundworm, *Ascaris lumbricoides* (unfertilized egg × 400). C. Same as B, fertilized egg (× 400). D. Whipworm, *Trichuris trichiura* (× 400). E. Pinworm, *Enterobius vermicularis* (× 400). F. Blood fluke, *Schistosoma mansoni* (× 400). G. Beef tapeworm, *Taeniarhyncus saginatus*. H. Dwarf tapeworm, *Hymenolepis nana* (× 400). I. *Echinococcus granulosus* (×1500). J. Fish tapeworm, *Diphyllobothrium latum* (× 400). (Adapted from T. J. Brooks, Essentials of Medical Parasitology, 1963.)

Shortly after, Nicoll (1911) made a definitive study of the ability of flies to ingest, retain, and excrete viable eggs of various tapeworms and nematodes. Most of his experiments were performed with *M. domestica*, a few with *F. canicularis* and *C. vicina*. He made the striking observation that flies are particularly attracted to tapeworms, upon which they feed with avidity. When an intact proglottid of *Taenia pisiformis*, *Bothriocephalus marginatus*, or *Dipylidium caninum* was mixed with feces, flies concentrated on the proglottid. The worm's attractiveness lasted several days, and proglottids remained a good source of eggs for flies for as long as two weeks. The nematodes, *Toxascaris leonina* and *Parascaris equorum*, were much less attractive, and flies were unable to penetrate the tougher cuticle of these worms. Nicoll observed that the crop invariably contained liquids only, whereas the midgut contents included particles of various dimensions up to about 40 μ. This led to experiments in which flies were fed helminth eggs of various sizes and their gut and feces were then systematically examined.

Some of Nicoll's data on the house fly are summarized in Table 6; the maximum size of an ingested egg appears to be approximately 40 μ. The few experiments with the lesser house fly and the blow fly suggest that larger flies ingest larger eggs. This was later confirmed by Pipkin (1943). Nicoll felt that the proportion of positive flies would have been much greater had he discovered earlier that the number of eggs taken up by a fly also depends on the nature of the infective meal. Thus, a larger proportion of flies became infected when feeding upon eggs suspended in water than when fed upon intact segments. Eggs were never found in the crop, though Zmeev (1936) found undeformed eggs of *T. saginatus* in various gut regions of wild *M. domestica vicina*, especially in the crop.

Calandruccio (1906) demonstrated the viable passage of eggs of *H. nana* through the gut of the house fly, and Joyeux (1920) confirmed his work with *Vampirolepis fraterna*, a related species with a similar egg. Joyeux fed his flies eggs suspended in syrup, and 12 hours later observed large numbers in the gut. The great majority of the eggs were still in their shells undeformed, and warming them elicited typical hexacanth movements. Aleksander and Dansker (1935) reported extensive tests in which *M. domestica* swallowed and subsequently excreted viable eggs of a number of helminths up to about 27 hours after ingestion. The worms he used were *A. lumbricoides*, *T. trichiura*, *E. vermicularis*, *D. latum*, *T. saginatus*, and *H. nana*. Opposing evidence was obtained by Pod'iapol'skaiā and Gnedina (1934), who fed *M. domestica* and *C. vicina* on feces that contained eggs of *A. lumbricoides*, *E. vermicularis*, and *D. latum* and found all three types in the excreta of the blow fly, but

TABLE 6

SUMMARY OF NICOLL'S DATA (1911)* ON THE INGESTION OF
HELMINTH EGGS BY HOUSE FLIES

Species of helminth	Size of egg (μ)	Number of positive flies over total	Maximum persistence in fly
Hymenolepis diminuta	70 × 65	0/35	—
Toxascaris leonina	80 × 70	0/20	—
Ancylostoma caninum	60 × 40	0/8	—
Trichuris trichiura	50 × 25	1/12	—
Dipylidium caninum	40 × 40	4/10	> 43 hrs.
Bothriocephalus marginatus	35 × 35	2/9	72 hrs.
Taenia pisiformis	35 × 35	17/45	14 days

*In addition to those listed in the table, the house fly was also shown to be capable of carrying eggs of the following helminths, both internally and externally: *Taenia solium, Hymenolepis nana, Diphyllobothrium latum* (?), and *Enterobius vermicularis*; and externally only: *Necator americanus* and *Strongylus equinus*.

none in the excreta of the house fly. The maximum number of eggs in a single fecal spot was 53, with all three egg types present. In Australia, unspecified flies were shown to be capable of ingesting eggs of the tapeworm, *Echinococcus granulosus*, and the much larger egg (65-80 μ × 45-50 μ) of the hookworm, *Uncinaria stenocephala* (Ross, 1929).

Two other studies bear on the ingestion and safe passage of eggs in adult flies. Gutberlet (1916) found difficulty in demonstrating eggs of the poultry tapeworms, *Raillietina cesticillus* and *R. tetragona*, in house flies that had fed on proglottids. He attributed this to his use of immature segments, for when he fed mature ones of *C. infundibulum*, whose eggs are larger than those of *Raillietina* species, he found cysticerci in the flies about two weeks later. Heinz and Brauns (1955) fed *Sarcophaga tibialis* on dog feces contaminated with ova of *E. granulosus* and recovered them from the surface and intestines of a number of flies. Infectivity of the ova was unaffected by passage through the gut, shown by animal tests.

Nicoll (1911) found that eggs were generally carried in the gut of the house fly for several days, and, in one case, viable eggs of *T. pisiformis* were excreted for at least 14 days. Round (1961), in studies on the role of filth flies in the epizootiology of bovine cysticercosis in Kenya, fed ova of *T. saginatus* to carrion flies and recovered them from the feces of *Chrysomya chloropyga* 258 hours later, from *C. albiceps* 18 hours later, and from *Sarcophaga* sp. 24 hours later. Viable ova were excreted by *C. chloropyga* for three days after the infective meal.

Eggs adhering to the surface of the fly are removed as the fly cleans itself, and this beneficial act of autosterilization usually occurs shortly after the fly quits the surface on which it has been feeding. Tao (1936-1937) studied the carriage of helminth eggs by flies in Shanghai and found eggs only inside the flies he caught. He concluded that flies had rid themselves of external ova in the neighborhood of the source and that more remote contamination of food occurs from the gut after the fly has flown away. Pipkin (1943) confirmed this by contaminating *M. domestica* with eggs of various nematodes. The flies remained externally contaminated for an average maximum period of only 3.47 hours.

Although external phoresy may be of shorter duration, it would be a mistake to conclude it is therefore unimportant. It is by means of external phoresy that larvae of *Rhabditis pellio* are thought to be transported from one breeding site to another by *Drosophila* (Aubertot, 1923) and by *M. domestica* (Menzel, 1924). Harada (1954) showed experimentally that larvae of the hookworm, *A. caninum*, can be carried on the bristles and pulvilli of *Calliphora* sp. and the house fly. Zmeev (1936) trapped 100 *M. domestica vicina* in toilets and found eggs, probably of *T. saginatus*, strongly attached to the legs of two of the four positive flies. In a study of the helminth carrier frequency of wild flies in the Volgograd area, Pokrovskiĭ and Zima (1938) found that, among 18 positive flies, 12 carried eggs of several species of helminths on their legs, one on the wing, and five in the gut.

Under certain conditions, such as infested poultry farms where high fly densities prevail, it probably makes little difference whether ova are located on or within flies. Such a tight ecosystem provides ample opportunity for cycling of parasites between chickens and flies. More tapeworm infestations have been observed among younger chickens (Gutberlet, 1916), and this may be correlated with their greater interest in catching flies as compared with adults, which seem little interested in them. Gutberlet (1920) was able, on several occasions, to infest chicks with the tapeworm, *Hymenolepis carioca*, by feeding them large numbers of *S. calcitrans*, which he caught around the chicken yards. During November and December, stable flies were very abundant and somewhat sluggish and therefore easy prey for chickens. This cestode was not common until this time, when the chickens became heavily infested. Clapham (1939) noted that adult *M. domestica* and *Phaenicia sericata*, infested from the maggot stage with the gape worm, *Syngamus trachea*, were also sluggish and would likely be easy prey for birds in the natural state, thus facilitating transmission; infested maggots appeared to have normal vitality. Ackert (1920) could not recover eggs or onchospheres of *R. tetragona* from the vomit and feces of house flies that had fed on egg masses. Nevertheless, house flies captured around chicken houses where there was a known infestation with this tapeworm produced the disease when fed to worm-free chickens. The previous year, he reported successful experimental transmission of *R. cesticillus* from house flies to chickens. Schiller (1954) experimentally transmitted *Echinococcus* sp. eggs to three of nine voles caged with *Phormia regina*, also without specifying location of the eggs on (or in) the fly. Experiments such as the last would be worth repeating on a larger spatial scale.

HELMINTHS IN MAGGOTS AND PUPAE

The maggot and pupal stages complicate our consideration of the internal carriage of eggs. One would assume from direct observation that the mouth of a maggot is too small to accommodate the egg of a worm. Indeed, there are statements in the literature to the effect that larvae only absorb liquid food, yet Stiles (in Nuttal, 1909) fed house fly maggots upon gravid *A. lumbricoides* and found eggs in their guts. More significantly, he also found eggs and larval worms in the guts of newly emerged flies bred from these maggots. Nuttal (1909) confirmed the former but not the latter observation, and he and Nicoll (1911) showed that, if anything, maggots are destructive of helminths and their eggs, ingested ova being destroyed in the gut. Nicoll found that eggs of *T. pisiformis*, *T. leonina*, and *P. equorum* were not transmitted from maggots to adults. This was extended by Gutberlet (1916) to include *R. cesticillus*, *R. tetragona*, and *C. infundibulum*. He bred 30 house fly and 50 stable fly maggots in chicken manure that contained proglottids of these cestodes. Histological sections of pupae and adults of both species of fly disclosed no suggestion of eggs or a worm stage. Additional support for the idea of a pupal barrier is found in the more recent work of Perez Fontana and Severino-Brea (1961). Eggs of *A. lumbricoides* (?) and *Trichurus* sp. and eggs and onchospheres of *E. granulosus* were found undamaged in the gut of *Lucilia* and *Calliphora* maggots after they had fed in feces containing the worms. Although eggs of *Trichurus* were present in the pupal gut, no helminth was recovered from adult flies. It is significant, nevertheless, that the onchospheres in the gut of the maggots underwent distinct molts. The authors believe that a more rapid summer development of the flies would have improved the survival of onchospheres through metamorphosis, but they did not investigate the temperature factor.

Supporting Stiles' thesis of an open passage of organisms from the maggot to the adult is the work of Roberts (1934), who found that eggs of *A. lumbricoides* may be ingested by *M. domestica* and *Phaenicia cuprina* and not only persist in a viable state through metamorphosis, but may also be deposited in the excreta of the adult flies.

FLY-NEMATODE ASSOCIATIONS

Available evidence suggests that cestodes, with the exception of *Raillietina* and a few others, are not especially at home in flies. Some may molt to become cysticercoids, or the eggs may endure much as do inert particles in the gut of the maggot and adult. This is not so true of *Ascaris*, which does occasionally survive pupation and may even hatch in the gut of the insect. However, the *Ascaris* larva is not invasive and is confined to the gut, which is unsuitable for its further development. Other nematodes have evolved larvae that are able to develop in the gut or to penetrate the gut wall and to develop to an infective stage in the tissues of maggot and pupa, ready for passive or active transfer from the adult fly to the definitive host.

There are examples of such fly-nematode stages, and these may possibly illustrate an evolutionary progression. Mitzmain (1914) observed a limited development in *S. calcitrans* of microfilariae of a species he called *M. sanguinus equi africano*. After being swallowed, the worm larvae escaped from the fly's gut and settled in the flight muscles, but they could not reach maturity. Parasitized flies suffered an increased rate of death during the first ten days of infection, when worm larvae were actively penetrating the fly's body. The increased fly mortality, phagocytosis of invading larvae, and the failure of such flies to infect well horses, all indicate that this fly-nematode relationship has not advanced beyond the stage of mutual destruction.

Though little is known concerning details of the life cycles of the parasites belonging to the genus *Stephanofilaria*, it is clear they have evolved a more specialized relationship to flies. Stoffolano (1970) has reviewed the pertinent literature and points out that these parasites occur in Asia and North America, where they attack bovids, horses, goats, and pigs, causing a cutaneous disease variously referred to as humpsore, Cascado, equine dhobie, and Krian sore. Fly involvement has been suspected for some time, though positive evidence has not always been forthcoming. The work of Ivashkin and his co-workers (1963) has demonstrated the life cycle of *S. stilesi* in *Siphona irritans*. The egg hatches in the gut of the fly, the larva invades the hemocele and eventually lodges in the proboscis. Studies with *S. assamensis* in *Musca conducens* show the parasite to be confined to the gut of the insect (Patnaik and Roy, 1966; Patnaik, personal communication). When a number of starved, three- to four-day old female flies were allowed to feed on cattle with chronic dermatitis due to *S. assamensis*, about 40% of the flies became infected. A few hours after the infective meal, the organism was recovered from the crop, and, four days later, "sausage" stages of the filaria were present in the midgut of some flies. The larvae migrate as they grow, and by the tenth day are present in the hindgut. By days 18 to 20, pre-infective-stage larvae have invaded the head and from the 22nd day on, mature infective larvae are seen in the proboscis. The developmental cycle takes place entirely in the digestive tract. Patnaik finds it puzzling that, since other filaria take only a week or so to attain infective stages in mosquitoes, etc., *S. assamensis* should require 20 to 25 days in the fly. He finds that very few *M. conducens* carry the infection in nature—less than 1%—and this agrees with Srivastava and Dutt (1963), who found infective larvae in only seven out of 787 captured specimens of the fly. This raises a question about the suitability of this fly as a vector. It is also puzzling that the rate of spread of the disease is rather slow when compared to that of other filarial infections, e.g. *Oncocerca*, *Dirofilaria*, and even Bancroftian filariasis in man. Workers in this field are agreed that much basic work needs to be done.

Other interesting associations are those between the nematodes *Parabronema skrjabini* and *Siphona irritans*, and *Syngamus trachea*, the gape worm, and the house fly. *P. skrjabini* is reported by Ivashkin (1959) to be widely distributed in ruminants and is of considerable importance in the southern semisteppe regions of the U.S.S.R. and Mongolia. The usual habitat of the parasite is the pyloric region of the abomasum, with the larvae settling in the wall itself and the sexually mature forms being found in the mucous membrane. Ivashkin found a concurrence between the peak invasion of ruminants and peak populations of *S. irritans*. Infection of the final host is by the oral route. Details of the life cycle in the fly are yet to be elucidated.

The gape worm has evolved several alternate routes for its development, including earthworms, snails, and various coprophagous insects, which is a rather wide range of intermediate hosts. The final host can be the chicken, pheasant, or other birds. There appears to be some selection operating in the bird for, according to Clapham (1939), it is not as easy to parasitize chickens with worms from starlings as it is with worms from other chickens. But passage through an invertebrate seems to neutralize these differences, and perhaps keeps the parasites from becoming over-selective. Clapham (1939) demonstrated transmetamorphic survival of worm larvae ingested as eggs by maggots of *M. domestica* and *P. sericata*. Adult flies of these maggots were infective when fed to chickens. In the maggot, the eggs hatched in the gut and the worm larvae invaded the hemocele. They were generally found loosely coiled in the fat body that enveloped the gut. All 12 maggots and ten of the 12 adults that were examined were found to be infested each, interestingly, with a single larva. Parasitized flies are sluggish and therefore easy prey for birds; infective larvae are transferred passively to the chicken when the fly is swallowed.

Though this method has been retained by *Habronema* species, and horses may become infested by swallowing flies, the worms have improved their chances by developing a more active and precise method of transfer. As shown in Figures 51 and 53, infective larvae migrate into the proboscis and enter the definitive host while the fly feeds. There are further refinements among *Habronema*. For example, larvae of *H. micro-stoma* plug the proboscis of the stable fly causing it to abandon its bite, which would generally result in a dermal dead end for the larvae. Instead, the fly is forced to adopt the house fly type of feeding, which, when it occurs around the mouth and nostrils, provides the larvae with a direct route to their destination in the stomach. Another adaptation, and one that seems to have not yet been perfected, is the emergence of the larvae in the proximity of warmed horse blood or saliva. These adaptations tend to synchronize the exodus of the delicate worms from the proboscis at a suitable time and place on the proper host.

Sukhacheva (1963) has expanded our knowledge of the destructive interaction between insect larvae and parasitic worm eggs in soil, dung, and water. He fed eggs of *A. lumbricoides* to various fly larvae and also to odonate and ephemerid naiads. After 24 hours in contact with the eggs, the insects were carefully washed to remove externally adhering eggs, and several hours later their guts and excreta were examined for viable and destroyed ova. In the gut of *Protophormia terraenovae* and *Themira putris* the eggs continued to develop. However, continued observation of the ascarid embryos after they had passed through the gut of these maggots showed that they soon lost their viability. Eggs that passed through the digestive tracts of *C. vicina, Fannia* sp., and *Eristalis* sp. developed at a normal rate, and worm larvae were normally viable. Maggots of *Ophyra* sp. and *M. domestica* were not capable of ingesting the eggs and therefore had no interaction except that the presence of house fly maggots in the medium seemed to stimulate emergence of ascarid larvae, which soon died for want of a host. If this occurs under natural conditions, it would help to rid dung of its parasite population. It is interesting to note in passing that mayfly naiads ate ascarid eggs with destructive effect whereas odonates had no perceptible effect.

Beliaeva (1960) found that encapsulated *Trichinella* larvae were decapsulated in the gut of green bottle maggots. The author did not state whether any further development occurred, and the significance of this observation is not clear.

Helminths in wild flies

Reports from various regions have shown flies to be frequent carriers of helminths. Galli-Valerio (1905b), in Italy, found eggs of *T. trichiura* and *A. lumbricoides* on captured flies, but he did not describe the circumstances. In some instances, the contamination of flies has been unusually great; for example, 26 out of 100 *Borborus punctipennis* carried eggs of *A. lumbricoides, N. americanus,* or *T. trichiura,* in St. Lucia (Nicholls, 1912). This is a coprophagous fly, but it is questionable whether its habits are of medical or veterinary importance.

Other examples of natural fly contamination are given below, but it is well to remember that carriage does not constitute transmission. On pig farms, for example, although flies may be demonstrably contaminated with ova of *A. lumbricoides,* there are much more likely avenues of dissemination, e.g. soil, food, and water (Roberts, 1934). On poultry farms, however, flies undoubtedly play a significant role as vectors of tapeworms. In addition to the supporting evidence cited previously, there is the remarkable finding of a house fly taken on a poultry farm in Kansas that harbored no less than 91 cysticercoids (Reid and Ackert, 1937).

It is generally not difficult to capture contaminated flies in other situations than animal farms. Of 206 flies, chiefly *Chrysomya* and *Lucilia,* caught near latrines in the Shanghai area, Tao (1936-1937) found 12 that carried ova of *Ascaris, Trichuris,* and *Ancylostoma* in their gut. In Tadzhikistan, Zmeev (1936) captured 100 *M. domestica vicina* in toilets and found four with eggs of *T. saginatus,* two to five eggs per fly. On Guam, the pooled excreta of trapped flies, predominantly *Chrysomya megacephala,* frequently contained ova of *T. trichiura, A. lumbricoides,* hookworm, and, in one case, an ovum of *H. diminuta* (Harris and Down, 1946). On the other hand, a ten-month collection of 5781 specimens of *M. domestica* in São Paulo, Brazil, yielded only one fly, taken near an incinerator plant, with a viable egg of *Hymenolepis* sp., and another with a *Habronema* larva (Coutinho et al., 1957).

The presence of contaminated flies in poverty areas where sanitation is low, food is unduly exposed, and human infection rates are high provides us with compelling though still not conclusive evidence of fly transmission. Wenyon and O'Connor (1917a) examined 200 house flies captured at random in various localities in Alexandria, Egypt, and, in the fecal spots of 15, they found protozoan cysts and also the eggs of *T.*

saginatus, Heterophyes heterophyes, T. trichiura, and *A. duodenale.* Also present was a large, lateral-spined egg probably of *Bilharzia mansoni,* which would measure 115 to 170 μ × 45 to 65 μ. It could only have been swallowed lengthwise by the gluttonous fly.

A similar degree of fly contamination was reported by Shircore (1916) in Mombasa, East Africa. Out of a total of 275 flies collected in hospital wards and meat markets, 29 (10%) were found to harbor ova of *A. duodenale* (5), *A. lumbricoides* (1), *T. saginatus* (8), *T. trichiura* (16), and *Bilharzia mansoni* (1). Parenthetic numbers indicate frequency. Shircore noted that the ova of *T. trichiura* and *T. saginatus* were less common in human fecal samples than ova of *A. duodenale* and *A. lumbricoides,* although they were more common in flies. There was no obvious explanation for this difference.

Buxton (1920) in Amara, Iraq, found that 4% of the house flies he collected in latrines, cookhouses, and compounds contained *Taenia* sp., *Trichuris* sp., and *Ancylostoma* sp.

Russian investigators have contributed much to our knowledge through their emphasis on medical ecology. Their studies have often shown a close relationship between contaminated flies and human food. Pletneva (1937) emphasized the role of flies as carriers of tapeworm eggs in Ashkhabad. Of 279 wild flies, predominantly *M. domestica vicina,* 13, or 4.65%, were found to be infested. The distribution by location and species of worm were as follows: Toilets—8 *H. nana* and 1 *E. vermicularis*; food store—2 *H. nana*; restaurants—2 *H. nana.*

Pokrovskiĭ and Zima (1938) examined a total of 2650 flies collected in the Volgograd region, 2531 of these were *M. domestica* (11) and the rest were *Calliphora* sp. (2). Eighteen flies carried eggs of *E. vermicularis, Hymenolepis, Diphyllobothrium,* and *Ascaris*; the number of positive flies is indicated parenthetically. A particularly significant finding is that half the infested flies were taken in food shops, where they were carrying four species of helminths.

Studies by Pod'īapol'skaīa and Gnedina (1934) among other Russian scientists further underscore the danger of fly contamination of human food. They examined 1600 fly specks at a restaurant and 900 at an abattoir and found an egg of *T. trichiura* at the former and two eggs of the fluke, *Dicrocoelium lanceatum,* at the latter. It was not surprising to find eggs of this fluke in abattoir flies, considering that the livers of 72% of cattle and 85% of sheep at Makhach Kala, the study area, were infested with the parasite. Sukhova and Talyzin in 1948 (see Sukhova, 1951) reported 13% of the blow fly, *C. uralensis,* caught in a fruit drink processing plant were carrying ascarid and other helminth eggs.

In Bukhara, Central Asia, the prevalence of *T. saginatus* infestations in bovids was reflected in both the number of taeniarhyncotic patients who were being treated at the clinic, and in a 12.9% infestation rate in flies caught near the clinic (Sychevskaĭa and Petrova, 1958). Flies were also taken in the market, mainly on melons and melon rind in fall, and on sweets in spring. Of 333 synanthropic flies caught in the market in 1953, 58 flies (six species) were found to be carrying taeniid onchospheres and three flies (two species) carried eggs of *A. lumbricoides*. The following year, 5% of the flies carried onchospheres. The authors were able to assign the onchospheres to the beef tapeworm with considerable confidence because infestations with the pork tapeworm are rare in Bukhara. The fly species involved were *Eristalis aeneus*, *M. domestica vicina*, *M. sorbens*, *M. stabulans*, *Dasyphora asiatica*, *C. vicina*, *Chrysomya albiceps*, *Pollenia rudis*, and *Sarcophaga haemorrhoidalis*. It is noteworthy that in 1955-1956, after sanitation and fly control programs were initiated in Bukhara, no flies were found contaminated with onchospheres.

Nadzhafov (1967) has also encountered high levels of fly infestation with onchospheres of the beef tapeworm. Onchospheres were found in 233 (4.8%) of 4,372 flies belonging to 20 species. The percentage of infested flies varied in different villages from 0 to 8.8 and depended on accessibility of *Taenia* proglottids in fecal masses. In another study of 10,845 flies caught on the premises of the Helminthology Laboratory in Fergana, 0.28 to 2.2% of *M. stabulans* and 0.26 to 3.22% of *M. domestica vicina* carried ova of *H. nana* both externally and internally. The onchospheres of a taeniid were also found in one house fly (Sychevskaĭa et al., 1959). In Beloruss SSR, 984 synanthropic flies of 11 species were examined by washings and by dissections and 8.3% were found to carry ascarid eggs, 1.4% *T. trichiura* eggs, and 0.6% eggs of *E. vermicularis* (Slavinskiĭ, 1960).

The winter danger of fly dissemination was underscored by Fedorov's (1962) finding of helminth ova in house flies taken during November and December in Omsk. The body surfaces of five flies (1.4%) had six ova of four worms, and the gut contents of 11 flies (3.17%) contained 16 ova of six helminths.

Investigations like the one conducted by Sychevskaĭa and Petrova (1958) prove that effective programs of sanitation and fly control at the community level can eliminate a specific helminth from flies. Is the same or a different level of control necessary to eliminate the fly risk in other diseases? We know almost nothing about the danger level of a fly population nor do we have well-tested criteria, as there are for *Anopheles* and malaria, for evaluating the vector role of the fly. The complexities and uniqueness of enzootic and endemic situations involving flies make generalization hazardous and often of limited value. The biggest gap in the logical development of incriminating evidence against flies is that which exists between our knowledge of the ability of flies to transmit and the actuality that they do. Two examples will illustrate how this impasse operates.

Noè's (1903) studies provided evidence that *S. calcitrans* is a carrier of *Filaria labiato-papillosa* (= ?*Setaria cervi*) of the ox. In the laboratory he found that worm larvae pass through the gut wall of the adult fly and settle in the proboscis. He also found that flies that fed on an infested ox never swallowed more than three larvae. Field studies in a highly enzootic area near Rome disclosed an infestation rate in stable flies of only 3 to 4%. These figures led Noè to discount the importance of the stable fly as a vector of the parasite. Wetzel (1936) fed proglottids of the chicken tapeworm, *C. infundibulum*, to house flies, only 20% of which became infested. He, too, considered this figure as evidence that flies are not involved in natural transmission of the tapeworm. One could argue that the infestation of one in every five flies is probably more than sufficient to maintain adequate parasite levels in poultry, especially if there is a high fly density and also a high proportion of younger, more insectivorous birds in the flock. As with Noè's research data, several interpretations can reasonably be made. Adequate evaluations are difficult to make on infection rates alone and would be aided by baseline data on the following: fly factors—species composition, density, communicativeness, experimental infection potential vs. natural infection rates; host factors—density, susceptibility, and sanitary standards.

Manipulation at community levels of the fly and other factors has been applied to epidemiologic studies of trachoma, bacillary dysentery, and other diseases, sometimes yielding unique data obtainable in no other way. However, the expense and complexities involved in the management of such extensive programs have discouraged their wider use. Model experiments that make use of an experimental animal community scaled down to the level of a flock, herd, or farm would be easier to manipulate than the human situation, more economical to conduct, and could be quite fruitful.

5

Flies and Animal Diseases

Far from looking upon them as dipterous angels dancing attendance on Hygeia, regard them rather in the light of winged sponges speeding hither and thither to carry out the foul behests of Contagion.

SIR JOHN LUBBOCK, 1871

FOOT AND MOUTH DISEASE

THIS IS an acute, highly communicable viral disease limited chiefly to cloven-footed animals. It occurs mostly in cattle, pigs, sheep, and goats and more rarely in other domestic animals. Man rarely becomes infected despite frequent exposure in some countries. The disease is characterized by vesicular eruption in the mucosa of the mouth and the delicate skin of the hooves and udders; there may be increased temperature, refusal of food, and general depression. It is usually mild, resulting in low mortality, although highly malignant forms have been seen; economic losses relate to loss in weight, milk production, and to malnutrition.

The disease is caused by a filterable virus 23 mμ in diameter and classified with the picornaviruses. This virus is considered to be the most infective among animal disease viruses, sometimes being infective in dilutions of more than 1 in 5 million.

Natural transmission commonly occurs by direct or indirect contact between infected and healthy animals or between susceptible ones and objects contaminated by infected discharges. The disease is characterized by fulminating outbreaks spread by direct contact. The reservoirs of the disease between outbreaks and the basis for recurrence of epizootics are not well understood. Various elements have been implicated, such as birds, rodents, man, wild animals, and chronic carriers among cattle. The disease is enzootic in certain parts of Europe, Asia, Africa, and South and Central America. In other regions where strict measures of control and eradication have been implemented, it has not become established.

Hecker (1899) stated that flies often spread foot and mouth disease, possibly by feeding on blood containing virus or by contaminating their bodies. He could not transmit the infection to experimental animals when they were fed flies that had sucked virulent material two hours earlier, but he did obtain transmission by feeding to the animals flies that had been immersed in infected saliva. He presumed that the virus was destroyed by the flies' digestive system.

Roch Marra (1908) conducted biting transmission experiments with *Tabanus bovinus* and successfully infected calves with flies collected from a region rampant with the disease and with flies that had been placed in contact with infected material. Schuberg (cited by Titze 1921) was not successful in transmitting the disease with *Stomoxys calcitrans* that had sucked infected material from guinea pigs and piglets. LeBailly (1924) inferred from one experiment that flies do not play a role in spreading foot and mouth disease. He transferred *Muscina stabulans* and *Musca domestica* from a stable with infected cattle to another stable holding healthy cows, and the latter did not become ill.

The year 1927 produced several reports on this subject. Wilhelmi (1927) in conjunction with Kunike's (1927) experiments published the results of his unsuccessful experiments at Riems in 1916 (1918). He had made 13 attempts to transmit the infection to piglets or steers using *S. calcitrans*, which were fed on the blood of sick pigs or on infected lymph from pigs, or were bred from lymph-drenched dung. Infections were attempted by fly bites and by injections of emulsions of flies infected in each way. All results were negative, except for one light case believed to be due to extraneous infection. He further noted that this disease could be spread widely without the aid of insects. Kunike's (1927) investigations, which Wilhelmi felt were done far more meticulously than his own and were not contradicted by him, were carried out on guinea pigs with *M. domestica* and *S. calcitrans*. The various experimental series were designed to cover all the theoretical or practical possibilities regarding fly transmission. In the first series, house flies were bathed in fresh lymph and then at daily intervals were rubbed, either dry or moistened, into the scarified plantae of guinea pigs; thus, it was found that the virus was infective in the body of a fly for up to 2 days. Further, the somewhat diluted lymph was placed in NaCl solution in petri dishes and numerous flies were allowed to tread through it, wetting their feet and proboscides. These members were removed and macerated in NaCl and inoculated cutaneously into the plantae. The results showed that 150 legs and 25 proboscides inoculated after a 24-hour interval did not produce symptoms, but 300 tarsi and 50 proboscides did do so when inoculated after 6 hours. Similar experiments using infected integuments macerated in NaCl did not result in transmission, nor did similar use of infected bovine spittle. The virulence of these materials was checked by direct injection into guinea pigs and was found positive.

In the second series, the flies were fed on virulent lymph and then, at hourly intervals, their intestines were dissected out and rubbed in guinea pigs' plantar cutis; it was found that the virus was infective up to 18 hours in the fly gut. Attempts with fresh and older fly feces were unsuccessful. In the third series, using biting flies, it was found that if the proboscis of a stable fly were macerated in NaCl solution 5 minutes after sucking on fresh vesicles and injected into guinea pig planta, infection resulted, but not if this were done after 7 hours. In blood-sucking experiments from 2 to 12 flies were used at intervals of 1 hour to 3 days; these flies were allowed to bite healthy animals on the plantae, the toes, or on the shaved thigh, after they had fed on virulent material. No transmissions resulted via fly bites.

Series four included biting, gut inoculation, and feces inoculation of experimental animals, using stable flies that had sucked blood from sick animals between the time of the appearance of primary and secondary vesicles; no transmissions were produced by these methods. In the last series of experiments, stable fly bites were substituted for artificial scarification of guinea pig plantae. In only one case was one inoculation-vesicle produced by the rubbing in of virulent lymph directly after a planta had been bitten by four flies. Kunike felt that these results showed that flies play a fortuitous role and only as occasional short-term carriers of the infection. Schmit-Jensen (1927) found that the virus did not persist in stable flies for more than 26 hours and was not excreted in the feces.

The use of cell culture promises a more sensitive system for detecting the presence of virus. Rosov and Prokhorova (1968) fed flies (*M. domestica*?) a solution with a titer of 10^{-7} virus of foot and mouth disease. The flies were washed or homogenized at various intervals afterwards, and these preparations were used to infect rabbit kidney cell cultures. By this method it was found that virus survived 96 hours on the surface of flies and 48 to 72 hours internally. Animal inoculation, as we have seen, showed a maximum survival time of two days.

Other authors have implicated ticks in transmission. Mohler (1926) stated that adult *Margaropus annulatus* taken from infected cows during periodic fever contain virulent blood, however, he was not able to demonstrate the passage of the virus through the life stages of this tick. Galloway (1937) observed the persistence of the virus for 10-14 days in *Argas persicus*, but transmission experiments with guinea pigs were unsuccessful. Dhennin et al. (1961) reported that *Ixodes ricinus* taken from a sick cow can be virulent, and also *Melophagus ovinus* (the sheep ked) taken from an infected sheep. They also demonstrated the presence of the virus in *Phaenicia sericata* and *M. domestica* after the flies fed on fragments of infected guinea pig epithelium. Homogenates of these flies produced disease in other animals when injected.

Hirschfelder and Wolf (1938) and Waldman and Hirschfelder (1938), in reviewing the literature and from their own observations, concluded that flies are a very minor factor in foot and mouth disease; the epidemiology shows that man and hoofed animals themselves are the principal vectors. At Riems it was observed that, when winds were strong, the flies did not use them for transport, but rather clung to footholds on walls or hid in sheltered places, thus failing to maximize migration; in fact, cattle on the mainland 800-1500 meters from the island were free from the disease, despite favoring northwest winds. This distance is readily negotiated by flies (see Table 3). Additional arguments against fly transmission were that the infection of biting flies was only possible during the approximately 48-hour period of viremia, but the period between the fly's meals (*S. calcitrans*, 2-3 days; tabanids, 3-12 days) is longer than survival of the virus in the fly. I think the interval between meals can be shorter, besides which interrupted feeding can circumvent this difficulty. Further, they argue, there is insufficient volume of virus in one fly, hence disease transmission is possible only during fly peak periods, especially that of tabanids, and then probably only for short distances; also disease spread is known to be independent of fly seasonality.

The highly contagious nature of foot and mouth disease facilitates transmission without fly intervention. The difficulty encountered by various investigators in transmitting the disease by means of flies, sometimes under highly artificial conditions, suggests that the virus occurs in low titer or is fairly rapidly destroyed in the fly. The availability of a cell system for titration of virus should encourage quantitative studies of virus survival in flies. In any case, available evidence on the house fly and stable fly indicate that at most they play a very subordinate role.

AFRICAN HORSE SICKNESS

This is an infectious viral disease of equines in which the mortality rate is high. It is characterized by marked edema of the subcutaneous tissues and lungs, haemorrhages in some of the internal organs, and accumulations of serous fluid in the body cavities. The causal agent is a small, viscerotropic virus with nine immunologically distinct types. It is widely distributed in southern and equatorial Africa, but recently it has spread into the Middle East and into Asian countries. According to Huq (1961), the countries stretching from Turkey to India lost between 200,000 and 300,000 horses and mules from this disease in 1960.

Horse sickness is not transmissible by contact. It is seasonal, with most outbreaks occurring in summer and especially during the rainy season. This fact is related largely to the seasonal variations in insect populations since the disease is insect-borne (Khera and Sharma, 1967). In the Belgian Congo, Van Saceghem (1918) thought that *Culicoides* spp. and a *Tabanus* sp. were likely vectors. Du Toit (1944) demonstrated the virus in wild *Culicoides* spp. in South Africa and transmitted the disease experimentally via these gnats. Alexander (1948) believed that the virus was introduced into the Middle East via infected equines or infected vectors carried in aircraft and that, once introduced, it was rapidly disseminated by the swarms of *Culicoides* that infest the valleys.

Other flies have been incriminated in horse sickness. Williams (1913) implicated *Lyperosia minuta* as a likely vector since it was prevalent in a Sudan desert outbreak where no other blood-sucking arthropods could be found. Van Saceghem (1918), without evidence, also mentioned *Lyperosia* sp. and *Stomoxys* sp. as possibly playing a role in transmission. In fact, Schuberg and Kuhn (1912) had previously reported successful mechanical transmission of horse sickness by *S. calcitrans*. We know too little to judge whether and to what extent biting muscids and tabanids transmit African horse sickness in nature. But the evidence strongly incriminates *Culicoides* and indicates that the virus is at home in other nematocera. Ozawa et al. (1966) report persistence of virus for more than a month in *Aedes aegypti* and transmission two to three weeks after an infective meal by this mosquito, *Anopheles stephansi* and *Culex pipiens* (Ozawa and Nakata, 1965; see also Mirchamsy et al., 1970).

BLUETONGUE

Bluetongue is a viral disease of ruminants, occurring principally in sheep. It is characterized by hyperaemia and ulcerative inflammation of the mucous membranes of the mouth, nose, and gastrointestinal tract. The tongue and mucous membranes of the mouth may become swollen and assume a purple or dirty blue color. The causal agent is a virus that is highly resistant to desiccation and antiseptics. The infection is not transmitted by direct contact but by the bites of *Culicoides* gnats, which accounts for its summer-fall incidence. It is endemic to Africa, from which it spread to Portugal, Spain, Cyprus, Turkey, other Near Eastern countries, and the United States (Neitz, 1966).

Commercial vaccines are available, but there is no specific therapy for bluetongue. Its spread in recent years has caused heavy economic losses, and it is considered an important emerging disease.

In South Africa, the involvement of *Culicoides* spp. as natural vectors was shown by Du Toit (1944); he obtained successful transmission of bluetongue by the bites of gnats that fed on infected sheep 10 days earlier. Wild *Culicoides* were also infected. The role of *Culicoides* in Kenya has been evaluated by Walker and Davies (1971). Price and Hardy (1954) in Texas, produced clinical bluetongue in sheep by injecting an emulsion of trapped *Culicoides variipennis*. Foster et al. (1963) obtained five transmissions of bluetongue in the laboratory when *C.*

variipennis were held 10 to 15 days after ingesting the virus and then allowed to bite sheep; no transmissions occurred with flies held less than 10 days. This suggests that multiplication may occur in the insect host. Virus isolation trials were generally positive for *C. variipennis*, but both transmission and isolation trials were negative for the stable fly, *Stomoxys calcitrans*.

Gray and Bannister (1961) have shown that the sheep ked, *Melophagus ovinus*, might harbor this virus and suggested a possible role for this ectoparasite in the epidemiology of the disease. Recently, Luedke et al. (1965) conducted experiments in which keds transmitted infection among sheep.

ARBOVIRUSES

To determine whether insects other than mosquitoes can transmit yellow fever, Hoskins (1933) allowed *S. calcitrans* to feed upon infected and then healthy rhesus monkeys, with the following results. In an interrupted feeding experiment, 11 flies collectively transmitted the virus by biting monkeys 6 hours, but not 16 hours, after the infective meal. By means of injections of fly homogenates into monkeys, Hoskins showed that virus remains alive in the fly for 42 hours, but not for 48 hours.

Maggot parasites of nestling birds were taken from birds' nests in Trinidad and tested as possible virus carriers (Aitken et al., 1958). The maggots belonged to several unidentified species of *Philornis*. They were able to become infected with St. Louis encephalitis virus by rasping the skin and sucking the blood of an infected bird. Furthermore, St. Louis or Ilhéus virus inoculated into maggots was recovered from adult flies 10 or 11 days later, at titers that leave little doubt that multiplication had taken place. Since the adult flies are not bloodsuckers, it is not known whether birds can become infected by feeding upon infected flies. Naturally infected flies have not been found, but even if they do occur, it is possible they are simply a blind alley for the virus.

Epizootics of eastern equine encephalitis (EEE) among pheasants in Connecticut and the presence of the black blow fly, *Phormia regina*, led to a study of the fate of virus in this fly. Bourke (1964) found that virus is destroyed in maggots and adults in less than a day. He mentions that Robert Wallis of Yale failed to transmit EEE to susceptible pheasants when he fed them larvae of *P. regina* taken from carcasses of infected pheasants. Neither was the virus recoverable from flies caught around pheasant pens during the epizootic. It therefore appears unlikely that this fly plays a role in transmission of the virus.

This is an infectious disease of cattle, horses, and swine characterized by fever and the development of vesicles on the tongue and other oral mucosa and sometimes on the skin of the feet. The symptoms are similar to foot and mouth disease. It is communicable to man, especially in enzootic areas and laboratories. The disease has been reported in Mexico, Panama, South America, and in the United States. It causes considerable economic loss due to temporary unproductiveness, loss of condition, reduction in milk production, and complications through secondary bacterial invasions.

The agent is a rod-like virus 60 mμ in diameter and 210 mμ in length, which infects a wide range of laboratory animals including guinea pigs, ferrets, hamsters, chinchillas, mice, and rats; its host range extends to sheep, dogs, bobcats, many rodents, and even cold-blooded animals.

The exact mode of transmission is not known. It does not spread readily by direct contact, and abrasions are necessary for entry of the virus. In the United States, outbreaks of the disease occur seasonally, but, in more tropical parts of Mexico, it occurs at all times of the year.

Radeleff (1949), in reporting an outbreak on a thoroughbred farm near Kerrville, Texas, stated that, if a vector existed, the most likely one would be *Stomoxys calcitrans*, which was present in hordes in and around the stables in question. It appears that the virus can survive only for a few days in barns and must be kept alive by continuous passage from one susceptible animal to another (Hanson, 1952). Hanson felt that the seasonal incidence, the ecological limitations, and the manner of spread of the disease suggest vector transmission. The disease usually occurs in the late summer or early fall and disappears shortly after insect-killing frosts. Disease spread sometimes follows lines of commerce such as roads and often follows natural waterways, sparing adjacent farms away from the water and infecting more distant herds along the waterway. Epizootics generally occur in regions in which cattle are pastured in woodlands where water is nearby rather than in open plains. Hanson pointed out that this provides conditions characteristic of the habitat requirements of a number of dipterans. Stable flies and biting flies with aquatic larvae, e.g. horse flies, black flies, and mosquitoes, are the most suspect.

Ferris et al. (1955) investigated Hanson's suggestions and attempted to transmit the disease with various dipterans by using the embryonated chicken egg as host. Twenty-four hours after inoculation with virus, the various fly and mosquito species were given access to the air sac chamber of the egg and left for one hour to one day until the egg shell membrane was found to be punctured by bites. Then the insects were removed and held for 12 to 24 hours and subsequently transferred to susceptible 8- to 9-day old eggs for 24 hours and similarly transferred every 24 hours until they died.

Their results showed that 14 species (and other unidentified species) of biting Diptera, which included the stable fly, six species of *Tabanus*, three species of *Chrysops*, four species of mosquitoes, and simuliids, were capable of mechanical transfer of the virus for 1-3 days; horn flies were negative. An attempt to transmit the virus to cattle by dipterans did not produce the clinical disease. The disappearance of tabanids before epizootics get under way in Wisconsin and the geographic limitation of epizootics argued against the importance of tabanids, stable flies, and some mosquitoes as vectors, but the abundance of certain mosquitoes and black flies (latter species not given), which could transmit the virus experimentally and which attacked cattle in late summer in regions where outbreaks had occurred, made them the most likely vectors. Field observations by Roberts et al. (1956) verified part of this opinion by suggesting that mosquitoes were more likely vectors than tabanids in Wisconsin. A definitive study would include the isolation of the virus from dipterans in the vicinity of an infected herd during an outbreak of the disease and transmission to healthy animals by a sample of that fly population.

Ferris et al. (1955) found no ovarian transmission or proliferation of the virus in the larval stages of the species studied; however, it is interesting to note that Périès et al. (1966) found that this virus is capable of multiplying in *Drosophila melanogaster*, and Mussgay and Suarez (1962) found that it grew in *Aedes aegypti* following ingestion or intrathoracic injection of the virus. It is not likely that *D. melanogaster* is involved (it does not bite), but Ferris et al. have reported transmission via three other species of *Aedes*. Recently, Chaverri (1970) has found the virus in *Phlebotomus* sp. in association with human cases of the disease. The wide spectrum of potential dipteran hosts invites further virus isolations from wild flies, and laboratory studies of virus propagation in flies and fly tissue culture.

RINDERPEST (Cattle Plague)

This is an acute, febrile, highly contagious, and highly fatal disease, primarily of cattle and buffalo. It is rare in sheep and goats. It is characterized by erosion and necrosis of the mucous membranes, conjunctival infection, profuse diarrhea, and rapid emaciation. Death occurs in 10 to 11 days, and mortality ranges from 15 to 75%. The agent is a highly infective virus, which is present in all tissues and fluids of infected animals and can be isolated in blood and nasal secretions 1 or 2 days before the initial rise in temperature to 104°F; it has the morphologic attributes of the myxoviruses.

Transmission is effected most commonly by direct contact, but infection by ingestion also occurs. The incubation period is 3 to 9 days. The virus is heat-labile and does not persist very long in food or water but remains infectious for weeks in the cold or in frozen animal products.

This disease is enzootic in parts of Asia, eastern Europe, and Africa, where it has been a serious economic scourge for centuries. Other parts of the world, including the Western Hemisphere and western Europe, are virtually free of the disease.

Experiments by Sen (1925-1926) in India tested the possibility of mechanical transmission of rinderpest by the house fly, *Musca domestica*. This study was indicated by the peculiar habit of these flies of sucking up buccal and nasal discharges of infected animals and at the next moment crawling right up the nostrils of the animals and sometimes even regurgitating material streaked with blood. He attempted transmission by introducing infected flies in various ways into susceptible bulls. Positive results were obtained in 3 out of 6 flies inserted subcutaneously up to 31 hours after flies were fed on virulent blood; by intravenous injection of saline suspensions of flies so fed (7 of 19), with a maximum interval of 12 hours; similarly, of flies fed nasal discharge (4 of 8), maximum 3 hours; of flies fed fresh feces (4 of 9), maximum 3 hours. Wild-caught flies were used in all these trials. Most important, all other trials using 15 different noninjection methods with laboratory-bred flies were negative. Thus, all simulations of natural conditions of fly contact gave negative results.

Hornby (1926) in East Africa reported success in mechanically transmitting the disease to a susceptible animal by *Glossina morsitans* under natural conditions in one out of two experiments. Crawford (1933) in Ceylon noted the resting habits of buffaloes in closely packed groups in swampy areas and the frequently numerous biting flies. Under these circumstances, *Stomoxys* and tabanids might convey the infection, spreading it from animal to animal in a herd once it is introduced. Taking his due from Crawford, Bhatia (1935) in India carried out a series of transmission experiments with *Tabanus orientis* and *S. calcitrans* via the interrupted feeding method. Wild-caught starved flies were singly induced to bite the shaved backs of bulls experimentally inoculated with the virus, usually at the height of the temperature reaction, and then transferred to a healthy animal for completion of feeding, usually ½ to 2 hours later. Under these circumstances, *S. calcitrans* was unable to transmit the disease. With *T. orientis*, successful transmission resulted only when a minimum of 36 infected flies was used. The author urged larger scale studies with other biting and bloodsucking flies.

Sen and Salam (1937) attempted further extensive experiments with *S. calcitrans.* One experimental bull received approximately 1081 interrupted stable fly bites, *en masse* and *singly,* up to 29 days after these flies had fed on bulls artificially infected with a virulent strain of the virus. The animal remained normal for 45 days after the last bite, and, when inoculated with 5 cc of virulent rinderpest blood, it died of the disease.

Attempts by various investigators to transmit infection via ticks, lice, and mosquitoes have also been unsuccessful. On the basis of the evidence, the part played by biting arthropods in the natural transmission of rinderpest is probably negligible. Before the door is finally closed on flies, the possibility of conjunctival transmission by nonbiting, eye-frequenting muscids should be investigated.

RABIES

The possibility that maggots may pick up rabies virus while feeding upon a rabid cadaver was considered by Fermi (1912). He found that maggots and pupae (undisclosed species) from a rabid rabbit did not produce rabies when injected subdurally or subcutaneously into susceptible animals. This is worthy of renewed interest since it has recently been discovered that animals can become infected via the oral route (Bell and Moore, 1971).

HOG CHOLERA (Swine Fever)

This acute, highly infectious, septicemic disease is limited to swine, generally characterized by sudden onset with high fever, morbidity, and mortality. It is caused by a filterable virus of about 35 mμ. The virus of swine fever occurs in Europe, America, Australia, and Asia; the virus of African swine fever is immunologically distinct, and animals immunized against either are fully susceptible to the other (Walker, 1930); however, they are related in that they have a common antigenic factor (DeKock et al., 1940).

Transmission is usually via the enteric route; raw garbage containing pork scraps from pigs in enzootic areas is a common source of infection. Outbreaks are precipitated by the introduction of pigs exposed in transit, in sales barns, or in stockyards to an unvaccinated herd. It is probably the most serious disease of swine in North America. It occurs throughout the year, but, in some regions of the United States, it tends to be seasonal in occurrence, with the highest incidence in late summer and fall (Udall, 1954).

Geiger (1937) stated that biting flies and lice are possible but unlikely carriers of this disease. Mohler (1929) noted that the greatest prevalence of hog cholera in Iowa in 1927-1928 corresponded with those periods during which the stable fly, *Stomoxys calcitrans*, was most prevalent; he reported experiments with screened and unscreened pens placed in proximity to hogs affected with the disease that indicated to him that flies or other insects that were excluded by the ordinary screen may be factors in the spread of the disease. However, Montgomery (1921) found that susceptible pigs separated by wire gauze for about one foot from affected ones also failed to develop the disease.

In earlier experiments by Mohler (1919) on the possibility of the spread of this disease by flies, *Musca domestica* and *S. calcitrans* were studied since both were commonly found about hog pens, and both are capable of traveling considerable distances. It was found that individual house flies that fed on the eye secretions or bloody discharges of sick pigs may harbor the virus for some days at least and may convey the disease to healthy pigs by feeding on their eyes or on fresh wounds on their skins. Yet other experiments with nonimmune pigs, which were placed in screened pens into which considerable numbers of infected house flies were liberated, failed to produce any infections, leaving doubt as to whether the house fly is involved with hog cholera under natural conditions. In similar studies infected stable flies conveyed infection either by feeding upon susceptible pigs or by being eaten by them. But it remains to be determined whether this fly is a factor of practical importance in the spread of the disease.

The very effective national hog cholera eradication programs in the United States will probably wipe out this disease soon, leaving this account largely historical for investigators in this country, but hopefully research on the fly question will be continued in other countries.

FOWL POX (Fowl Diphtheria, Avian Molluscum)

Different strains of the fowl pox virus attack a wide range of domestic and wild birds on a worldwide basis. The virus is epitheliotropic, causing wartlike nodules on the skin and necrotic lesions of the upper digestive and respiratory tracts. It is spread by direct contact, or mechanically by pecking and mosquito bites. In 1932, Bos showed that two quite unrelated mosquitoes, *Anopheles maculipennis* and *Theobaldia annulata*, are infective for 70 days; two years later, the infective period was extended to 210 days. This and subsequent isolations of fowl pox virus from wild mosquitoes leave little doubt of their importance as vectors and reservoirs. Bos also experimented with *S. calcitrans* and obtained positive transmission in 7 out of 11 tests, for a maximum of 15 days. The stable fly must therefore be regarded as a possible vector of the disease.

MINK ENTERITIS VIRUS INFECTION

Many bacteria have been incriminated in mink enteritis, a disease that causes losses wherever mink are raised, but, except for virus enteritis, the etiology is largely unknown. The disease often occurs in the warm season of the year on farms where sanitation is poor. Symptoms include a black, tarry diarrhea, anorexia, and lassitude.

Schofield (1949) pinpointed flies and poor sanitation as factors in outbreaks of mink enteritis virus (MEV) on Canadian ranches. Bouillant and Hanson (1965a) presented evidence that MEV excreted in feces from infected mink remains infective in the natural environment for periods extending beyond nine months. Since *M. domestica* is usually abundant in the vicinity of mink pens, the authors investigated its possible role in mink enteritis (Bouillant et al., 1965b).

Their first experiment involved the release of flies into a fly-proof enclosure in which MEV infected and uninfected mink were housed. Three groups were housed within the enclosure. The first group of mink was inoculated with the virus; the second group was not inoculated but was exposed to flies having contact with feces of group 1; group 3 was not inoculated and was kept isolated from flies by gauze. A fourth group was placed outside the enclosure, uninoculated, and served as the virus challenge control. Three thousand uninfected *M. domestica* flies were allowed to emerge from pupae into the enclosure; 3 days later, 3000 more pupae were added. The results showed that 3 out of 4 mink in group 1 became infected. All four mink in group 2 showed varying degrees of anorexia. None of the group 3 or group 4 mink showed signs of MEV until later challenged, though group 3 mink were within the enclosure and exposed to the air. It was concluded that the group 2 mink were infected by flies, which had unlimited contact with infected mink and feces.

In the second experiment flies that had been exposed to MEV infected feces were fed to susceptible mink. Each of two mink became ill after ingesting a pool of 75 flies collected 5-7 and 8-9 days after group 1 developed clinical signs of MEV; pools of 50 and 25 infected flies fed to mink appeared to produce less pronounced signs of infection. Thus it seems that contaminated flies can readily close the MEV cycle between diseased and healthy mink directly by being swallowed or indirectly by contaminating the food.

EQUINE INFECTIOUS ANEMIA (Swamp Fever)

This viral disease of equines is usually associated with intermittent fever, depression, progressive weakness, loss of weight, edema, and anemia. The virus is present in the blood, body secretions, and excretions in all stages of the disease, including the milk, semen, nasal and lacrimal secretions, urine, and feces containing blood. It has been known to persist for 14 years in an infected horse. The virus is fairly resistant to disinfectants, heat, drying, and putrefaction.

The infection can be transmitted mechanically to susceptible animals by contaminated surgical instruments or by insect bites. The disease spreads slowly and usually occurs sporadically. However, conditions become favorable for epizootics during the transportation and congregation of large numbers of horses, some of which are carriers; transmission seems to occur more readily among horses in pasture than among stabled horses. The disease has a seasonal incidence and is most prevalent during late summer and autumn. It is now worldwide and a problem in most countries where there are sizable populations of horses.

Mohler (1908) described the disease in the United States as more prevalent during wet years, usually making its appearance in June and increasing in frequency until October. These characteristics leveled suspicion against flies and mosquitoes. Kinsley (1909), pointing out that the disease has been transmitted from infected to healthy animals by blood inoculation, wrote that flies and mosquitoes appeared to have no relation to the transmission of the disease, since only a single case may occur in a large barn where these insects swarm around diseased and healthy animals. Norton (1911) agreed and felt that poor sanitary conditions had much to do with the development of the infection, as well as over-working of the animals. However, Van Es (1910) recommended that horses with swamp fever be protected from biting insects on the suspicion that they may be agents in transmission.

A report by the Japanese Commission of the Horse Administration Bureau (Tokyo, 1914) described its experiments and conclusions: 1. The infection could be propagated without the aid of insects among horses living together, but this mode of infection was very weak. 2. Infected and healthy horses were kept in separate enclosures during the insect season, allowing free access of insects to the animals. Three trials clearly demonstrated transmission of the disease without contact between the horses and with free access of insects; the infections were to the degree as found in the mixed pasturing. *Stomoxys calcitrans* was abundant in the stables, but experimentally, transmission of the disease by this fly was not so marked as in the mixed pasturing and enclosures. Therefore, it was not thought that stable flies were the true transmitters of the disease. 3. On the other hand, the spread of the disease coincided with the appearance of horse flies in the pastures, and, although it could not be demonstrated experimentally, it was concluded that various tabanids accomplished the significant transmissions.

Experiments in which various kinds of infected mosquitoes were placed in screened cages with five healthy horses all proved negative (Scott, 1915). When a considerable number of flies—house flies, stable flies, tabanids, and other wild flies—were introduced to infected horses in the cage, only house flies and stable flies thrived. Stable flies attacked the horses viciously (house flies, of course, did not attack the horses), and when three healthy horses were introduced into the cages, one of the horses contracted and died of the infection (proved by subinjection) and another developed a chronic case. Scott tentatively concluded that stable flies were responsible for transmission. Confirmatory results were published by Howard (1917) and Lührs (1919). Later, Scott (1920) reported experiments in which infected horses were rotated through two cages, one of which held two healthy horses and was fly-free and the other of which held three healthy horses with stable flies. Two of the three healthy horses became infected, proving that flies and not contamination was the means of transmission. When stable flies confined in small, boxlike screen cages were exposed to and allowed to bite the backs of infected and then healthy horses, at least two of four horses developed the disease. *Tabanus septentrionalis* also transmitted infection to two of three horses by the interrupted feeding method. Stein et al. (1942) showed that the horse fly, *Tabanus sulcifrons*, as well as the stable fly might act as vectors. Scott (1922), in reviewing all his own experiments and those of other investigators, stated that he was convinced that certain biting flies, particularly *Stomoxys* and *Tabanus*, furnish the most important means of natural transmission of this disease (see also Wilhelmi, 1922).

The work of Rohrer and Möhlmann (1950) has reinforced this opinion. They kept susceptible horses in insect-proof stables under very hygienic conditions in close contact with infected horses for six months and these horses remained healthy and proved to be still susceptible after the lapse of the experiment. Lührs (1919) was able to transmit the infection by means of the mosquito *Anopheles maculipennis* collected in stables housing infected horses, and Stein et al. (1943) demonstrated that another mosquito *Psorophora columbiae* might also carry the infection. It is now generally accepted that infectious equine anemia is spread mainly through the agency of various bloodsucking dipterans.

PASTEURELLOSIS
(Fowl Cholera, Rabbit Septicemia, Buffalo Epidemic)

Fowl cholera is an acute or chronic, generalized or local, infectious disease of domestic poultry and wild birds. It often shows a sudden onset with high mortality; symptoms include enteritis. The agent is *Pasteurella multocida*, a small, nonmotile, nonsporogenous, gram-negative, aerobic rod, showing bipolar staining. A wide variety of birds and mammals, including man, are susceptible. In mammals, the disease is generally termed hemorrhagic septicemia.

The organism survives for at least three months in contaminated soil or in a decaying carcass; drying and direct sunlight kill it. Carrier animals contaminate food, air, water, and soil, from which the infection can be picked up.

Scott (1917) isolated *P. multocida* (= *Bacillus cuniculicida*) from *Musca domestica* caught in his laboratory under circumstances that linked them to outbreaks of rabbit septicemia among his rabbits and guinea pigs. The organism isolated from the flies was very pathogenic for rabbits; via intravenous inoculation, it killed them in 36 to 48 hours. Skidmore (1932) conducted a series of experiments on the possibility that house flies, which often fed on fowl cholera infected material, transmit the disease to turkeys, which eat such flies. Fifty flies were caught outdoors, fed for one hour on infected rabbit heart blood, removed and allowed to dry and clean themselves for four hours, and then 30 of these were washed in sterile water; when two rabbits were injected intravenously with 0.5 cc each of this wash water, both died within 16 hours, showing characteristic bipolar organisms in their blood. These same 30 flies were then surface sterilized in a 1:200 solution of corrosive sublimate, rinsed, and homogenized. Two rabbits injected with the fly homogenate died of septicemia within 16 hours. In other experiments, 50 wild-caught flies, fed for 12 hours on infected rabbit blood, were killed 2 hours later and fed to a turkey, which died of fowl cholera 48 hours later; 200 flies were allowed to feed on infected rabbit blood for a half hour and then fed to a group of 10 (including 2 controls) 14-week-old turkeys, three of which died of fowl cholera, with 4, 4, and 23 hours between the time the flies were removed and the time they were fed to them; 175 flies were allowed to feed on infected rabbit blood for 2 hours, removed and allowed to dry themselves for 6 hours, and then they were killed and fed to 3 turkeys, which all died of the disease. Thus, house flies that have fed on infected fowl cholera blood can transmit the disease at least one day later if they are swallowed by turkeys.

In cattle, buffalo, and sheep, a related type of infection is variously termed buffalo epidemic, hemorrhagic septicemia, or shipping fever. It is characterized by pneumonia and septicemia. The etiology may be viral, complicated with certain bacterial agents, namely, *Pasteurella* spp. In Java, Nieschulz and Kraneveld (1929) experimented on the possible role of biting dipterans in hemorrhagic septicemia of buffaloes. Rabbits were used as the experimental animals, and the following species were tested: *Tabanus rubidus, T. striatus, Chrysops flaviventris, Stomoxys calcitrans, Lyperosia exigua, Musca inferior, Anopheles fuliginosus, Aedes (Stegomyia) fasciatus,* and *Armigeres obturans.* The insects were allowed to bite the infected animals individually, and microscopic and cultural examinations were made of the blood of the rabbits infected. The results indicated that, by far, the tabanids were the best transmitters. Transmissions were still successful after 4 and 6 days with just a few specimens of *T. striatus* and *T. rubidus,* respectively, and it appears that they can be infective for even longer than 6 days. Nor did the tabanids lose the infection in the first sucking act; a single fly could infect three experimental animals one after the other. *C. dispar* was only a little less effective than the *Tabanus* spp. *Stomoxys, Lyperosia,* and *M. inferior* appeared similarly to be good transmitters. Transmissions with *Stomoxys* and *M. inferior* were successful after 24 hours and with *Lyperosia,* after 6 hours. For all three muscid species, the transmission expectations were 1:8 after from 3 to 24 hours. For the mosquito species, the expectations were about 1:20 after ½ to 6 hours and 1:30 after 1 to 2 days. Transmission to several experimental animals one after the other was relatively easy with *T. rubidus, Stomoxys,* and *Armigeres.* Thus, from these results it appears that under suitable circumstances various species of flies can be of significance in the spread of buffalo epidemic.

PLAGUE

Plague has been known from ancient times and has had great impact on the course of history. It is a severe and often fatal disease caused by the bacterium *Pasteurella pestis.* The great epidemics of human plague have been preceded by epizootics in the rat. As the rat dies, the infected and hungry fleas are forced to seek man for food; they usually bite on the lower extremities, and the bacilli spread rapidly through the lymphatics and enlarge the lymph nodes in the groin, causing buboes.

The rodent-flea sylvatic and urban cycles of plague are well understood as the primary pathways. In addition, bed bugs, human lice, mosquitoes, ants, beetles, cockroaches, *Triatoma*, ticks and lice and mites of rodents and other animals have been proved or suspected as plague vectors. In most cases, the plague bacillus can be found in or on the arthropods in question and, by biting or contamination, they can transmit the infection to laboratory animals.

We find this situation with flies.

Since the 14th century, observers have associated an increase in the number of flies or fly deaths with plague. Yersin (1894) remarked on the unusual number of dead flies lying about his laboratory when he made autopsies on plague animals. He was the first to show by inoculation of guinea pigs that such flies contained virulent plague bacilli, and he concluded that they are capable of spreading the infection. Matignon (1898) was also struck by the large number of fly deaths during plague times in Mongolia.

These observations were confirmed by several investigators. Nuttall (1897) fed *Musca domestica* on the organs of animals dead of plague and found that they survived 8 days when kept cool (12° to 14°C) and retained virulent bacilli for more than two days under these conditions. At higher temperatures, however, infected flies died more rapidly than controls.

Hunter (1905) examined a large number of house flies during the plague epidemics in Hong Kong in 1903-1904; they were caught in the Public Mortuary and in the Infectious Disease Hospital when both institutions housed many plague victims. Though he observed no unusual fly mortality, he succeeded in isolating *P. pestis* from emulsions of crushed flies from the hospital and from the mortuary even before post-mortems were performed. Small pieces of sugar, known to be free of the plague bacillus, introduced into sterilized test tubes containing plague-infected flies, were found to contain plague bacilli when examined. At the same time, Herzog's (1905) failure to transmit plague to guinea pigs by means of infected house flies that fed upon the syrup-smeared backs of these animals is not surprising. By what means was the organism expected to gain entry into the host?

Wayson (1914) demonstrated that mechanical transmission by the bite of *Stomoxys calcitrans* was possible. He allowed stable flies individually in glass vials to bite an infected guinea pig and reported that eight bites by one or more flies (not over two were used at the same time) effectively transmitted the disease to a second healthy animal (which died in 5 to 9 days) if the application was made within an hour after the first feeding; washings of the flies in normal saline and of flies slightly crushed and injected subcutaneously produced similar results.

The question of transmetamorphic infection in flies that normally feed as maggots in cadavers of plague-infected rats has also been considered. Gosio (1925) showed that *M. domestica, Calliphora vomitoria,* and *Cochliomyia macellaria,* emerging from infected larvae, were still infected with plague bacilli, though with fewer numbers. He noted that such flies were subject to septicemia and death within 24 hours after emergence. Russo (1930) reported that the organisms were also carried over in *Sarcophaga carnaria* and *Lucilia caesar* and the bacilli were recovered in the excreta of these flies.

The most recent study fails to confirm the Italian work. Lang (1940) placed larvae of the "green blow fly" on a cadaver of a Siberian marmot dead of experimental plague septicemia. In smears made from these larvae recovered after one and two days of feeding, the specific bipolar rods were usually observed, but in smears made from larvae close to pupation, from pupae, and from young adults such rods were not observed.

Pollitzer and Meyer (1961), reviewing various aspects of plague, state that it is not at all likely that *M. domestica* takes any part in its propagation. From available evidence, it seems reasonable to suppose that the various stages of nonbiting flies probably do harbor the pathogen under natural plague conditions. And this sojourn in flies leaves the virulence of the organism and the development of the fly essentially unaltered, although it seems to shorten the adult's life. Nevertheless, plague in these flies appears to be a dead end for the organism. On the other hand, incidental mechanical transmission by biting flies has not been ruled out.

TULAREMIA

This is a disease of wild rodents and especially of rabbits, hares, and birds. Nidi of the disease exist in populations of ground squirrels, wild rats, meadow mice, opossums, cats, dogs, foxes, sheep, skunks, and snakes. Man is an accidental host; in acute cases, the disease may assume the form of a septicemia, resulting in death between the first and second weeks. In animals, it is characterized by small foci of infection in the viscera; there may be glandular enlargement, acute necrosis of the spleen, liver, lymph nodes, bone marrow, and lungs and frequently a rapidly fatal septicemia. The disease occurs in the United States, Canada, northern Europe, Russia, Japan, and many other parts of the world.

The causative agent is *Francisella* (= *Pasteurella*) *tularensis*, a nonmotile, nonspore-forming, gram-negative, aerobic rod. The disease is transmitted from animal to animal by the bites of blood-sucking arthropods such as flies, fleas, lice, and ticks, predation and cannibalism, and by contaminated drinking water. Rabbits are among the chief reservoirs of infection in this country, and many human cases occur from the bites of flies, ticks, etc., that had direct contact with the diseased animal. Hunters and other animal handlers are apt to acquire the infection through contact with rabbits or hares; other avenues are consumption of uncooked, infected tissues or contaminated water and perhaps through the respiratory passages.

Steinhaus (1946) gives an extensive list of the insect and tick species that may be concerned in this disease. Regarding dipterans specifically, Wayson (1914) first called attention to the possibility of *Stomoxys calcitrans* and *Musca domestica* as mechanical vectors. He transmitted a fatal tularemia from sick to healthy guinea pigs with one or two stable flies biting at least eight times within an hour. Washings of the flies, and of flies slightly crushed, produced similar results when injected under the skin of test animals. He also allowed house flies to crawl over and feed on the infected viscera of an animal dead of the disease and immediately transferred them to conjunctivae, causing a severe purulent conjunctivitis after 48 hours and death in 5-9 days with typical visceral lesions; contaminated house fly feces produced similar results. Transmission by biting occurred, apparently, only from animals with advanced bacteremia, close to death. The organism did not persist, and flies were no longer infective 24 hours after feeding.

Francis (1919) described the disease as occurring in rural populations in summer months coincident with the prevalence of the deer fly, *Chrysops discalis*. Francis and Wayne (1921) reported experimental transmission of tularemia by this fly, which is found especially in Utah and vicinity. It is probably a mechanical vector since it loses its ability to transmit after about five days. Jellison (1950) pointed out that the geographical distribution of this fly roughly corresponds to that of the disease in the western United States. *C. discalis* is undoubtedly the most important fly involved in transmission, but certain ticks are of major importance in maintaining the disease in animals and transmitting it to man. Francis (1937) reported that of 1824 human infections in the United States, 68 were caused by fly bites (*C. discalis*) and 115 by ticks. Parker (1933) lists other species of tabanids and a black fly (*Similium decorum katmai*) that mechanically transmitted the disease to guinea pigs. Also, Philip et al. (1932) showed experimentally that several mosquito species may be significant in transmission to humans, by biting, being crushed into the skin, and by depositing excrement on the skin. Olin (1938) believes that the infection can be transmitted to man by mosquitoes in Sweden.

Several Russian authors provide additional perspectives on the transmission of tularemia in other ecosystems. Somov et al. (1937) reported that laboratory-bred *S. calcitrans* transmitted the infection to rabbits and guinea pigs 15 minutes to 53 hours after an infective meal but not after 68-96 hours. Infected animals developed papules on the bite sites that turned into abscesses. The pathogens remained viable in the fly gut for 52 hours after feeding on sick rats. Fly feces were infective 1-4 hours after feeding, but not after 20 hours. *S. calcitrans* was considered a likely transmitter in summer epizootics, but this did not exclude other insects such as mosquitoes and horse flies.

Olsufiev (1940) confirmed these results by transmitting the infection from bacteremic water rats to healthy guinea pigs, via *S. calcitrans*. Transmission occurred between two minutes and 24 hours after infection of the flies. The organisms were found in the flies for 8½ days and in the feces for up to 5 days. Despite its evident potential, the synanthropy of the stable fly would tend to minimize extensive contact with wild rodents and specifically with water rats, which makes *S. calcitrans* an unlikely transmitter of tularemia compared with mosquitoes and horse flies.

Demonstration of the causal organism in wild *Chrysops relictus* and *Haematopota pluvialis*, and the identification of water rat and human blood albumin (precipitin test) in the gut of *C. relictus* caught in endemic foci, provide clearcut evidence for the existence of this cycle in certain regions of the U.S.S.R. (Romanova, 1947).

Bozhenko and Kniavskiĭ (1948) confirmed previous work with *S. calcitrans*, adding the observation that infectivity can be extended by keeping the experimental flies at lower temperatures. At 14° to 15°C and below, the bacteria persisted in the internal organs of flies for up to 31 days and were excreted in the feces for up to eight days. The bite of these flies was infective for 5 days. The organism was not transferred to the progeny of the infected fly as emulsions of eggs and larvae did not cause symptoms in test animals. Interestingly, a group of 9 stable flies was periodically fed on the blood of one of the authors (Bozhenko) while he was ill with tularemia, but the organism could not be demonstrated in the flies.

Available evidence suggests that we assign to the stable fly and house fly—the only muscids studied—a relatively minor place among the many arthropods known to be involved in the transmission of tularemia.

BRUCELLOSIS

This disease primarily affects goats, cows, swine, and sheep. Humans may contract the infection (undulant or Malta fever) either by direct contact with these animals or by the consumption of milk and milk products. It is caused by bacteria of the *Brucella* group, and, in cows, it is characterized by inflammatory lesions of the endometrium, the fetal membranes, and of certain parts of the fetus; abortion occurs in otherwise healthy cows. To a lesser extent there may be infection in the accessory sex glands of the male and infertility in both sexes. Abortion in human cases is much rarer than in cows. Bacteriologically, it is characterized by the presence in the blood and the organs, especially the spleen (in humans), of *Br. melitensis* or one of its varieties, *Br. abortus* (contagious abortion in cattle) or *Br. suis*. Except for abortion, the infection in animals usually produces few symptoms and almost never causes death. The infection occurs worldwide and is prevalent in the U.S.S.R. and countries of the Mediterranean and Near East.

Studies of the possible involvement of flies in this disease began in the early 1900s, and the most recent reports on this subject appeared in the 1960s. Kennedy (1906), in his studies of undulant fever in Malta, noted the ubiquitous *Stomoxys calcitrans* in this area viciously attacking both man and animals and collected these flies from infected goat quarters. He fed them on healthy monkeys (they bit well), but no infection occurred. Also, flies were fed on infected monkeys and then on healthy monkeys with no transfer of the disease and a negative serology. *Br. melitensis* was not recovered in cultures of the flies' organs. Eyre et al. (1907) allowed *S. calcitrans* to bite on the denuded skin of infected guinea pigs. At varying intervals several flies were killed and the gut contents plated out. It was found that the excrement contained numerous living *Br. melitensis* for many hours after feeding on the infected animal and that, in some cases, the bacteria persisted up to five days in the alimentary tract. In transmission experiments, the monkeys were replaced by a highly susceptible goat, but again infected flies failed to transmit in seven attempts.

Suspicion of flies and other insects persisted because brucellosis appeared in noninfected herds kept under ideal conditions. In the United States, Ruhland and Huddleson (1941) investigated *Musca domestica*, *Muscina stabulans*, *S. calcitrans*, *Calliphora* sp., and *Lucilia* sp. because they were often found in and around dairy barns feeding on the excretions of animals. The flies were starved and then fed freely for two hours on the growth of a virulent strain of *Br. abortus* in solid medium. At intervals, a small droplet was squeezed from the anus of a few flies and streaked over the surface of a tryptose-agar plate containing crystal violet. Their results agreed with those of Eyre et al., who had used *Br. melitensis*. House flies, blow flies, and stable flies excreted the bacteria for at least four days. Wollman (1927) had previously reported that

Br. abortus was eliminated from axenic house flies in a few hours and persisted in *L. caesar* at least 48 hours.

In Germany, Wellmann (1950) reported that, experimentally, *Br. abortus* types *bovis* and *suis*, and *Br. melitensis* were transmitted to guinea pigs by the bites of *S. calcitrans* and various species of tabanids. He used contaminated milk and naturally infected fecal material as sources of infection in the insects. The experimental animals were regularly infected by a single fly several hours after the fly's infective meal.

In the Soviet Union, Galuzo and Rementsova (1956) obtained artificial brucellosis infection in *M. domestica* under laboratory conditions. Teskey (1960) stated that there is evidence to suspect that *Musca autumnalis* might be involved in the transmission of infectious abortion in cattle.

Wellmann (1959), speaking of *S. calcitrans* as a transmitter of zoonoses, states that these insects are not the principal mode of infection. But, he feels, since biting flies are often present in extraordinary numbers in the warm season on animals in pasture and in barns, the prospects for disease transmission are increased by the summation of their efforts, and, further, these flies may also suck on material containing a much greater concentration of infective agent than could be present in blood.

A number of argasids and ixodids, *Culex pipiens*, and other arthropods have been found naturally infected with *Brucella* organisms, and transovarian transfer has been demonstrated in ticks. In the U.S.S.R., at least 15 species of rodents, 5 species of birds, and 5 species of reptiles are known to be susceptible, and many of these retain the organism for longer than a year. Considering the spectrum of vertebrate reservoirs and potential arthropod vectors, we can assume that the cycles of *Brucella* among wild animals are extremely diversified. The involvement of muscoid flies in sylvatic and domestic animal cycles is unknown. Despite their capability, flies probably have a subordinate role in the domestic animal cycle wherein direct contact with sick animals and ingestion of contaminated milk and milk products are the major pathways for the spread of *Brucella* organisms.

INFECTIOUS KERATOCONJUNCTIVITIS
(Pink-eye, Infectious Ophthalmia)

This is an infectious disease of domestic ruminants characterized by inflammation of the superficial layers of the eye, photophobia, lacrimation, and conjunctivitis. It is common in cattle and sheep, but may also affect goats, pigs, and even fowls.

Various agents have been incriminated as causal agents of the disease in ruminants. In cattle, *Moraxella bovis*, a short, thick, gram-negative diplobacillus, infectious bovine rhinotracheitis virus and other viral agents are considered capable of causing pink-eye, and it appears probable that all may cause clinical disease singly as well as concurrently. According to Henning (1956), symptoms are associated with the presence of rickettsia-like bodies in the conjunctival epithelium, and ocular and nasal discharges. He considers that bovine infectious keratitis is caused by a large virus or *Rickettsia*. Other workers in a number of different countries have made a strong case for the etiological significance of the bodies described as *Rickettsia conjunctivae*.

The disease occurs suddenly in the initially affected animals and tends to spread rapidly. Dry, dusty environmental conditions, bright sunlight, feeding in tall grass, or the presence of large numbers of flies tend to propagate the disease. More younger animals than older ones are affected, but all ages may be involved. The clinical course varies from a few days to several weeks.

Natural transmission generally takes place when susceptible animals are kept in close contact with others that have the clinical disease or are carriers. A number of observers believe that vectors play a part in transmission because the infection can spread from infected to healthy animals separated by wire netting and because of the seasonal incidence of the disease in summer and autumn, when flies are plentiful.

Cheng (1967) studied the frequency of pink-eye incidence in cattle in relation to face fly abundance by taking a fly census in different parts of Pennsylvania, where there were outbreaks of the disease. Analysis of the results showed that cases of pink-eye consistently increased in pastures where cows were more heavily burdened with face flies.

Mitscherlich (1943) reported success in transmitting ophthalmia (*R. conjunctivae*) from the eyelids of infected to healthy cattle by means of *Musca domestica* and *Stomoxys calcitrans*. The infection may persist in the house fly for at least 24 hours. It is not surprising that he found no transfer from female flies to their offspring.

In the United States, Jones and Little (1923) described the symptomology of the disease in dairy cows and found a characteristic diplobacillus in smears of exudates. The disease did not necessarily attack neighboring cows; the cases occurred irregularly throughout the barn and finally spread to other barns. They thought it possible to account for these facts by assuming transmission by flies, which were numerous during the outbreak and were frequently observed feeding on the exudate. The disease subsided after treatment and the onset of cold weather, but 5 of 10 cows re-examined after 3 to 4 months were still carrying diplobacilli and a mild infection in one cow occurred in late January. The next year, Jones and Little (1924) attempted experiments with house flies caught feeding upon the exudate about the eyes of spontaneous cases. Six house flies caught feeding on the eyes of spontaneous cases were caged over the eye of a cow within a half-hour; the flies did not feed but some came in contact with the eyeball. The eye remained normal, and diplobacilli were not found in the lacrymal fluid. In the next series of experiments, flies were captured in laboratory rooms, starved, and then fed on exudate obtained on sterile cotton swabs from spontaneous cases and placed in a little sterile bouillon. Many flies were caged over the eye of a cow and came into contact with the eye and fluids and a few flies were even stunned or killed and placed directly on the nictitating membrane, but again the outcome was negative. Other experiments demonstrated that the organism cannot remain viable in the gut of flies for an interval as short as five minutes and suggested that organisms may be carried on the external surfaces of flies for only relatively brief periods. Although their experiments failed to show that the disease could be transmitted by flies, Jones and Little thought it possible that their methods were at fault.

Moutia (1929) concluded from his observations that *Stomoxys nigra* was the vector of ophthalmia of bovines in Mauritius. The infection occurred yearly from December to March, when this fly was in greatest abundance. These flies have a strong predilection for the eyelids, and disease symptoms were seen to follow fly bites.

Steve and Lilly (1965) suspected that *Musca autumnalis* contributed to the increase of bovine infectious keratitis (pink-eye) in cattle in the United States. Their more significant findings were that *Moraxella bovis* may remain viable in laboratory environs up to at least 3 days (instead of the 3 hours reported by Jones and Little in 1924), that the bacterium was readily recovered from the exudate from infected eyes, that it was readily recovered from wings and legs of face flies up to three days after exposure to laboratory cultures, that it is rapidly destroyed in the digestive tract of this fly, and that *M. bovis* was recovered from laboratory-reared flies exposed to the lacrymal exudates on infected cattle in the field. They felt that the face fly may play a role as a mechanical carrier in the dissemination of pink-eye, and that further intensive investigation is warranted.

Brown (1965) conducted preliminary investigations on the transmission of *M. bovis* by the face fly. Six thousand newly emerged face flies were exposed to 36-hour cultures of *M. bovis* on blood agar for 4 hours and then confined for 11 days within a calf pen containing 4 animals. Attempted cultures from the eyes previous to fly exposure dem-

onstrated no growth of *M. bovis*. All cultures made from the eyes following exposure to the contaminated flies were positive for *M. bovis*; cultures made from face flies on the second day following release were
positive for *M. bovis*. The symptomology and lesions produced in the
animals were suggestive of mild to moderate pink-eye. Thus, face flies,
under the conditions of the experiments, were able to transmit *M. bovis*
from artificial culture media to the eyes of susceptible cattle. The pathological lesions obtained in eyes of calves exposed to noncontaminated
flies indicated that the face fly may also cause traumatic inflammation,
thus predisposing the tissue to invasion by infectious agents. It must be
remembered that pink-eye was prevalent in the United States many years
before the face fly appeared and that outbreaks do occur in winter when
flies are absent.

BOVINE MASTITIS

This is a disease complex that refers in this account to an invasion
of the bovine udder by pathogenic organisms. Sheep and goats are also
attacked. Among the etiologic agents associated with this disease are
coliforms, yeasts, PPLA, *Pseudomonas*, and *Clostridium perfringens*.
Those bacteria in which we are especially interested as causative agents
involved with flies are: *Streptococcus agalactiae*, *Corynebacterium pyogenes* (mastitis), and *Staphylococcus aureus* (= *Micrococcus pyogenes*).

Mastitis is of greatest economic importance in dairy cows; it results
from an interplay between infectious agents and poor hygienic practices.
Most cases of chronic mastitis are due to streptococci or staphylococci
and are characterized by repeated mild mammary swelling, with the milk
containing clots or flakes. *S. agalactiae* requires the mammary gland for
its perpetuation in nature. It enters the gland through the teat opening
and resides in the milk and on the surface of the milk channels. It does
not penetrate the tissue, and its action on the parenchyma is through an
irritant that forms in the milk. Streptococcal mastitis spreads from cow
to cow during milking, and calves fed on milk containing the pathogen
may transmit it to the immature glands of their young penmates. *S.
aureus* is common in the milk of larger herds of dairy cows and may be
found in more than 50% of the milk of the cows in a herd. However,
penetration of the organism into the tissues appears necessary before a
true infection can produce mastitis; milking practices causing tissue
stresses may enhance tissue penetration. *C. pyogenes* produces a characteristic mastitis in dry cows and is occasionally observed in mastitis of
the lactating udder, but may be a secondary invader. Pyogenes mastitis
may occur in epizootic form among dry cows kept in small enclosures

during a long wet period. At worst, the various forms of this infection can cause an intense inflammatory reaction that produces a deterioration of the functional structures and stimulates the proliferation of the connective tissues or results in varying degrees of palpable fibrosis or atrophy.

Solution of mastitis problems involves critical diagnosis of the specific microbial agent involved, the correction of faulty hygienic practices, and judicious use of intramammary therapy. Indicated managerial practices include milking mastitis cows last, disinfecting hands and teat cups regularly and thoroughly, properly functioning milking machines, complete evacuation of glands, and avoiding teat injuries. Well-established cases of *C. pyogenes* infection have been satisfactorily treated by surgical removal of the teat to establish drainage; animals with multiple udder abscesses should be slaughtered. Antibiotics against the specific agent(s) injected into the teat canal have been successful in treatment; animals not responding to treatment should be sold.

Those species of flies closely associated with cows and that may come into contact with the lactating udder are suspected of transmitting mastitis. In Schleswig-Holstein, Pfeiler et al. (1927) stated that flies appeared to play a role as carriers because they were present in abundance in the hot summer when the disease became especially severe. In addition, they were attracted by the foul secretions of the udder, and the bacteria were found present in drops of liquid around the teat. However, the kinds of flies involved were not mentioned.

Karmann (1928) in Ostfriesland conducted five experiments with "*Bact. pyogenes*" to test transmission by flies. He easily collected flies from severely infected animals by slipping a glass tube over a teat on which flies were sucking. He then taped the tubes containing "infected" flies over the teats of five cows. None of the animals became infected, but, admittedly, the number of trials was small. Again, the species of flies were not determined, and no bacteriological examinations of flies were made. Karmann concluded that it was unlikely that flies play any important role in transmission for the additional reason that cows energetically defend their udders against flies.

In reviewing the literature on bovine mastitis through 1935, Munch-Peterson (1938) stated that once this disease got a foothold in a herd, the infection could spread in several ways: by the milkers' hands; milking machines; infected bedding; or by flies. Sanders (1940a) pursued these speculations in light of the prevalence of mastitis in several dairy herds in Florida. He noted that house flies were abundant on the floors of barns feeding on spilled milk, that they crawled over the surface of the teats of cattle in the milking line to persistently feed at the teat opening, and they crawled or flew from one teat opening to another on the same and on different cows in the corrals and milking sheds.

For his transmission experiments, Sanders confined and isolated infection-free cows in a screened, sanitary building. House flies were captured in contaminated premises, caged, and fed heavily contaminated milk for 3-7 days. The flies were then given no food for a short time and were exposed to the teat opening by means of glass mantles. The teat was pressed to bring out small droplets of lactic secretion, on which the flies fed readily. Multiple exposures were made in this manner on each cow. Flies were also released in healthy cow cages and given infected milk to feed upon. Mastitis developed in each of several experimental animals. Sanders felt that this showed conclusively that *M. domestica* is a natural vector of the disease. He did not report the species of bacteria involved.

Sanders (1940b) also investigated the possible role of *Hippelates* gnats (species not given). These flies were observed in abundance on the lower abdominal region in the vicinity of udders and feeding on lacrymal fluid, sebaceous material, wound secretion, milk droplets spilled on the legs and feet of cows, and at the tips of the teats of cows known to be carriers of mastitis organisms. He, again, did not diagnose the causative organism but did make stained smear preparations of *Hippelates* from infected cows, which showed the presence of organisms resembling those observed in cases of the disease, and found that milk incubated with these flies and filtered to remove the insects, produced purulent, bloody, ascending mastitis when injected into the milk cistern of healthy cows through the teat opening; other cases were produced by gnats brought into contact with the opening via rubber sacks containing lactic secretion. He repeated the techniques used in transmission studies with the house fly, and multiple exposures again produced mastitis in healthy animals. He concluded that *Hippelates* spp. also serve as vectors of bovine mastitis.

Ewing (1942) went much further in his investigation of *M. domestica* and mastitis. In one experiment, he placed house fly pupae in a screen box, and the emerging adults were fed skim milk containing *Streptococcus agalactiae* for 4 days. Bacteriological examination showed that 6% of the flies examined carried the bacteria externally for 7 days; and some were internally contaminated for at least 10 days, but all had eliminated the pathogen by the fourteenth day. Schumann (1961) showed that viable *S. agalactiae* were present throughout the house fly gut and could be recovered for 5 days after the flies fed on milk from infected cows. The organism was also found in the fly's feces. A second experiment by Ewing showed how readily flies came in contact with the pathogen by feeding at the external surface of the teat sphincter of mastitic cows. However, mastitis bacteria were not present on the external surface of the teat following the milking and stripping of cows by machine. In a third experiment, *M. domestica* were trapped in a barn containing cows of which 46% had mastitis. None of the flies contained *S. agalactiae*, and only a few carried pathogenic staphylococci externally. Ewing concluded that it was very unlikely that flies would have been important agents under naturally infective conditions in the dairy barn.

In his reports on "Summer Mastitis" (S.M.) in Jutland, Denmark, Bahr (1952-1953) criticized Sander's (1940a, b) work, especially the lack of quantitative analyses of the infective potential of flies and the lack of a real bacteriological study of the problem. He discussed his studies of 15 cases of S.M. in 14 rural districts and the insects found on the tips of the teats of the infected mammary glands. All but one (*Staphylococcus*) of these infections were verified as *C. pyogenes* mastitis. In all the cases female *Hydrotaea irritans* was collected from the tip of the teats. In flies from six of the districts, gram-positive corynebacteria were found in the lumen of the foregut and crop. In only 3 cases, however, was subculturing possible, and in one case *C. pyogenes* was confirmed. Bahr postulated the existence of other types of *C. pyogenes*, each perhaps differing in its capacity for exotoxin production. The observation that this fly's maximum population in the field (it is an exophilic fly) coincided very closely with the appearance of S.M. in Jutland lends further support to the theory that *H. irritans* is important as a conveyor of *C. pyogenes*.

In a later report, Bahr (1955) presented further data on this issue. From 1950 to 1954 he obtained a total of 22 fly samples from 22 Danish and Swedish herds (dry animals) suffering from S.M.; all samples contained *H. irritans*. He pointed out the importance of taking the fly samples from the teat tips of the affected udder and not from the udder itself or from other parts of the animal. Transmission experiments were not possible because this fly could not be successfully bred in captivity. However, a series of practical preventive measures were carried out. An ointment was prepared to kill the flies and protect the cows from infection by that route. Three or four times during the summer at intervals of three to four weeks, it was rubbed on all four teats and, in particular, the teat tips of all the dry healthy animals among the herds under study. In an exploratory experiment, 8 of the 24 untreated animals were attacked by S.M. and only 2 of the 48 treated animals were attacked. In one experiment in June-July, 1954, 18-20 cases occurred in untreated animals in 10 herds, and in the same 10 herds and 4 other herds, only one of 96 treated animals was attacked. In another experiment with 22 herds in 1954, among a total of 375 treated animals only 9 to 10 in 7 herds were attacked and the remaining 15 herds were free of S.M.; in the previous three years, the disease had occurred in 15% of untreated animals annually. Even though the last two studies did not have controls, Bahr felt that they offered strong indirect proof of the significance of *Hydrotaea irritans* as a carrier of summer mastitis infection.

Skovgaard (1968) investigated the incidence of *C. pyogenes* in cattle and found it constantly present and often in high numbers in those places (nasal cavity, conjunctiva, and vagina) frequented by flies. Thus he pointed out the origin of the S.M. bacteria in the natural orifices of the cow and the possibilities for transmission by flies to the teats. Elliott (1965) isolated *Staphylococcus aureus* from the vagina, nose, and rectum of dairy cows in a mastitic herd in Wisconsin. The organism was widely distributed in the barn environment and among the dairymen and their children, and dogs, cats, and flies. Of 241 house flies sampled during a period of 14 months, 19.9% harbored coagulase positive *S. aureus* in their digestive tract, including flies taken in winter. Coagulase-negative staphylococci, possibly *S. saprophyticus*, were even more widespread, also occurring in *Phaenicia sericata, Fannia canicularis* and *Stomoxys calcitrans*, and in fly specks throughout the barn. Evidence was not developed to show whether the pathogenic agent in cows and in flies was indeed the same.

There are numerous symbovine flies whose habits qualify them as possible vectors of mastitis. These include *M. autumnalis, Muscina stabulans, Morellia hortorum* and *M. simplex, Hydrotaea meteorica*, and the biting species, *S. calcitrans* and *Haematobia irritans*. As far as we have been able to ascertain, nothing has been published concerning the relationship of these flies to mastitis.

SWINE ERYSIPELAS

This is an infectious disease, primarily of swine, that shows a wide variety of clinical forms. The acute form is generally characterized by septicemia, fever, the development of diffuse, red patches on the skin, and rapid death. The urticarial form is shown by the appearance of large, generally diamond-shaped haemorrhagic patches on the skin. Quite often swine erysipelas does not cause severe death losses, but the greatest economic losses come from the mild chronic forms, which cause general unthriftiness.

The disease is caused by *Erysipelothrix insidiosa* ($=$ *rhusiopathiae*), which is a gram-positive, nonsporing, bacillus. It is readily isolated from tissues of sick pigs and also from the tonsils of apparently normal ones. The organism is widely distributed in nature, occurring on the surface slime of both fresh and salt water fish. It has also been found in polyarthritis in sheep, joint ill in lambs, and infections in horses, cattle, turkeys, and peacocks. Man becomes infected by contact with fish, infected animals, or animal products. In swine erysipelas, the infection may be spread by direct or indirect contact between infected and healthy swine and by ingestion or contact with the abraded skin of material contaminated with infected feces and urine. The disease is fairly common in the swine-growing areas of the United States, Europe, and elsewhere. It vies with hog cholera in causing the greatest economic losses in swine around the world.

Wellmann (1948, 1949a, 1949b, 1950) reported a long series of transmission experiments with *Stomoxys calcitrans*. Using swine blood inoculated and incubated with swine erysipelas bacilli and a mixture of swine blood and meat broth as an infective meal for the flies, he then allowed them to feed on swine. Three hogs developed the disease via the bites, and the places infected produced a large number of distinct pustules (1948). Subsequently (1949a), his experiments showed that the disease was not transmitted by flies that had sucked pustules of artificially infected animals. However, studies with mice demonstrated that the stable fly can carry erysipelas from infected to healthy mice by their bites. Greater transmission occurred when a very large number of flies was used; it was diminished when less than eight were used and was rather rare with only one fly. Infection was also rare when fly bites on healthy animals did not immediately follow bites on the sick ones. However, healthy mice were infected by flies that had bitten a sick mouse 24 hours earlier. In the fly gut, the bacteria remained alive for at least 48 hours and in their feces for less than 24 hours, but feeding mice with infected flies full of blood rarely produced an illness. In further studies (1949b, 1950), Wellman experimentally transmitted the disease by stable fly bites from mouse to pigeon and pigeon to pigeon. The infections were verified by an examination of bird organs and by cultures. The flies were confined in a plucked breast area and fed thereon. In the mouse-to-pigeon experiments, all 5 birds were positive, including one after a 15-minute interval. In pigeon-to-pigeon experiments using 9 birds with immediate transfer of flies, 2 were positive and 7 were negative. Also, 5 birds were used at intervals of 5 to 15 minutes between transfer, and 2 positive cases resulted. Wellman obtained successful transmissions with a variety of blood-sucking Diptera, *Haematopota pluvialis*, *Tabanus bromius*, *Heptatoma pellucens*, *Aedes* spp., *Theo-*

baldia spp., and *Anopheles* spp. From the results, it appeared to him that the *C. pluvialis, T. bromius,* and *H. pellucens* were even more efficient than *S. calcitrans.* In some cases, a single bite sufficed to transmit infection. Wellman (1954) experimentally demonstrated that *Musca domestica* is also able to transmit swine erysipelas. In a series of tests, house flies, which were allowed to fly into a stable after having fed upon an erysipelas culture in another room, transmitted clinical illness to at least 3 out of 14 pigs and subclinical infections to four others.

Tolstiak (1956) in the Soviet Union undertook experimental infection of white mice and swine via the bites of stable flies that had fed on the blood of mice and swine suffering from the disease, and also on material from mice dead of the infection. Flies were caught 2 to 5 hours in advance and placed in small flasks closed with gauze, which were placed against the skin of a sick animal and then similarly against that of a healthy one; most of the flies actively bit. The sick animals used to feed the flies and those subsequently infected by fly bites were microscopically and bacteriologically examined. In three experiments involving 27 mice, he was able to transmit erysipelas infection from sick to healthy mice by *S. calcitrans* bites up to 68 hours after fly infection, and also by the bites of flies fed on corpse material from infected dead mice. To study fly bite transmission between swine, 5 healthy piglets (5-6 months of age) were used. The unshaved skin of piglet number 43 was exposed to flies that had been fed on the blood of a white mouse dead from experimental infection. That of number 40 was exposed to flies that had sucked blood from piglet number 43 on the day when, after its infection, a high fever appeared. Piglet number 40 obtained the disease by fly-borne transmission from number 43. In a second swine-infection experiment, piglet number 21 was similarly infected via flies that fed on the mouse blood culture, and piglets numbers 5 and 7 were fed on by flies that fed on the lesions of number 21 after it had become infected. Numbers 5 and 7 both developed light attacks of the disease, possibly because of some immunity acquired at the farm whence they came. These experiments showed how easily the infection can be transferred by the bites of *S. calcitrans.* Tolstiak suggested insect control as erysipelas control.

In 1956, Ostashev reported a four-year study of the microflora of flies caught on swine and in piggeries in the Molotovsk region. Besides *Salmonella,* in the intestine of one of 57 flies caught on a farm where erysipelas was present, he bacteriologically and biologically confirmed the presence of erysipelas bacilli. Unfortunately, the species of fly involved was not given.

Wellman (1959), reviewing the research on biting flies as zoonoses transmitters, states that, with erysipelas just as with brucellosis, transmission by insects is not the principal mode of infection. However, his and Tolstiak's work surely indicates the relative ease with which biting flies can transmit the disease. Khera and Sharma (1967) consider biting insects probable factors in the transmission of infection among swine because of the usual summer incidence of the disease. The characteristic periodic rise and fall of this disease in various years and its seasonality, however, are still not well understood. Further research such as Ostashev's on the bacterial flora of flies associated with swine, and studies on the effect of fly suppression where erysipelas is prevalent are clearly needed.

ANTHRAX (Charbon, Milzbrand, Malignant Pustule)

This is a cosmopolitan disease to which nearly all domestic mammals and man are susceptible. Cattle, sheep, horses, goats, and wild herbivores are most commonly affected. It is caused by *Bacillus anthracis*, a gram-positive, nonmotile, spore-forming, rod. The bacilli occur singly or in pairs but, in advanced cases, are seen in the blood as chains of rods. Spores probably do not form in an unopened carcass, but sporulation occurs when organisms are discharged from the body of an infected animal. The spores are highly resistant to heat, low temperature, chemicals, and desiccation, and may retain viability for 20 to 30 or more years in pastures. In endemic areas, the disease has a marked tendency to be seasonal in character, occurring in epizootic form during summer and early fall, but sporadic outbreaks occur at any time.

The average incubation period ranges from 1 to 5 days. The acute and subacute forms common in cattle, horses, and sheep are accompanied by high body temperature and a distress syndrome. Infectious, bloody discharges may emanate from the mouth, nose, and anus. Death occurs in a day or two in the acute form, while the subacute form may lead to death in a few days or to complete recovery. The chronic form is observed mostly in swine and is marked by swelling of the throat and tongue, which in some cases causes death by suffocation; some animals may die without obvious symptoms. Cutaneous anthrax occurs in cattle, horses, and in man when anthrax organisms lodge in wounds and abrasions, and is characterized by swellings on various parts of the body.

This is an old disease, relatively uncommon in England and the United States, but a scourge in Austria, Hungary, Germany, France, and eastern countries. Animals can be actively immunized by various vaccines, and more recently several antibiotics are in use. Preventive measures include:

burying diseased carcasses and sterilizing wool and other materials from anthrax areas.

In evaluating the role of flies in this disease, it is well to keep in mind the three ways in which the disease may enter an animal's body; 1) through pricks or lesions (inoculation), producing local anthrax, 2) inhalation of spores, producing pulmonary anthrax, and 3) ingestion with food, producing an intestinal form. Consideration must also be given to conveyance from bloody discharges and watery secretions from the eyes of sick animals.

The pre-scientific incrimination of flies as disseminators of anthrax has an interesting history. The first recorded association with flies of which we are aware was by Gilbert, who wrote in 1795: "In the district of Argenton, Department of L'Indre, where I have been combatting this epizootic which inflicts such fearful ravages, the malady attacks all animals indiscriminately and was communicated to persons solely by the bite of flies which had pumped the blood of dead animals." The idea continued into the latter decades of the 19th century that anthrax results from a fly bite, and this idea was shared by laymen and professionals alike. Walz (1803) noted that skinners were particularly afraid of being bitten by flies while skinning carcasses. Wuttge (1828) reported the death from anthrax of a shepherdess two days after having allegedly been bitten by a fly, and Schwab reported a similar case in 1832. Seiderer (1839) associated two cases with fly bites; both individuals developed malignant pustules from fly bites on the cheek, and in both instances sheep that had died of anthrax lay open nearby. William Budd (1862) considered conveyance of malignant pustule from flies to man more important than direct contact or contaminated food. He noted that flies can transmit the disease after contact with the carcasses of diseased cattle and that the disease usually appears on exposed parts of the body. Additional cases of anthrax attributed to fly bites were reported in 1869 (Weiss) and again in 1872 (see Nuttall).

Bridging the historical and modern approaches are separate papers by Ricque and Déclat, both published in 1865. Ricque describes fly prevalence associated with sewage pollution of a river that passed by a city near an army camp. He mentions animal corpses in the river on which flies "suck the putrid juices and then by their stinger or their mandibles deposit septic virus in tissues of living creatures," causing carbuncle and malignant pustule. He gives the detailed case histories of seven soldiers who, after being bitten by insects (some cases were clearly flies; all were so assumed), developed carbuncles, edema, anorexia, and fever and were hospitalized for a week or more. He describes and illustrates two flies he believes responsible for these cases: ". . . only one, *Musca phasia* (var. *maculata*) has a piercing mouth; the other, *Musca vomitoria* (var. *calliphora*) has only a simple infundibular proboscis, like the common fly." He concludes that the "majority, if not almost all, of septic bites (of humans) are produced by contact with the proboscis of *M. calliphora*." Déclat also emphasized the dangers of the nonbiting fly and urged burial of diseased carcasses in order to prevent the contamination of flies. Neither man performed any experiments to support his thesis. The latter's position was buttressed by the work of Davaine (1868, 1870) and Raimbert (1869, 1870), who pioneered the experimental approach to the fly-anthrax question. Davaine considered both biting and nonbiting flies transmitters of the disease and showed that a millionth of a drop of blood from a diseased animal—comparable to that delivered by a fly bite—could kill a guinea pig. He felt that biting flies, including tabanids, transmitted the disease in open country, and nonbiters were responsible for wound to wound transmission in stables. Raimbert discounted the importance of *S. calcitrans* for the curious reason that he could not get captive flies to feed on blood. On the other hand, nonbiters fed readily on infected cadavers, and Raimbert's experiments convinced him that bacilli could penetrate the skin of healthy animals without a fly bite. Both investigators obtained simple mechanical transmission by injecting parts of artificially contaminated house flies and *C. vomitoria* into guinea pigs, and both emphasized the importance of flies in the dissemination of the disease.

Mégnin (1874) and Bollinger (1875) tempered this view, the former pointing out that the disease also attacks animals in winter in the absence of flies. From personal experience, he concluded that only biting flies such as *Stomoxys* were able on occasion to be agents of anthrax. Bollinger cited anthrax cases that occurred in a 20-year-anthrax-free high pasture at an 8-10 hour traveling distance from a known focus of infection, and that, he stated, could hardly be explained other than by flies. However, he did not think that flies were usually the exclusive mode of infection. In reviewing the subject, Koch (1886) concluded that not enough was known about "the really dangerous" insect species.

The literature before and since that time has been replete with case histories that offer circumstantial but highly plausible evidence of fly involvement. Many of these were discussed in the historical chapter. A typical more recent report (Beyer, 1920) concerns a patient in a hospital at Jackson Barracks, Louisiana, who had chills and fever and a reddish, abraded area on his face, such as produced by an insect bite. Soon a pustule developed proving positive for anthrax. This occurred in March, when *S. calcitrans* was one of the early spring flies present in abundance in the nearby stockyards. In this instance, transmission probably could have occurred from an undetected case among the transient cattle.

After Davaine and Raimbert, the only worthwhile work of an experimental nature for the remainder of the century was that of Bollinger (1874). He found that most flies (including the "dangerous" tabanids) are unharmed by blood infected with anthrax. He fed maggots of a carrion species for two days on large quantities of infected blood, and they were unaffected after 14 days. In a significant, earlier experiment, he produced anthrax in rabbits with flies caught on anthrax cadavers.

With the turn of the new century, research interest quickened. Partly, it was stimulated by the influence of Nuttall (1899), who perceived the state of the problem as too many positive opinions about flies and too little experimental evidence. Partly, it was a generally low public opinion of flies that rekindled the attack.

B. ANTHRACIS IN FLIES

Buchanan (1907) let *M. domestica* and *Musca* (= *Calliphora*) *vomitoria* run over an agar culture of anthrax bacilli and demonstrated that large numbers of organisms adhered to their feet and were deposited in gradually diminishing numbers on surfaces wherever the flies settled. In Australia, Cleland (1913) noted that *Musca vetustissima* (= *sorbens*) readily ingested films of anthrax blood that had been dried for days, and virulent anthrax bacilli were then cultivated from the fly's excreta and body. Anthrax colonies grew out on agar visited by *Calliphora* sp. and *Lucilia* sp. that had sucked the juices of infected meat (Osins'kiĭ, 1938).

More thorough and sophisticated experiments were performed by Graham-Smith (1910, 1911, 1912). In one series (1910), he placed 24 *M. domestica* in a cage for one hour with an opened carcass of a mouse just dead of anthrax. The flies were seen to feed on the blood. The resulting vomit and excreta were cultured and examined microscopically, and at intervals flies were dissected and cultures made from their legs, wings, head, crop, and intestinal contents. The results showed that nonspore-bearing anthrax bacilli survive on the exterior of flies for no more than 24 hours. However, they may remain alive in the intestine for 3 days and in the crop for 5 days, especially when partially coagulated blood remains in that organ. No spore-bearing forms were found. The bacilli were present in the feces 48 hours after infection.

In experiments with anthrax spores, flies were fed an emulsion of an old culture and were dissected at intervals. Again cultures were made from external and internal parts of the fly and from fecal deposits. Numerous smears were made from gut contents at various times and no bacilli were seen microscopically, indicating that the spores do not develop into bacilli in the fly. Cultures were also made from drops of syrup after the flies had been allowed to feed on them, and they were positive for up to 10 days, but not later. Spores were carried on legs and wings for at least 12 days and were present in considerable numbers in the gut for at least 7 days; vomit and fecal spots contained viable spores for 6 days or longer. Later experiments showed that spores remained viable externally and in the gut for at least 20 days and were passed in the feces for 14 days.

In 1912, Graham-Smith studied the problem of bacterial survival through metamorphosis of flies. His conclusions were: 1) that adults of *L. caesar* and *C. vicina* bred from larvae allowed to feed on the bodies of guinea pigs dead of anthrax are not infected with *B. anthracis* (in his 1911 experiments, it was likely that the emerged flies had fed on the infected meat; this possibility was avoided this time); and 2) that a large proportion of house flies that develop from larvae contaminated with anthrax spores are infected. Morris (1920) observed that house flies and "blue bottle" flies that bred out of anthrax carcasses rid themselves of the pathogen, but flies bred in the presence of spores retained the organism. Using modern methods, Radvan (1956) has re-confirmed these observations on house flies.

TRANSMISSION EXPERIMENTS

Schuberg and Böing (1913) investigated the transmission of anthrax by biting flies. In earlier experiments, they had obtained successful transmission from mice to guinea pigs via fly bite and then extended these to larger animals. But when infected *S. calcitrans* were put on a shaved area of a goat's throat no transmission occurred. When a sheep was bitten by 12 flies, 6 of which had fed on guinea pig anthrax-infected spleen and 6 on a live infected animal, it died of the disease on the third day. However, attempts to infect sheep with a single bite failed. Mitzmain (1914) reported that the stable fly does not normally attack carcasses of animals but could be induced to do so in the laboratory. In his experiments, *S. calcitrans* (interrupted feeding) and *Tabanus striatus* were allowed to bite guinea pigs dying of anthrax and then were transferred to healthy animals; this resulted in fatal infection. Negative results were obtained when animals recently dead of the disease were used. Nieschulz (1928) obtained similar results.

Morris (1918) fed *Lyperosia* ($=$ *Haematobia*) *irritans* for one minute on infected guinea pigs and then on a healthy one for 1-3 minutes. Transmission was most successful when flies were fed a short time before the death of the animal. However, *H. irritans*, like *S. calcitrans*, would not feed on anthrax carcasses in nature.

304   FLIES AND ANIMAL DISEASES

Nieschulz is convinced that once an outbreak has begun, tabanids play an important role in the further spread of the disease because of their high experimental "capacity index," their great numbers, and, particularly, because they remain infective for several days. Next to them, if in a much smaller degree, the muscids play a role. The work of Sen and Mivett (1944) also somewhat fits this view. They reported that *S. calcitrans* failed to transmit anthrax to goats by its bites or by defecating on the scarified or cauterized skin of goats. There was no evidence of anthrax bacilli in the mouthparts of the stable fly except in two cases immediately after feeding. However, the bacilli were present in the bodies of such flies up to 72 hours and in their feces 21 to 72 hours after an infective meal. Both *M. domestica* and *C. vicina* mechanically transmitted the disease when brought into contact with the cauterized skin of goats after having fed in incisions on infected carcasses of goats (see also, Rinonapoli, 1930).

In summary, muscids are capable of harboring anthrax bacilli for a sufficient period of time to be potentially important vectors. We do not know whether the organism multiplies in flies, nor do we know the dosage delivery of the stable fly, house fly, and other suspect muscids. Transmission experiments suggest that the smaller biting flies generally pick up too few organisms from a sick animal to transmit with much success or consistency. Tabanids, on the other hand, ingest a greater volume of blood and presumably deliver a larger inoculum. Factors that would favor transmission by small flies are the cumulative effects of a large number of bites and numerous interrupted feedings, with the source, an animal near death at the height of bacteremia. Van Ness (1971) minimizes insect propagation of anthrax among animals in the United States, but cautions against human infection from bloodsucking insects that are disturbed while feeding on an animal that has died of the disease.

LIMBERNECK (Botulism in Birds)

This is a type of intoxication due to ingestion of toxins of *Clostridium botulinum*. Types A and C toxins of *Cl. botulinum* are the most common causes of botulism in birds. Various toxin types also affect man and many other animals. In all cases, this organism occurs commonly in the soil and may contaminate food stuffs. In order to produce botulism, the bacteria must have multiplied and formed toxin in the food before it is ingested by the host. The most important symptoms in birds are loss of appetite, uneasiness, ruffling of the feathers, and weakness of legs, which rapidly progresses to complete paralysis.

In the United States, this disease in poultry has been known for a long time as "limberneck." Outbreaks have been caused by feeding poultry on spoiled canned beans, peas, corn, or asparagus that in some cases had already produced the disease in man. This disease also occurs in South Africa, where, Henning (1956) notes, as early as 1893 a reference was made in the *Agricultural Journal of the Cape* to "Lamseikte" (botulism) in ducks due to the ingestion of maggots or decomposed meat.

Saunders et al. (1914) mention that it is a well-known fact that on large poultry farms, young turkeys, pheasants and chickens are fed maggots, which are obtained by hanging meat over boxes of corn meal, and which they devour in large numbers; on these same farms, there are large losses of birds from time to time due undoubtedly to a toxic larva being introduced by the green fly *Lucilia caesar* (= *illustris*). They experimentally produced botulism in fowls by feeding them on the larvae of this fly, obtained from limberneck carcasses. Following this work, Bengston (1922) isolated *Cl. botulinum* type C from *L. illustris* larvae and produced botulism symptoms in laboratory animals by inoculation or feeding them the toxin. Later, Graham and Boughton (1924) isolated an apparently identical organism from chickens and ducks.

The experiments of Wilkens and Dutcher (1920) corroborated Saunders's work in part. They found that *L. illustris* larvae removed from a limberneck fowl carcass or the carcass of a hog that died of paralysis could produce the disease in healthy fowls when fed to them. *Musca domestica* and *Calliphora vomitoria* were also tested and gave negative results. Bishopp (1922-1923) reported that *Cochliomyia macellaria* larvae reared in carcasses of limberneck fowls were also capable of producing limberneck when fed to healthy fowls. In another experiment, adults of *C. macellaria* bred from a limberneck hen carcass were placed on healthy beef, and, when the larvae reared from these flies were full grown, they were fed to a normal hen, which exhibited a typical case of limberneck the next morning and died that same day. Bacteriological examinations were not made. Bishopp felt that other species of blow fly larvae may be involved under similar conditions and that the causative agent of limberneck may be carried by larvae in a limberneck carcass through the pupal and adult stages to the larvae of the next generation reared in beef. Larvae and adults of the cheese skipper, *Piophila casei*, are readily contaminated with *Cl. botulinum*, but limited tests have shown that the organism is destroyed during the first few days of metamorphosis (Legroux and Second, 1945). Disposal of all carcasses by burning, especially those that have died of limberneck, has been stressed.

Cheatum et al. (1957) also described an outbreak of botulinum type

C poisoning in game farm pheasants in New York state, where the primary source of toxin was blow fly larvae. They stated that the spores of this bacteria, commonly present in the soil, are probably being continually ingested and excreted by the birds, and only when stasis of the digestive tract occurs in sick or dead birds, or when dead birds lie in contact with soil-borne spores, are conditions suitable for germination of spores and the production of toxin. More recently, Lee et al. (1962) found *Cl. botulinum* type C toxin in *Lucilia illustris* larvae taken from carcasses of pheasants on a game farm in northern Wisconsin during an epizootic in which about 4,000 birds died between July 25 and August 8. A suspension of 1.0 gram of larvae was found to contain $2 \times 10^{4.9}$ mouse LD_{50}. Blow flies are usually abundant from late spring to late fall and are constantly searching for food and oviposition sites, such as dead or dying birds. Maggots were found in injured living birds as well. The investigators estimated, on the basis of two carcasses examined, that up to 5,000 or more maggots in various stages of development may be found on a single carcass. In their experiments, it was found that 8 third-instar larvae contained a lethal dose of toxin for a 10- to 12-week-old pheasant. The role of adult flies was not clear, but they possibly spread the epizootic into adjoining pens if they had been feeding on contaminated carcasses.

Fish et al. (1967) report an outbreak of botulism among pheasants on an island in Lake Ontario in which *L. illustris* was the prime fly. It was found breeding in large numbers in pheasant carcasses, and pheasants, in turn, fed voraciously on the maggots, consuming hundreds of them in one day. As noted above, this is far in excess of the number required to kill a bird. *C. botulinum* toxin was obtained from washings of maggots and from homogenates of thoroughly washed specimens. Type C toxin was obtained from *C. botulinum* cultured from maggots. Maggots that were held at $-23.3\,°C$ for nine months were still lethal for mice and pheasants. We do not yet know the effect of hibernation or metamorphosis on the infectivity of the various fly stages. Fly control is economically feasible where birds are confined to restricted areas, otherwise, vaccination seems the only hope.

BLACKLEG

Blackleg is a fatal disease of cattle caused by *Clostridium chauvoei* and preventable by vaccination. Sauer (1908) showed that *M. domestica* and other flies become infected by feeding on an animal that has died of the disease, and cadavers of such flies are still infective by inoculation one year later. Flies are also suspect because the organism gains entry through small punctures or minute abrasions in the skin and mucous membranes.

Human intestinal infections transmitted by flies are discussed in Chapter 4. Here, we mention only in passing that flies may also be involved in cases of clostridial food poisoning. In addition to the usual enteric pathogens, blow flies collected in butcher shops, a fish shop, a food store, etc., at Slough, England, were frequently contaminated with *Clostridium welchii*, a heat-resistant anaerobe responsible for poisoning of cold and re-heated meats (Green, 1953).

CUTANEOUS ACTINOMYCOSIS (Streptothricosis),
Dermatophilus congolensis Infection

This is a disease of wild and domestic herbivores including cattle, horses, sheep, goats, and deer, and was first recognized in cattle in 1915. It is characterized by the formation of horny crusts, which adhere firmly to the infected skin. It is caused by *D. congolensis*, an aerobic actinomycete, which forms a branched mycelium that divides transversely and then longitudinally to produce thick bundles of very small cocci; these enlarge and mature into flagellated ovoid "zoospores." When the crusted lesions are wetted, the zoospores emerge to the surface, where they are available for the transmission of the infection. There is no resistant stage, and transmission is presumed to be direct. This disease is of economic importance in many countries, causing damage to animal hides, debility, and sometimes death. There is evidence that plants, weather conditions, and arthropods may serve as mechanical means of transfer. Damage to hides by thorns favors the introduction of infection (Zlotnik, 1955), and the eradication of thorny shrubs from pastures greatly reduced the incidence of leg lesions on cattle (Macadam, 1964). Heavy rains may cause skin damage leading to infection. All observers found that the disease was widespread during the rainy season and uncommon during the dry season. High humidity may or may not have an effect on transmission, but, in any case, the tick and biting fly populations increased greatly with onset of the rains and decreased suddenly at the end of this season (Macadam, 1964). Plowright (1956) has demonstrated the importance of ticks in the spread of this infection. Roberts (1963) suggested that flies were involved in the transmission of the zoospores of a closely related species, *Dermatophilus dermatonomus*, among sheep in Australia.

In Nigeria, Macadam (1964) studied the possibility that biting flies might be vectors of this disease. Four naturally infected cattle were kept fly-free at a mean relative humidity of 92%. The lesions healed in two cases and increased slightly in the third case; new lesions developed in the fourth, and it died. Two other naturally infected cattle were exposed to fly attack. One of these had a mild infection and was exposed to a light biting fly attack for six weeks in moderate (mean 52%) humidity and recovered. The other, a severe case that was kept at a mean 86% relative humidity, was exposed to a very heavy attack of *Stomoxys calcitrans* for eight weeks; new lesions arose, and it was shot while dying from the disease. His results are inconclusive.

However, Macadam points out that many of the lesions on the fly-attacked cattle were not on tick inhabiting sites, such as the axilla, scrotum, groin, tail, and ears, but rather on the back especially just behind the hump and spread over the back, sides, and rump, where ticks were seldom found. Such lesions may have been caused by *S. calcitrans* since the fly populations were highest at this time. He cautioned that experimental cattle should be kept free from ectoparasites for five weeks before beginning observations so that any incubating lesions may develop, and concluded that high humidity alone had little, if any, effect on the lesions.

In the laboratory, Richard and Pier (1966) carried out fly transmission studies with *S. calcitrans* and *M. domestica*. *S. calcitrans* transmitted the infection to a normal rabbit up to 24 hours after feeding on an infected rabbit. Transmission attempts were unsuccessful when the flies were permitted to feed on a dry lesion of a donor rabbit. When flies were permitted to feed on moist lesions, 17 of 20 attempts were successful, and moistening of the feeding sites on recipient rabbits facilitated transmission. Isolation of *D. congolensis* from flies proved difficult due to overgrowth of other microorganisms and was demonstrated only on the feet of *S. calcitrans*. Transmission attempts with *M. domestica* were successful, and this is important for it shows that mechanical disruption of the host's skin by the fly is not necessary for transmission of *D. congolensis*.

The extent that biting and nonbiting flies are involved and under what conditions they may spread the disease, whether by bite or direct contamination of sores, are some of the questions that need further detailed study.

LEPTOSPIROSIS

This is an infectious, febrile disease affecting animals and man. Clinical symptoms vary according to the species and strain of *Leptospira*, and depend on the species of affected animal. Leptospiral infections are known to occur throughout the world in cattle, pigs, goats, dogs, horses, and sheep. Muridae are reservoir hosts, and they, as well as convalescent animals, can become renal carriers, shedding leptospires in the urine. Invasion of the host usually takes place through breaks in the skin, through the nasopharynx, and the gut. It may also occur through animal or possibly insect or tick bites.

The possibility that insects transmit leptospirosis was considered even before the first species of *Leptospira* was discovered in 1915. The increased incidence of the disease associated with marshy zones, summer, and rain and humidity hinted at a malaria type of epidemiology. Mosquitoes and horse flies were suspected (Hecker and Otto, 1911). After discovery of the causative agent, the theory was largely abandoned, though several workers attempted to demonstrate its validity even up until fairly recently.

Reiter and Raum (1916) showed that *Haematopota pluvialis* may transmit leptospires to guinea pigs mechanically. Uhlenhuth and Fromme (1916) believed that insects, especially biting flies, played a role as intermediates in this disease, but when they injected homogenized flies from an endemic region into an experimental animal, the results were negative. Later, however, Uhlenhuth and Kuhn (1917) infected guinea pigs with *L. icterohaemorrhagiae* through the bites of *Stomoxys calcitrans*. More recently Kunert and Schmidtke (1952) infected nonbiting flies with leptospires (*L. icterohaemorrhagiae, grippotyphosa, canicola*) in order to test their role as transmitters. Adults and larvae of *Calliphora vicina, Musca domestica*, and *Phaenicia sericata* were used; infection was accomplished by feeding or rectal injection. The causative organisms were viable for up to 26 hours in the intestinal tracts of adult flies. A diet of carbohydrates seemed to favor the survival of leptospires in the intestinal tract. Larvae either died or freed themselves of leptospires before undergoing metamorphosis. It was concluded that, while larvae are unable to carry the infection, non-blood-sucking flies might be responsible for sporadic cases by transmitting leptospires to food and water.

Numerous studies have been carried out with various species of mosquitoes with generally negative results. An exception is Noguchi (1925), who maintained that *Aedes aegypti* may be a carrier and transmitter of leptospires. Other workers reported success with lice and bedbugs, but their results could not be confirmed. Experiments with fleas and ticks have been negative, but recently it has been demonstrated that some *Ornithodoros turicata* ticks may become infected with leptospires for as long as 518 days and may emit them with their coxal liquid and pass them on to their offspring (Burgdorfer, 1956; Schlossberger et al. 1952, 1954).

Babudieri (1958) in his review of the animal reservoirs of leptospires considers flies and other insects without practical importance and transmission by ticks as a rare occurrence that happens only under very particular circumstances or in well-organized experiments.

SPORADIC BOVINE ENCEPHALOMYELITIS

This is an acute infectious disease of cattle that is probably worldwide. It is caused by *Miyagawanella pecoris*, which also attacks sheep. Outbreaks are sporadic and often occur in closed herds, and for these reasons carriers and vectors are suspected. At Cornell University, a herd of cattle was kept free of the agent for five years by means of scrupulous sanitation, despite the presence of infected animals less than 300 yards away and accessibility of flies to both groups (Lindsay and Scudder, 1956).

"Q" FEVER

Normal guinea pigs were injected with a supernate of homogenized *M. domestica* that came from a room where guinea pigs with "Q" fever were quartered. The injected animals developed typical symptoms of the disease and also proved to be refractory to subsequent challenge with *Coxiella burneti*, the causative agent. Homogenates of flies that developed from maggots fed on infected spleen, and homogenates of maggots removed overnight from infected material, were not infective when injected into susceptible guinea pigs. Tests to transmit the agent between guinea pigs by means of the house fly were also negative (Philip, 1948).

ANAPLASMOSIS

This is an infectious disease of cattle caused by *Anaplasma marginale*, which destroys the red blood cells and causes typical anemic changes. The infection is frequently fatal, but, when recovery takes place, the organism usually resides in the host for a long time while the animal is immune. The causative agent is observed in a stained blood smear as round purplish bodies 0.3 to 0.8 microns in diameter, one to three in number, located within some of the red blood cells near the margins. Soulsby (1968) states that there is little doubt that *A. marginale* can no longer be regarded as protozoan, and that there is every evidence of its rickettsial nature.

The disease is enzootic in the warmer parts of the world. It has been reported from most states of the United States, but is most prevalent in the Gulf states, lower plains states, and in California. Piercy (1956) noted that in the Texas gulf coast the disease followed a typical seasonal pattern in which the greatest clinical incidence was from June to September or October with peak incidence during August and early September; cases in other months were minimal and sporadic. The infection is limited to cattle and related ruminants; wild species such as deer and antelope harbor latent infections and are important factors in maintaining the pathogen in enzootic areas.

The infection is easily transmitted by mechanical transfer of infected blood, and major and minor operations in cattle husbandry such as dehorning, castration, vaccination, blood sampling, and ear tagging have been traced as responsible for outbreaks of considerable proportions both in and out of season.

Ticks are probably the primary vectors of anaplasmosis around the world. Piercy (1956) listed 19 species in seven genera of ticks (*Argas*, *Boophilus*, *Dermacentor*, *Hyalomma*, *Ornithodoros*, *Rhipicephalus*, and *Ixodes*) that are concerned in experimental transmission and may serve as vectors in nature. Some features of tick transmission are of a biological rather than a mechanical nature. The causative agent remains alive during the relatively long periods of time required for the stage-to-stage development of ticks and, in some cases, is transmitted transovarially to the larva. Sanborn and Moe (1934) obtained negative transmission results with a number of species of ticks that infect cattle in the southern United States. This led to the incrimination of other arthropods, namely flies, as vectors in these areas.

Transmission by flies has been considered in detail (Piercy 1956), and this method appears to be common in the southern United States. With fly and mosquito vectors, the transmission is accomplished mechanically by the proboscis, which carries blood from infected to healthy cattle when feeding is interrupted, and, therefore, transfer must be immediate to be effective.

Previously, on the dipteran transmission question, Parodi (1917) found that *Stomoxys* and tabanids did not appear to transmit anaplasmosis since mixing sick and healthy cattle, all deprived of ticks, resulted in no transmission, although the animals were attacked and bitten by flies. However, Sanborn et al. (1931) succeeded in transmitting the infection to a susceptible cow by a total of 41 composite transfer feedings with *Tabanus gracilis*, *T. sulcifrons*, and *Chrysops sequax*. Howell et al. (1941a) also reported positive transmission with several species of *Tabanus* and believed that any species of *Tabanus* is capable of transmitting the disease by its natural feeding habits if it is active in sufficient numbers. Transmissions failed when transfer feedings were interrupted for five minutes or more; but undelayed transfer feedings from sick ani-

312   FLIES AND ANIMAL DISEASES

mals by only 13 *T. oklahomensis* transmitted the infection. Dikmans (1933) tabulated the experimental results in tabanid transmission by various workers and found that, in 29 experiments involving undelayed transfer feedings from animals with clinical cases, the disease was transmitted in 59% of the trials; in 14 experiments using immediate transfer feedings from recovered carrier animals, transmission occurred in only 7% of the susceptible cattle. Deferred feedings by flies from either clinical cases or recovered carriers did not produce the disease. Thus, there appears to be sufficient experimental and epidemiological evidence to identify tabanids as important vectors in nature.

However, Taylor (1935), in anaplasmosis-free Great Britain, failed to transmit the disease via biting flies. He described experiments in which a large number of bites was inflicted by flies that had every opportunity of transmitting the disease (46 bites by *Haematopota pluvialis* and over 6,000 by *S. calcitrans*) and in which the strain of anaplasm used was sufficiently virulent for 0.1 cc of infected blood to transmit the disease. The negative results strongly suggested to him that the danger of transmission of anaplasmosis from one animal to another via biting flies was unlikely under natural conditions in Great Britain. However, he felt that the small number of trials and the different species used did not disprove the American investigators' mechanical transmissions. Similar results have been reported from Australia (Anon., 1936).

Other dipterans have been investigated. Sanders (1933) in Florida reported that *Stomoxys calcitrans* transmitted the disease from a clinically sick animal in one experiment that involved direct and undelayed transfer feedings of several hundred flies. Howell et al. (1941b) produced positive transmission to a susceptible animal after composite undelayed transfer feedings from a clinical case with approximately 1500 mosquitoes, *Psorophora columbiae* and *P. ciliata*, and with 241 of these species and *Aedes aegypti*. The evidence is probably too meager to cite the specific role of mosquitoes or stable flies as vectors in nature, but, from all indications, they are much less important than either ticks or tabanids.

DERMATOMYCOSES

The dermatophytes are fungal parasites of keratinized parts of the body—skin, hair, and nails—and are responsible for such cosmopolitan afflictions as athlete's foot, jockey itch, and ringworm. The pathogenic species thus far associated with flies include *Trichophyton tonsurans*, *T. acuminatum*, *T. mentagrophytes* var. *granulosum*, *T. rubrum*, *T. verrucosum*, and *Microsporum canis*. Dermatophytes are less fastidious about the species of host than they are about the specific requirement for keratin-containing ectoderm. They infect various domestic and laboratory animals and man, but always superficially. If the parasites are injected intravenously into a guinea pig, systemic infection does not occur, but skin infection appears wherever the skin is traumatized.

Dermatomycoses are highly contagious and are spread primarily by contact. But as early as 1879, Aubert suspected that flies and other insects may play a part. *M. canis* has since been isolated from fleas on cats, and from ants found in the vicinity of infected animals. Lice on children suffering from favus have been shown to harbor *T. schoenleinii*, one of the causative organisms.

Koch (1963, 1964) isolated *T. verrucosum*, an agent of ringworm, from house flies taken in a barn where cattle infected with the disease were quartered. The flies were often observed to settle on the skin lesions of the animals. The following year, he succeeded, in a limited trial, in transmitting ringworm from an infected guinea pig to a calf by means of flies. He also induced infection in a small number of guinea pigs by means of house flies that had been in contact with guinea pigs infected with *T. mentagrophytes* and *Microsporum gypseum*. On the other hand, Richard (1963) did not succeed in transmitting *T. equinum* through the agency of *S. calcitrans* between ponies or between guinea pigs. Although viable organisms were present in the skin lesions upon which the flies fed, they were apparently not picked up by the flies.

Working on the west coast of Sweden, Gip and Svensson (1968) isolated *T. mentagrophytes* var. *granulosum* from 3 house flies out of 59 collected in a cattle barn. The organisms were apparently carried on the exterior of the flies. It is interesting to note that there was no known occurrence of dermatomycosis in either man or animals (cows, pigs, and poultry) in the environment in which the contaminated flies were found. A word about the method of isolation. Wings, legs, and body were separated and pressed onto the surface of Sabouraud's glucose agar, which contained 300,000 units of penicillin, 300,000 μg streptomycin, and 0.5g cyclohexamide per liter.

Kamyszek's (1965a, b, 1967) thesis is that factors that increase fly density, e.g. high humidity and a temperature between 29° and 35°C, favor the spread of dermatophytes among animals. In a personal communication, he mentions a successful transmission of pathogenic fungi by house flies to guinea pigs, particularly young, weak ones. Bolton and Hansens (1970) have shown in the laboratory that house flies ingest and pass cultivable spores of various fungi; they gave no data on the length of time spores may persist in the gut and be passed in the feces.

314 FLIES AND ANIMAL DISEASES

This is a chronic, nodular, suppurative disease of the skin, lymphatics, and mucous membranes of horses in Europe, Africa, and Asia. The agent is a fungus, *Cryptococcus farciminosus*, which is about 4 μ in diameter and reproduces by budding. Mild cases recover in a month, but the majority of cases are incurable. Jewell (1904) called attention to the possibility that the disease is spread by flies that pass from infected to healthy animals and contaminate sores and abrasions while feeding. Since the organisms are present in skin lesions, contaminated hands, harnesses, direct contact and flies may disseminate the organism. Harber (1913) reported that the use of Stockholm tar and oil on the wounds of horses repelled flies and reduced the incidence of infection. Chatelain (1917) studied the disease on the Moroccan coast and noted that cases appeared in healthy horses at considerable distances from sick ones. He cited Vehr, who found cryptococci in the proboscis of flies that had just fed on an abscess, and Teppaz, in Senegal, who reported that the disease occurred almost exclusively in fly regions. Jarvis (1918), implicates ticks, tabanids, and stomoxyds as vectors.

TRYPANOSOMIASES

The protozoan genus *Trypanosoma* contains a large number of species, which parasitize a range of vertebrates from fish to man. In most hosts infection is benign, but in man and domestic animals it may be severe. In Africa, human sleeping sickness has been recognized since the 14th century, and animal trypanosomiasis has dominated and seriously interfered with the economic development of about one-fourth of the continent. According to the World Health Organization, it is the greatest single impediment to the development of the stock-raising industry and blocks efforts to improve the deficient diet of African populations. As an example, Zebu cattle are bred north of the tsetse line and supply 90% of West Africa's beef. They have to be trekked through fly belts to the consuming areas. Along one such route, from Mopti to Accra, losses from trypanosomiasis have been estimated at $2.5 million annually. In tsetse-free regions, mechanical transmission by *Stomoxys* and other flies can occur, resulting in widespread epizootics. In 1946 and 1947, in one country alone, cattle died at the rate of 10,000 a month, and a total of a million and a half were attacked by the disease. WHO points out that the cost of treatment is high. In one West African country, the nine-year cost to cattle owners was $450,000, while the government spent $1.5 million on its treatment program.

Trypanosomes infect primarily the blood and tissue fluids of their vertebrate hosts, producing anemia and emaciation, and causing enlargement of the lymph nodes, liver, and spleen. In the Gambian more than in other kinds of trypanosomiasis, and in man more often than in domestic animals, invasion of the central nervous system may produce characteristic symptoms of drowsiness and stupor. In animals, drug therapy is less effective than in man, and infection is frequently fatal. However, much depends on the species of trypanosome and host, and one of the serious problems is that infections may be inapparent, with such hosts serving as reservoirs.

It is not known whether the genus *Trypanosoma* originated in Africa, but it appears to have undergone considerable speciation there. One hypothesis is that the *brucei* subgroup arose in Africa and then spread to India and the New World, where further speciation occurred. Trypanosome diseases are not confined to Africa but are prevalent in South and Central America and in Asia, as well. Two trypanosomes, *T. theileri* and *T. uniforme*, were recently found in the blood of Holstein cows in Ontario, Canada (Woo et al., 1970). Trypanosomiasis is endemic in regions where tsetse are absent and *Stomoxys*, tabanids, and other biting flies are common. Although a great deal of work has been done with these two groups of flies in Africa, Asia, and elsewhere, there has been little in the way of a concerted review of the results. Buxton's (1955) excellent monograph and the papers of Willett (1962, 1963), and Chinery (1965) provide a good perspective of tsetse-borne trypanosomiasis. Our purpose here is to examine the evidence for the involvement of flies other than tsetse.

Trypanosomes originated as parasites of the insect intestinal tract. It is possible that types such as *Herpetomonas muscarum*, a parasite of common flies such as the house fly, represent a line that evolved in blood-sucking insects and, through countless inoculations into vertebrates, finally became adapted to them. Some trypanosomes established beachheads in aquatic vertebrates through leeches and mosquitoes, and in marine fish through unknown intermediaries; related types became parasites of *Euphorbia* and other plants, through the agency of Hemiptera.

Based on Hoare's classification, Lapage (1968) has provided a concise summary of the two groups and the three major subgroups of *Trypanosoma*, with distinguishing features of the more important parasites. With the exception of *T. evansi* and *T. equinum*, trypanosomes normally undergo cyclical development in the insect, which may include changes of form. This is comparable to the cycle of *Plasmodium* in the mosquito except that most trypanosomes develop in the lumen of the gut rather than in the gut epithelium; also, there is no sexual stage in trypanosomes.

316 FLIES AND ANIMAL DISEASES

They are grouped according to their location in the gut of the insect. Anterior station forms develop in foregut, midgut, or salivary glands and are usually transmitted through the proboscis (inoculative). This type includes the *brucei*, *vivax*, and *congolense* groups with which we are concerned. Posterior station forms complete their development in the hindgut, are voided in the feces, and are transmitted through contamination of wounds and bites. They include *T. cruzi*, agent of Chagas disease, which develops in triatomids, and *T. lewisi*, which is disseminated by the rat flea. Whereas some anterior station types require up to five weeks to produce infective metacyclic forms in the fly, certain members of the *brucei* group, e.g. *T. evansi* and *T. equinum*, have abbreviated the insect phase to a very brief excursion in the fly, which momentarily interrupts its feeding on one host to resume quickly on another. It is important to emphasize that all species of trypanosomes that undergo cyclic development in tsetse are also infective when introduced mechanically by a fly's proboscis. *T. equiperdum* has eliminated the insect "middleman" by adopting the venereal route. Dogs and cats with abrasions of the digestive tract lining can acquire surra by eating the meat of horses infected with *T. evansi*, but this is an exceptional route since most cases of trypanosomiasis occur among herbivores.

SURVIVAL OF TRYPANOSOMES IN NON-TSETSE FLIES

A great many futile searches were made for evidence of cyclical development of trypanosomes in *Stomoxys*, *Haematobia*, and tabanids before a mechanical role was finally accepted as the only possible one. In the East Indies, Penning (1904) ruled out *Stomoxys* as a vector of surra on the grounds that the parasites are rapidly destroyed in the fly. A few hours after ingestion, the parasites begin to degenerate and are usually dead after about 24 hours (Currey, 1902; Nabarro and Greig, 1905; Minchin, 1908; Fraser, 1909; Baldrey, 1911; Fletcher, 1916; and Dieben, 1928). Taylor is one of the few who has reported multiplication in the gut of *Stomoxys*. He observed the production of a large number of short forms of *T. brucei* during the first day, but the population died out the next day. Baldrey's (1911) observation of a partial cyclic development of *T. evansi* in the gut of *S. calcitrans* is open to question since he admits to the presence of *H. muscarum* and the two species may have been confused. The only arthropods other than tsetse for which we have found evidence suggestive of a developmental cycle are ticks e.g. *Ornithodoros crossi*, with which a few successful transmissions were accomplished with feeding intervals of 17 days and a month (Cross, 1923); there appear to have been no follow-up studies.

Noteworthy is a report of successful transmissions of *T. evansi* by

S. calcitrans between dogs, with intervals of one or two days. This, as far as we know, has also not been confirmed and is of particular interest in light of the transient survival of trypanosomes in the proboscis of the fly. Infective *T. evansi* survive for only about 30 seconds in the proboscis of *S. calcitrans*, according to Mitzmain (1912b), and for ten minutes in *S. nigra* (Moutia, 1928). Beck observed viable *T. gambiense* in the proboscis of *Stomoxys* for ten to twenty minutes.

There is no question that mechanical transmission by means of the fly's proboscis is the prime if not the only mechanism. But we should be cautioned against too simplistic an acceptance of the "hypodermic needle" analogy. Van Saceghem (1922) dissected numbers of stable flies immediately after they had fed upon an animal infected with *T. congolense*. Trypanosomes were invariably present in the crop, sometimes in the pharynx, but never in the food canal of the proboscis. He suggested the following conditions as necessary or desirable for transmission to occur: interrupted feeding with immediate resumption on another animal; the crop should be full of blood to dilute the fly's toxic saliva; and the pharynx should be full if a good inoculum is to be delivered. Of the three conditions, the first has been amply confirmed, and the other two have not been investigated.

To demonstrate that trypanosomes are transferred from the surface of the proboscis, Sergent and Donatien (1922a, b) compared the number of successful transmissions of *Stomoxys* flies that bit guinea pigs normally with those that had to pierce a fine cloth in order to bite. In both cases, the technique of interrupted feeding and immediate transfer was used. There were 6 out of 12 transmissions in the first case compared with only 2 out of 11 in the second case. Nieschulz (1930) tried to transmit surra to pairs of rats that were bitten successively by the same flies, but only the first rat became infected. *Tabanus striatus* is also limited in this way (Mitzmain, 1916).

The rapid disappearance of trypanosomes from the proboscis of *Stomoxys*, *Haematobia*, and other flies suggests a toxic effect, possibly of the saliva, that kills on contact. If this is so, it is not evident in species of *Glossina* and *M. domestica*. Working with *T. brucei*, Taylor (1930) found that the proboscis of *G. tachinoides* favors survival of trypanosomes for up to 3 hours (though mechanical transmission ceases after 20 minutes); by contrast, only 3 out of 13 *Stomoxys* had trypanosomes in their proboscis immediately after an infective meal, and none after 5 minutes. Obviously, the salivary glands of tsetse are not toxic to trypanosomes since they proliferate in them. Working in Panama, Darling (1912) reasoned that, since viable *Herpetomonas* can be expressed from the proboscis of the house fly, the same environment should not be inimical to trypanosomes. In fact, he was able to demonstrate live forms of *T. equinum* in the house fly proboscis two hours after a meal. There is evidence that *Musca sorbens* probably behaves like the house fly in this regard. But this analogy cannot be carried too far since *Herpetomonas muscarum* thrives in the gut of various flies, whereas trypanosomes soon die out. Nevertheless, these observations suggest specific differences in the proboscis of flies that may be attributable to salivary factors. This key aspect of mechanical transmission deserves more study.

SURRA

The name derives from a Hindi word meaning "rotten." The disease is endemic in India and in the Orient generally, in parts of Africa, and in Central and South America. In Algeria it is known as el debab, in Sudan as gufar or mbori, in Panama as murrina, and in Venezuela as derrengadera. *T. evansi*, the causative agent, has acquired a number of aliases, e.g. *T. annamense*, *T. berberum*, *T. cameli*, *T. hippicum*, *T. soudanense*, and *T. venezuelense*. The disease was first described in 1885 by Evans in sick horses in India, and its host range includes mules, asses, camels, elephants, capybara, deer, dogs, and rats. It can also be transmitted by inoculation to cattle, goats, sheep, and laboratory animals. The blood of cattle, buffalo, and pigs sometimes swarms with trypanosomes, often with little effect.

In originally describing surra, Evans called attention to such biting flies as tabanids as being likely transmitters, since "it is a fact that the disease does spread mostly at those posts where the horses are closely packed and flies are in the greatest numbers." The ensuing years have witnessed a counterpoint of positive and negative evidence that still leaves the status of flies far from clear.

Experimental transmission

Unless stated otherwise, the technique employed in the following accounts is interrupted feeding on a sick animal with immediate resumption of feeding on a well animal. Rogers (1901) was the first to experimentally transmit surra between infected and healthy dogs by the agency of tabanids.

Compared with *Stomoxys* and *Haematobia*, tabanids have generally proved to be the most effective transmitters, at least in the laboratory. In Malaya, mechanical transmission of surra was successful with four species of *Tabanus*, but not with *Stomoxys*. Two experimental methods were used: naturally infected horses and cattle were put together with well animals and flies in specially constructed fly-proof enclosures; and parasitemic and healthy laboratory animals were alternately bitten (Fraser and Symonds, 1908; Fraser, 1909). In the Philippines, Mitzmain (1916) concluded that *Tabanus striatus* is the primary vector, *M. domestica* has to be given serious consideration, but *S. calcitrans* is not important. He succeeded once in transmitting surra via stable flies by repeated feedings of a large number of flies for six days (1912b). In India, *Tabanus albimedius* has been shown by Cross (1923) to be an efficient vector in the laboratory, even transmitting, though with fewer successes, when the donor animal did not have parasitemia. In the East Indies, experiments have demonstrated the superior vector capacity of *T. rubidus*, *T. striatus*, *T. minimus*, and *T. immanis* (Nieschulz, 1930).

The reader is cautioned against concluding that tabanids are sole disseminators of surra. Epizootics have occurred when tabanids were absent and *Stomoxys* and *Haematobia* were the predominant biting flies. We shall now examine the relevant evidence.

Musgrave and Clegg (1903) in the Philippines, succeeded in passing surra from sick to well monkeys and also to the horse, dog, rat, and guinea pig by means of biting flies (probably *S. calcitrans*). Working in Mohand, India, Leese (1909) got one positive result in three trials when he used 10 to 15 *Stomoxys* flies per trial; with *Tabanus*, several trials were successful when only 4 flies were used. Despite the ease of transmission with *Tabanus*, Leese implicates *Stomoxys*, pointing out that, during the surra season in the area he studied, there were 500 *Stomoxys* for every *Tabanus*. He suggests that infection is usually transmitted from camel to camel because *Stomoxys* flies prefer camels to horses. When camels are not in the picture, tabanids transmit infection from carrier bovids to equids, the reasoning being that the larger capacity of their proboscis compensates for the paucity of trypanosomes in bovid blood. *Stomoxys* then transmits among closely positioned horses, since horses suffer a more intense parasitemia.

A horse exposed to many *Stomoxys* will, by constant movement of the head, tail, feet, and skin, prevent the flies from completely engorging and cause them to finish their meal on another animal. Observers have reported that in North Africa *Stomoxys* seem to prefer horses to camels, the reverse of the situation in Mohand. *Stomoxys* also respond differently to various breeds of cattle. The smooth-skinned, long-tailed, active native cattle of Tanganyika seem to keep relatively free from attack, whereas imported bulls, especially of the heavier breeds, are victimized to a "truly distressing extent," according to McCall (1926). Calves, presumably because of their size, seem less prone to attack than larger animals in the herd.

Jowett (1911) succeeded once in transmitting trypanosomes by successive feedings of both *Stomoxys* and *Haematopota*. Since the results were negative in other experiments in which the flies were tried separately, it is not known which fly might have been responsible. In North Africa, Sergent and Donatien (1922a, b) transmitted surra among guinea pigs six out of twelve times with *Stomoxys* sp. Because of the seasonal occurrence of these flies in the animal quarters under investigation, they considered *Stomoxys* as the primary vector. However, earlier limited success of experimental transmissions with *Stomoxys* led the brothers Sergent (1905) to believe that tabanids are the more likely vectors in Algeria. Dieben (1928) also reported only partial success with *S. calcitrans*; however, his observations of the extent to which these flies harass water buffaloes and make them "sweat blood" inclines him to consider them important in the East Indies, together with *Haematopota* and other biting flies.

The first outbreak of surra on Mauritius occurred in 1901 and was traced to a consignment of cattle from India. The entomological fauna of the island lacks tsetses and is generally poor in biting flies, with the exception of *S. nigra*, which is present in great numbers. This fly constituted 99% of a total catch of 2000 insects taken in a fly trap (Moutia, 1928). Large numbers of trypanosomes were reported in the gut of presumably wild flies, and laboratory transmissions have been successful with this species (Edington and Couts, 1907; Moutia, 1928). In 1933, Adams and Lionnet speculatively traced an outbreak of surra among wild deer to nearby bovids, by way of *S. nigra*. The flies were seen feeding on deer cadavers four or five hours after they had been shot. Unfortunately, no effort was made to ascertain the presence of the pathogen in the gut of these flies.

Nieschulz can be credited with the most thorough contribution to the study of surra and flies (1926, 1927, 1928a, b, 1929, 1930, 1941); his 1930 paper is especially informative. Working in Java and Sumatra, he performed about 116 experiments with *S. calcitrans* and *S. brunnipes* on various mammals. The experiments are extensive, and only the briefest review can be given here. He obtained 44 successful transmissions using a total of 4008 flies, 10 to 25 flies per trial. Nieschulz calculated the average value of the transmission expected per fly as 1:91. His experiments reveal important differences not only in the vector capacity of certain flies but also in the donor capacity and susceptibility of mammals. With a horse as a surra carrier (or donor), the calculated transmission expectation values are: to another horse, 0:1044; to a monkey, 1:25; to a laboratory rat, 1:33; and to a mouse, 0:100. The following expectation values reflect the susceptibility of the recipient animals, regardless of the donor: horse, 1:1197; monkey or rabbit, 1:25; guinea pig, 1:31, rat, 1:27; and mouse, 1:55. With *Stomoxys* as vector, the species of recipient animal has a decisive influence on the outcome. From the above, it is clear that horses are much more refractory to infection than small laboratory animals. Differences in donor capacity are due to the varying degree of parasitemia each animal experiences. The rat has an intense parasitemia, therefore more trypanosomes can be picked up by the fly, and the number of laboratory transmissions is expected to be greater than between horses, in view of the latters' poor carrier value (0:1044) and resistance to infection (1:1197, average other animal ⟶ horse). Nieschulz concludes that *S. calcitrans* can hardly play a part of any importance in the natural transmission of surra. In a previous trial, he had failed to transmit surra between horses when he used 700 stable flies. With guinea pigs, on the other hand, he had 2 out of 5 successes when he used a mere 10 flies per trial (1928b). Later, when he substituted a Bulgarian strain of *T. evansi*, his successes diminished to 2 out of 17 (1941), which suggests that differences between strains are also important (see also Taylor, 1930; Lucas, 1955). The results he obtained with *S. brunnipes*, although more limited, indicate that it behaves like its congener, *S. calcitrans*.

Chaudhuri et al. (1966) doubt that *S. calcitrans* is involved in the transmission of surra in India. Their experiments with guinea pigs, dogs, and ponies, which were maintained under fly-free conditions, yielded one positive in 13 trials with guinea pigs and one positive out of 6 with dogs. The investigators employed larger numbers of flies to compensate for the reluctance of the flies to feed successively. For example, in one positive trial, 30 of the 126 flies completed their meal on the susceptible animal after feeding briefly on the donor. In another experiment, numbers of stable flies were released into a fly-proof stable that housed an infected and a healthy pony, not in contact. This experiment was repeated three times; in the last trial several thousand flies were allowed to emerge from natural breeding sites within the enclosure. The period of exposure ranged from 11 to 17 days in their experiments, but none of the animals became sick, based on daily temperature records and blood tests. Although no experiments with tabanids were performed, the investigators are convinced of their importance in India. Yet, Holmes, a half-century before, observed sick and well horses stabled side by side with numerous tabanids flying about, and not a single case of surra occurred among the healthy horses in three years (see also Penning, 1904; Carment, 1916; Branford, 1919).

The puzzling inconsistencies in experimental and natural transmission of surra are typical of trypanosomiasis in general, and we will discuss this in the conclusion.

322   FLIES AND ANIMAL DISEASES

The first clue as to possible involvement of nonbiting flies in trypanosomiasis came from the work of Musgrave and Clegg (1903). In a successful experiment they performed in the Philippines, house flies passed from a sick to a well dog, each with an exposed wound, and transferred surra in the process. Mitzmain (1914) pursued this by simulating the "normal" feeding interaction between *S. calcitrans* and *M. domestica*. He placed 200 flies of each species in a bottle and inserted the tail of a monkey infected with surra. The house flies engorged on the blood that flowed from the bites of the stable flies. Next, the tail of a healthy monkey was inserted and the feeding interaction was again enacted. Five trials were made, each with a negative result. The house flies did not seem to be able to infect by contaminating the wounds made by stable flies. Two years later, however, Mitzmain reported 4 out of 5 successes in similar trials, without giving reasons for the reversal. A possible field confirmation of house fly involvement was an outbreak of surra among over 100 pack and saddle horses on the island of Luzon. Although the first horses contracted the disease when *Tabanus striatus* was present, the main outbreak occurred when gad flies were absent and house fly density was great. Almost all the animals had open sores, which may have provided house flies with the means for acquiring and spreading the infection.

Lamborn's investigations of *Musca sorbens* in Nyasaland qualify this insect as a potentially serious candidate. *T. brucei* passed down its digestive tract and was infective in the feces at least 6 hours after ingestion, and *T. rhodesiense* was infective from the proboscis for at least half an hour (Thomson and Lamborn, 1934). The synanthropy of this fly would tend to confine the risk to villages.

Lamborn's efforts with this fly were not uniformly successful for, in another report (1934), he mentions that he failed to transmit *T. brucei* and *T. congolense* by means of *Musca sorbens*, or by means of *M. tempestiva* and *Bdellolarynx latifrons*, which are also nonbiting, hematophagous flies. In a large series of experiments, the flies fed upon the scored ears of guinea pigs, rats, and puppies, after feeding upon the blood of an infected ox. Nieschulz (1930) obtained negative results in a limited trial with *Musca inferior*. Thus, the state of our knowledge concerning nonbiting flies continues in limbo.

Musca crassirostris, unlike its flabellate relatives, uses prestomal teeth to give it independent access to the tissue fluids of an animal (Fig. 18). During April and May in the Philippines, this fly builds up populations that may equal those of the stable fly. Mitzmain (1914) and Fletcher (1916) reported some successful transmissions of *T. evansi* with this fly, but Nieschulz's (1930) results were negative. Fletcher points out that successful transmissions were limited to a maximum interrupted feeding interval of seven minutes, even when viable organisms were present in the fly's gut for 28 hours.

We have found no reports of positive transmissions with *Haematobia* (= *Lyperosia*) *exigua* (Nieschulz, 1930), even when infective parasites were demonstrable in the gut (Mitzmain, 1914). Taylor (1930), in Africa, failed in eight trials with *T. brucei*, *Haematobia* sp., and guinea pigs. Mitzmain (1912a) noted that *Haematobia* was the predominant biting fly on horses, cattle, and carabao in haciendas where surra existed. But, when he put two infected carabao with a healthy one in a fly-proof enclosure together with 5000 *Haematobia*, nothing happened. There was a noteworthy phoretic relationship between the fly and the louse, *Haematopinus bituberculatus*, which persisted for six weeks, until *Haematobia* populations declined. The association may have been specific because the louse was never found on *Stomoxys* flies even when they predominated. The louse was infective to healthy animals two hours after removal from an infected animal. It is possible that *Haematobia* may not serve as a direct vector for *T. evansi*, but it may occasionally serve as a vehicle for infective lice.

The involvement of lice and other insects is by no means certain as counterclaims have muddied these waters. Leckie (1925) found trypanosomes in stable flies and camel lice after they fed on a sick camel, but the lice did not transmit infection to guinea pigs. He cites an instance where fleas may have been involved in an outbreak among foxhounds in Lahore, for no further cases occurred after the fleas were exterminated. With further regard to fleas, a dog infected with *T. brucei* was kept next to a healthy dog for six weeks in the presence of a heavy flea infestation. The dog remained well, as did another dog inoculated with a suspension of these fleas (Duke et al., 1934).

Leese (1909) in Mohand, India, could not transmit *T. evansi* through *Culex*, *Anopheles*, *Stegomyia*, or *Phlebotomous*, and Taylor's (1930) data on three species of *Aedes* and *T. brucei* are also negative. Nieschulz (1940a) reports a few successes in 48 trials with *Aedes aegypti* and *T. congolense* in guinea pigs. Infectivity of the parasites in the gut of the mosquitoes was limited to 3 hours. He had previously had some success with *Anopheles fuliginosus* and *Armigeres obturans*.

TRYPANOSOMA EQUINUM (= *T. hippicum*, MAL DE CADERAS)

The organism is considered by some to be synonomous with *T. evansi*. It attacks horses, mules, donkeys, dogs, cattle, sheep, and goats, in descending order of severity, in South and Central America. It appears to be enzootic in Argentina, Bolivia, and Paraguay. The capybara, a large rodent found near water, is considered to be a reservoir, and indeed, the nidus of the disease is in wet swampy areas where tabanids are also abundant. The incubation period of 4 to 10 days is followed by fever and parasitemia, with death following quickly or after a chronic course of several months. Weakness of the hindquarters results in a faltering gait (mal de caderas). Restricting the spread of disease is the fact that trypanosomes diminish in the peripheral blood in more advanced infections.

Sivori and Lecler (1902) claim to have transmitted fatal mal de caderas by means of tabanids and *S. calcitrans* from sick to healthy horses, but they have given no details. Elmassian and Migone (1903) counterclaim that winged insects are not important in transmission. For 6 months they kept 300 well horses close to 34 horses that were sick and dying of the disease. Separation was maintained by wires, which allowed free access of biting flies and other insects, yet no trypanosomiasis occurred in well animals.

In a series of papers (1911a, b, 1912, 1913), Darling discusses an outbreak of the disease among mules, which automatically excludes venereal transmission. The animals were never infested with ticks, tabanids were present only in the woods, and *S. calcitrans* was eliminated because it attacked saddle horses and mules alike, yet the horses, though equally susceptible, never suffered from the disease. Further, dissections of stable flies taken from an infected mule and from around the corrals were negative for trypanosomes. House flies were suspected because the sick animals had open sores that were constantly visited by these flies. Experimental transmission to one of three mules occurred when house flies were quickly transferred from a sick horse. As an additional test of his hypothesis, Darling instituted two methods of house fly control: the local house fly population was suppressed; and the flies were repulsed by applying creolin, etc., to the animals' wounds. Possibly as a result of these measures, no new cases occurred in the next two years, while elsewhere in Panama the disease remained endemic.

DOURINE (MAL DE COIT)

This venereal disease of horses, caused by *T. equiperdum*, is widespread in Europe, Asia, Africa, and parts of the Western Hemisphere. Biting flies may be involved in a minor way. In 1908, Sieber and Gonder declared that the disease may be spread by *S. calcitrans* in the absence of coitus. The following year, Schuberg and Kuhn (1909a) announced they had succeeded in transmitting the disease between rats by crushing flies on the rat's skin and by fly bite.

The trypanosomes we shall now discuss have an exclusively African distribution, with the exception of *T. vivax* (=? *T. viennei*), which is also found in South and Central America and the West Indies.

T. vivax (= *T. cazalboui*, *T. pecorum*) causes a disease known as souma in Africa. It attacks cattle, goats, sheep, horses, and camels, but dogs, pigs, rats, and mice are resistant. Reservoir hosts are antelope in Africa and deer in South America.

T. congolense infects a wide range of domestic and laboratory animals, with antelope, zebra, and other game animals serving as reservoirs. In cattle, the disease produced by this and other species of trypanosome is known as nagana. In camels, infection is less severe. *T. dimorphon*, which has been split from *T. congolense*, has a similar distribution. *T. simiae* also belongs to this group. It produces fulminating infections in pigs, monkeys, and camels, mild symptoms in sheep and goats, and usually no symptoms in horses, cattle, and dogs.

T. suis is highly pathogenic to pigs but appears to be innocuous to other domestic animals.

T. brucei is widely distributed in tropical Africa between the latitudes of 15° north and 25° south (Soulsby, 1968). It is most severe in horses, dogs, sheep, and goats, and relatively mild in cattle and pigs. Laboratory mice are highly susceptible, rabbits and guinea pigs less so.

In the tsetse belt, all the above agents are transmitted primarily by tsetse. In other regions, and sometimes within the belt, other biting flies are implicated in transmission. It is this aspect we shall now examine, looking first at vector capability revealed by transmission studies and then at the evidence from field observations.

Bouffard (1907a, b) in Sudan, placed two de-ticked heifers, one of which was infected with *T. vivax*, under netting and beyond physical contact. He then introduced 40 wild *Stomoxys*, which survived for three days and bit during this time. On the twelfth day the blood of the well animal contained trypanosomes. In another positive report (Bouet and Roubaud, 1912), wild *S. calcitrans*, *S. bouvieri*, and *S. boueti* were interrupted in their feeding on an infected sheep and then transferred to well sheep for several days. Also positive was a similar experiment between cats. A series of trials with *T. dimorphon* yielded negative results and led the investigators to conclude there is a gradient of difficulty in achieving transmission with various trypanosomes.

Bruce et al. (1910) consider *Tabanus*, not *G. palpalis*, to be the probable common carrier of *T. vivax* in Uganda. *Stomoxys* were numerous around a herd that contained healthy and infected cattle, yet there was no spread of infection. Instead, the presence of disease seemed to be closely timed with the appearance of tabanids in areas where tsetse were absent (see also Richardson, 1924, 1928, for similar observations with *T. vivax* and *T. congolense*). Even the case against tabanids does not rest on unshakable ground. In a recent outdoor study in northern Nigeria, where there are no tsetse (Folkers and Mohammed, 1965), 8 cattle infected with *T. vivax* or *T. congolense* were kept in close proximity to a herd of 22 well cattle for three months. No transmissions occurred despite intense parasitemias in the cattle and the presence of numbers of tabanids and *Stomoxys*.

Transmissions are limited when *T. brucei* is used as the agent and *Stomoxys* is the vehicle. On the positive side, are the results of Martin et al. (1908), Schuberg and Kuhn (1909), and Taylor (1930). The latter scored 3 out of 10 transmissions with *S. calcitrans*, but 10 out of 15 with *Glossina tachinoides*. In Uganda, cattle trypanosomiasis is often a very serious disease in districts where *Glossina* are not known to occur. Carmichael (1933) succeeded several times in transmitting the infection among cattle with *Tabanus congolense* and *T. thoracinus*, and once with *S. calcitrans*. He observed a rapid destruction of trypanosomes in the proboscis and felt that under Entebbe conditions desiccation would probably limit mechanical transmission to less than a minute. He notes a severe outbreak of nagana, which coincided with a population explosion of *S. calcitrans*, *nigra*, and *brunnipes*. Eight percent of these flies taken while feeding on a sick donkey proved to be positive in their gut and proboscis. There have been relatively few validated instances of natural transmission of *T. brucei* by flies. Carmichael (1948) did not observe it once in 16 years among animals quartered together in a stable where biting flies were abundant. The negative findings of Nuttall (1908) with *T. brucei* in the laboratory, and of Martini (1903) in the environment of a stable, should also be taken into account.

In Uganda, in the absence of tsetse, two animals in a small herd of five cattle contracted *T. congolense*, 17 and 19 days after an infected animal was introduced. Since the incubation periods were only a few days longer than they would be if the organisms were inoculated by syringe, all signs pointed to mechanical transmission. Experimental efforts could not corroborate this; numerous trials with about 200 *Stomoxys* in a fly-free enclosure with sick and well cattle for six weeks ended negatively, as did interrupted feeding experiments (Hall, 1927; McCall, 1926). In this instance, we cannot be certain that stray tsetse had not been involved. In another case, transmissions occurred at a field site in western Uganda, 20 miles from the nearest tsetse zone. There were numerous flies and ectoparasites on the cattle, but homogenates of *Stomoxys* and *Haematopota* from infected cattle were negative when inoculated into mice (Lucas, 1955).

With regard to *T. simiae*, even less concrete evidence exists. Hoare discusses epizootics of pig trypanosomiasis in tropical Africa caused by this organism and emphasizes that, in most reports, tsetse were considered to be scarce or absent, whereas *Stomoxys* and tabanids were numerous. According to Hoare, it is the consensus among investigators that these flies are responsible for intra-herd transmission. In Bevan's (1917) experience, the disease was not contracted when infected and healthy pigs were closely stied together in the presence of swarms of *Stomoxys*.

We have seen only scattered researches on *T. gambiense* and *T. rhodesiense* in relation to non-tsetse flies. Beck (1910) failed to transmit *T. gambiense* by means of *Stomoxys*. Lamborn and Howat (1936) fed *Musca sorbens* on the blood of a patient infected with *T. rhodesiense* and obtained a positive outcome when the flies were immediately transferred to an incision on the ear of a dog. In 1933, Lamborn demonstrated that stomoxyds and tabanids transmit this trypanosome provided a number of flies are transferred to the healthy host within five minutes after feeding on an infected animal, and that this procedure is continued for a period of 20 days. Strong (1944, p. 180) mentions that Duke transmitted *T. rhodesiense* between monkeys by using 7 to 10 wild *Stomoxys*.

CONCLUSION

At this juncture, there is need for clarification and perspective. In view of the welter of conflicting information, we would be wise to reserve judgment on the relationship of stomoxyds to the maintenance of trypanosomiasis in the absence of tsetse. Since we are not dealing with an all-or-none phenomenon, we ought to expect only partial success from any natural mechanical means of transmission. Admittedly, stomoxyds have sometimes won their reputation solely by default of other candidates, and often because they generally manage to be on the right host at the right place and time, and in very convincing numbers. Thus, in the absence of tsetse in the Sahara, and at times of the year when tabanids are also absent, the year-around abundance of *Stomoxys* spp. suggests that they maintain the surra cycle between naturally infected dogs and dromedaries (Donatien and Parrot, 1922; Donatien and Lestoquard, 1923). Jack (1917) has given numerous case histories to show that transmission of trypanosomiasis occurs in the absence of tsetses. But, in areas contiguous with tsetse zones, it is difficult to rule out the possibility that infected tsetses may stray miles from their haunts to introduce infection into herds (Buxton, 1955, pp. 650, 651).

Another knotty problem is that of assigning relative importance to stomoxyds and tabanids where they coexist in the absence of tsetse. There are, fortunately, a few interesting studies that bear on this question. On the island of Zanzibar the incidence of *T. congolense* in cattle is 16 to 18% where tabanids are abundant, and only 2 to 6% where *Stomoxys* and *Haematobia* predominate (Mansfield-Aders, 1923-1924). Furthermore, no trypanosomes were found in the blood of stall-fed cattle in town, despite the nearby presence of infected animals and the ubiquity of *Stomoxys*. This corroborates Khan's (1922) observations that in low-lying, well-watered places in Zanzibar, where tabanids are numerous, infection rates in cattle are high, whereas, in dry, rocky country where *Haematobia minuta* is the dominant biting fly, infection rates are much lower. It was discovered after these studies, that *Glossina austeni* is common in parts of the island. What light this sheds on previous findings remains to be seen.

Investigators early noted that in certain parts of Africa spring epizootics coincide with the appearance of *Tabanus*, while *Stomoxys* is abundant the year around (Cazalbou, 1907; Kinghorn and Yorke, 1912). For centuries, camel drivers in Algeria have driven their animals to winter or summer pasture along ancestral routes away from standing water and vegetation, where tabanids abound. They have also followed the practice of moving to pasture in early morning and returning before the heat of day brings forth the greatest activity of these flies. In scientific circles, the belief has long prevailed that tabanids transmit in swampy regions and between herds while *Stomoxys* transmit within the herd.

The limits placed on *Stomoxys*-type outbreaks would make it a markedly localized phenomenon that spreads slowly. Jones' (1915) description of an epizootic aptly illustrates this important feature: ". . . when the disease was raging on the eastern side of the railway the cattle to the west enjoyed good health, though at krall No. 18 there was nearly 500 head of cattle, and the distance between them and the infected ones was not great."

The success and failure experienced with *Stomoxys* in the laboratory and in the field are less disconcerting when we realize that tabanids and *Glossina* spp. also transmit with irregularity. A large part of the problem has to do with the threshold dose of an animal vis-à-vis the dose a fly is capable of delivering. We can expect the threshold dose to vary somewhat with the strain and species of trypanosome, but a major barrier to transmission is the host's resistance to infection. By inoculating graded doses of *T. congolense*, McCall (1928), in a small but suggestive study, found that the minimum infectious dose for rats is 6.5×10^4 and for cattle, 2×10^5. This is undoubtedly far beyond the capacity of single flies to deliver, but the cumulative inoculations of many flies might reach the required level. Within what time span must this cumulative inoculation be given to be most effective? Might not the intermittent injection of small numbers of trypanosomes over a period of time have a significant effect by altering an animal's immune response? Despite the volume of research on flies, the quantitative aspects of uptake and delivery of parasites by the feeding fly in relation to host susceptibility have scarcely been investigated. It is reasonable to suppose that, in terms of size alone, the larger proboscis of tabanids and tsetse can deliver more parasites than the proboscis of *Stomoxys*, *Haematobia*, mosquitoes, or fleas. But comparison and other kinds of studies are urgently needed if we expect to unravel the inconsistencies that have bedeviled mechanical transmission experiments.

Ford (1964) has raised the interesting possibility that *T. vivax* has a greater disposition for acyclical transmission by hematophagous (non-tsetse) flies, than does *T. congolense*. This would account for the predominance of *vivax* in certain non-tsetse areas, whereas in tsetse areas one commonly finds both trypanosomes. That species of trypanosomes behave differently in flies is a reasonable hypothesis that Bouet and Roubaud (1912) suggested earlier. It is another aspect worth exploring.

The state of the problem remains as it was summed up by the Expert Committee on Trypanosomiasis of the World Health Organization in 1962: ". . . apart from the problems of the cyclical transmission there is need for work on mechanical transmission, to what extent and under what circumstances it occurs and to define the species of biting flies responsible."

BOVINE VENEREAL TRICHOMONIASIS

The evidence pertaining to the possible involvement of flies in the transmission of this disease is discussed in Chapter 4 under Amebic Dysentery and Related Diseases.

COCCIDIOSIS

Coccidia are major protozoan parasites of domestic and game animals, but are of little importance as disease producers in man. Details of the life cycles, hosts, and symptomatology are too extensive for our consideration here. Metelkin (1935) studied *Eimeria exigua* (= *perforans*) and *Eimeria irresidua* in six species of flies: *M. domestica*, *S. calcitrans*, *C. vicina*, *L. caesar*, *C. mortuorum*, and *P. groenlandica*. Oocysts of both species of *Eimeria* are small enough to be swallowed, and they pass through the fly gut undamaged, beginning 5 to 10 minutes after feeding and continuing for at least 24 hours. Superficially, the legs of flies pick up the greatest number of oocytes, but these are rapidly killed by desiccation or removed as the fly cleans itself. Metelkin tested viability of the oocytes by incubating them in a 2.5% potassium dichromate solution at 25°C for several days. This is a better test than staining with eosin. Metelkin does not consider flies necessary for the spread of coccidiosis, particularly among rabbits, which he found 100% infected. Since the parasites are rarely pathogenic to man, and are readily spread among animals when sanitation is poor, there has been little incentive to study flies.

TOXOPLASMOSIS

This is a protozoan infection caused by *Toxoplasma gondii*. It is a wasting disease causing anorexia, weakness, depression, ocular and nasal discharges, vomiting, coughing, and finally encephalitic symptoms, tremor, incoordination, and paralysis. It may be acquired or congenital in all domestic animals, wild animals, and birds, and in man. The infection occurs in all parts of the world, but it is most prevalent in warm, humid regions. The pathogen is an intracellular, motile protozoan.

Until recently no special life cycle was known, but cyclic development has now been convincingly demonstrated in the cat by Frenkel et al. (1970), Galuzo and Krivkova (1970), and others. Enteric infection in this animal is revealed by the passing of oocysts in the feces and by the presence of schizogonic and gametogenic stages in the epithelium of the small intestine. Fecal forms become infective after undergoing exogenous development for 3 to 5 days during which the oocyst produces paired sporocysts, each in turn giving rise to 4 sporozoites. The invasive trophozoites exhibit catholic organotropy and are capable of multiplying in nucleated cells from all three germ layers, including the brain. Cysts occur in various tissues. There is evidence that the life cycle is obligately sequential since oocysts appear in the feces in 3 to 5 days when cysts are fed to kittens but do not appear for 8 to 10 days when trophozoites are fed, and not for 21 to 24 days when fecal oocysts are fed. Evidence for the coccidian nature of *Toxoplasma* is presented by Frenkel et al. (1970) and others, but Galuzo and Krivkova (1970) argue that aspects of the life cycle and the unusual organotropy are sufficiently distinctive to justify creation of a new order.

Transmission may take place via the ingestion of contaminated food (e.g. milk, raw meat, and eggs) from infected animals; besides feces, sputum, nasal secretions, and urine may also be infectious, and infection via the respiratory tract has been reported. The exogenous phase provides the link for transmission to herbivores through contamination of grass, water, and feed (Galuzo and Krivkova, 1970). Carnivores become infected by ingesting cysts and oocysts; trophozoites are less likely to infect because of their sensitivity to stomach acids. Recent studies (Sheffield and Melton, 1969; Frenkel et al., 1969) show that *Toxoplasma* is not dependent on the nematode *Toxocara cati* or its eggs for transmission, as was formerly thought.

There have been extensive studies over the years on the possibility of transmission of this disease by arthropods. These were probably suggested by the infective nature of the organism and possible analogies to dipteran-borne malaria and trypanosomiases, which have thus far not been borne out.

The first systematic study of the possible role of flies (van Thiel, 1949) concerned transmission of *Toxoplasma* by nonbiting *Calliphora vicina*. In the first series of experiments, flies were fed toxoplasma-infected mouse brain, and, after various intervals, their intestinal tracts were aseptically removed and injected subcutaneously into mice. Only three times—after 4 hours, 1 day, and 4 days—out of 19 trials could the presence of viable toxoplasma be traced in the intestinal tract of flies. In the second series, 61 whole, infected flies were triturated and injected into mice, but virulent toxoplasma were present in only one fly one day after the infectious meal. In the third series, 12 flies were individually fed on toxoplasma-brain (after fasting 6 hours). The intestines of 5 flies were inoculated into mice with negative results; 5 whole flies were ground and inoculated into mice, also with negative results; and the intestinal tracts of 2 flies were smeared serially on slides and stained, but did not show the organism 2 days after feeding. However, it was shown that fly vomit contained virulent toxoplasma 2 hours after the infectious meal. In another test, mice failed to become infected after feeding upon whole flies or the feces of toxoplasma-fed flies. In fact, van Thiel succeeded in infecting only 5 of 21 mice orally with infected mouse brain suspension, and, since this was so difficult, he was not inclined to attach to nonbloodsucking flies an important role in the transmission of toxoplasmosis to man.

Blanc et al. (1950) collected four *S. calcitrans* in a stable, kept them unfed for 24 hours, and then let them infect themselves on a squirrel, *Xerus getulus*. On the same day 2 of the flies were ground and injected into a mouse, which subsequently died from the disease; the other two flies were ground and injected into a mouse after 48 hours with negative results. Also, 44 stable flies fed for three days on infected *Xerus* or guinea pigs failed to infect healthy guinea pigs with their bite 1 to 6 days later.

Kunert and Schmidtke (1953) found that 2 to 48 hours after feeding on the intestinal contents of toxoplasmic mice, the feces of *S. calcitrans*, *C. vicina*, *M. domestica*, and *Phaenicia sericata* were negative when injected into healthy mice; an exception was the 24-hour, pooled feces of *S. calcitrans*, which caused one death. Fly foregut contents were infective as long as infectious food was in the crop. *Toxoplasma* in fly feces 48 hours after feeding did not infect mice. The intestines of *S. calcitrans* and *C. vicina*, 12 hours after feeding, caused infection when injected; 2 out of 63 mice died after eating *Stomoxys* intestines 4 hours after feeding.

Kåss (1954) allowed 24 flies (*M. domestica* and *Muscina stabulans*) to feed on material dissected from mice dead of the disease and failed to produce an infection when the flies were injected into healthy mice two days later.

Laarman (1956) studied in greater detail the role of *S. calcitrans* as a possible vector. In his first series of experiments, after the flies had fed on the blood of mice containing toxoplasma, they were injected subcutaneously into mice at certain intervals after the infectious meal. When fly-injected test animals died, the results were confirmed by examining two brain and one lung smears; when a test animal survived after one month, brain suspensions were injected into two mice (which were subsequently examined), and brain smears were examined for pseudocysts. The results demonstrated the presence of virulent *Toxoplasma* in the fly intestinal tract for up to 22 hours after the infectious meal, but not afterwards, although examinations were continued for 9 days. In the second series, which tested the possibility of transmission by fly bites with 64 flies, the infectious meal was interrupted and the insects were allowed to continue their meal on 42 healthy animals after less than 10, 20, and 30 minutes; results were negative with one possible exception in which 13 stable flies transmitted infection to a guinea pig 24 hours later. However, serology was not performed, and the guinea pig may have already had a latent infection. The results did not indicate that the pathogen passes through a developmental cycle in *Stomoxys*. Evidently, mechanical transmission is possible but difficult.

Varela and Zavala (1961) reported attempts to transmit toxoplasmosis to white mice via *M. domestica*. The flies were lab bred, and both the vegetative and encysted forms of *T. gondii* were used for experimental infection. Whole flies, legs, wings, abdomens, proboscis, guts, and feces were examined for the pathogen 1 to 24 hours after contact with or ingestion of contaminated food. There were 7/50 (whole fly and parts), 5/100 (gut), and 5/50 (feces) positives for intervals up to 24 hours. At 1 and 24 hours after the infective meal, inoculations with a "triturate" of 5 flies per mouse produced no (0/5) infections; and 2/10 mice became infected by eating food on which 25 flies (infected the previous day) had defecated.

The erratic data we have reviewed typify difficulties that have generally been encountered in transmitting toxoplasmosis, whether by food contamination, carnivorism, insect bites, or other means. Yet the extensive distribution and versatile host range of toxoplasma suggest that the pathogen is readily disseminated in nature. Although *T. gondii* does occur in the peripheral blood of man and animals and contaminant materials (e.g. feces and carcasses) are infectious, the evidence for dissemination by these routes via biting and nonbiting flies is far from convincing. Parallel studies with fleas, biting Hemiptera, ticks, and mosquitoes are on a par with the fly work (Raccuglia, 1952; Woke et al., 1953; Laarman, 1956; and Varela and Zavala, 1961), except for a longer period of retention of toxoplasma in some other insects e.g. six days in *Rhodnius* and *Triatoma*, and 13 days in *Pediculus humanus*. The most impressive record of persistence is 65 days in *Periplaneta americana* infected *per os* (Mayer, 1965).

Newer knowledge of the extrinsic development of sporozoites in the feces justifies re-study of fly transmission. Important in this context is that the time of appearance of oocysts in the cat's feces—5, 10, or 24 days after the infective meal—was found to depend on whether cysts, trophozoites, or oocysts were fed. Transmission experiments must be designed to bracket the host's "hot" period. It may not be amiss to recall that the introduction of such a chance item as a raisin into the diet of *Phlebotomus* flies provided the key factor in the conversion of the fly from a marginal to an excellent host for *Leishmania*. We cannot help wondering whether some cyclic, nutritional, or other factor has thus far been missed. In the meantime, should Mayer's work be confirmed, we may have found an important insect vector and reservoir in the cockroach.

HABRONEMIASIS

Adults of the large stomach worms *Habronema muscae*, *H. microstoma*, and *H. megastoma* (Aschelminthes; Spiruridae) cause a condition in horses known as gastric habronemiasis. *H. megastoma* produces large tumor-like swellings in the stomach wall, which may interfere mechanically with the function of the stomach, but which are otherwise not usually harmful. The other two species occur freely on the mucosa and may penetrate into it, producing an irritation that leads to a chronic catarrhal gastritis with the formation of much mucous; *H. microstoma* may also produce ulcers.

Cutaneous habronemiasis, also known as summer sores, granular dermatitis, bursati, swamp fever, etc., is caused by the larvae of these species (especially *H. megastoma*), which are deposited in existing wounds by infected flies, and which then migrate into and irritate the sore, resulting in a granulomatous reaction (Fig. 50). According to Nishiyama (1958), the larvae may also penetrate apparently intact skin. The larvae are also associated with a granular conjunctivitis, especially in areas where cutaneous habronemiasis is prevalent (Australia, Africa, the East, North America, and the U.S.S.R.); the wart-like lesion, which itches intensely, is on the inner canthus of the eye, and, rarely, the whole conjunctiva is a mass of granulation tissue.

The embryos or larvae of the worms are passed out in the feces of the host and are ingested by maggots of *M. domestica* (*H. muscae, H. megastoma*) and *S. calcitrans* (*H. microstoma*), which serve as true intermediate hosts. The worms reach the infective stage at about the time the maggots pupate; in the adult fly, the larvae occur freely in the hemocoele and pass forward into the proboscis (Fig. 51). Horses are infected by ingesting parasitized flies or by the worm larvae, which emerge from the flies' proboscis as they are feeding on the lips, nostrils, or wounds. In the case of *H. microstoma* in *S. calcitrans*, the infective larvae interfere with the fly's ability to pierce the skin, and it reverts to an imbibing method of feeding from the moist surfaces of the horse. The larvae are liberated in the stomach of horses, where they grow to maturity.

The chronology of discovery of the fly-borne nature of this disease and subsequent research is fairly long, and only the historical and important research will be considered. Carter (1861), in India, first described the presence of a nematode in the head of a house fly; Diesing (1861) took this worm as the type of a new genus (*Habronema*). Leidy (1874) reported that *H. muscae* occurred in 20% of the flies he examined at Philadelphia, Pa. (see also Generali, 1886). Von Linstow (1875) described a nematode larva in the head of *S. calcitrans*, which he named *Filaria stomoxeos*. Johnston (1912) found a larva resembling *H. muscae* in *S. calcitrans*, and a similar one in *Musca vetustissima* (= *sorbens*). Ransom (1913), in the United States, gave the first account of the life-history of *H. muscae* in the house fly and also expressed the opinion that the larvae found by Von Linstow and others in *S. calcitrans* might be that of *H. microstoma*. His work showed that the embryos of *H. muscae* are taken up by the larvae of *M. domestica*, that they develop through the larval stages in the fly larva and pupa (Fig. 52), and that the final stage of the worm is reached in the adult fly and is usually found in the head and proboscis.

Not much more than the above was known when Bull (1919) and Hill (1918) independently published their work on the life-histories of these three species. Bull (1919) found that *M. muscae* passes through its larval stages in *M. domestica*, but shows no development in *S. calcitrans*; *H. megastoma* was found to have a similar life-history. *H. microstoma* passes through its larval stages in *S. calcitrans* and sometimes showed an aberrant development in *M. domestica*. In the final stage, larvae migrate to the head and proboscis. The escape of the larvae from the proboscis of the flies was found to depend on the rupture of some portion of that organ, probably the thinner chitinous membrane on the inner surface of the labium. The larvae apparently do not possess the power to penetrate the structures in the proboscis, but rupture seems to depend upon the pressure exerted by the number and activity of the larvae present. The larvae did not appear to live in saline, horse serum, or water for longer than 2-3 days (rarely for as long as 7 days); longevity outside the body of the fly may depend to some extent on attainment of the final stage and escape or removal from the proboscis. Hill's (1918) work was largely confirmed by and agreed with Bull's findings, except that Hill concluded definitely that *M. domestica* occasionally acts as host for *H. microstoma*. Other experiments by Bull demonstrated that these species are the cause of the granulomatous reaction in horses. Both authors left open the possibility that other fly species may be involved.

Following this work, Johnston and Bancroft (1920) in Queensland investigated other fly species as transmitters of *Habronema* infection. They reported that *H. muscae* and *H. megastoma* may be transmitted by the muscids, *M. domestica*, *M. vetustissima* (= *sorbens*), *M. terraereginae*, *M. hilli*, *M. fergusoni*, and *Pseudopyrellia* sp. and also by *Sarcophaga misera* and no doubt by other sarcophagids. *Anastellorhina augur* could become infected, as also probably could other blow flies with similar habits. In Ceylon, Crawford (1926) demonstrated that drosophilid flies that developed fortuitously in a portion of a diseased stomach taken from a fatally infected horse contained larvae of *H. megastoma* upon emergence. Five percent of a large number of *S. calcitrans* and *M. domestica* taken in a Bulgarian village carried *Habronema* (Iwanoff, 1934).

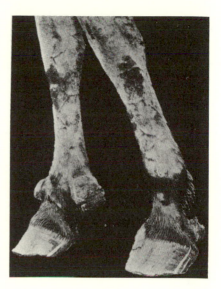

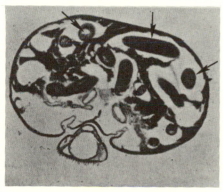

FIG. 50. Forelegs of horse with lesions due to cutaneous habronemiasis. (From Bull, 1919.)

FIG. 51. Cross-section of the proboscis of *S. calcitrans* at the level of the bulb, showing larvae of *H. microstoma* in situ. Note the food channel and salivary duct. (From Bull, 1919.)

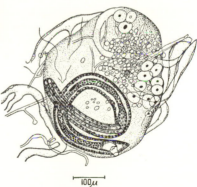

FIG. 52. Stage 2 larva of *H. muscae* encysted in pupa of *M. domestica*. (From Ransom, 1913.)

Johnston and Bancroft (1920) showed that the larvae can escape from the proboscis of infected flies (of several species) while the flies are feeding on warm, human saliva. Attempts by de Margarinos Torres (1925) to obtain spontaneous deposition of larvae of *H. muscae* by house flies were negative when the flies fed on sugar water, cow's milk, or human and horse saliva. But the same flies would then emit larvae into horse blood. Blood of other animals (man, guinea pig, rabbit) was also negative. This specificity of the hemotropism means that there is probably no danger of *Habronema* infection of human wounds. Hemotropism may also be influenced by conditions of temperature and humidity, for the emission experiments were most successful on hot, dry days. Later, Mello and Pereira (1946) showed that emission of larvae occurred only when the air temperature was above 22°C and the temperature of the fluid was above 23°C.

Other workers elaborated on some of the details of the fly-worm interactions. For example, Roubaud and Descazeaux (1921) reported that after the eggs of *H. megastoma* are ingested by *M. domestica* maggots, the embryos enter the lumen of the malpighian tubules, molt, and then invade the cells of the tubules causing hypertrophy. The spiny third-stage larvae invade the pupal abdomen, and finally locate between the labella and oral aperture. Descazeaux and Morel (1933) developed xenodiagnostic procedures for gastric habronemiasis in the horse. They diagnosed *Habronema* carriers by rearing flies (*M. domestica*) on suspected dung, and could even differentiate *H. megastoma* from *H. muscae* by their location in the infected flies. The former localized at the level of the malpighian tubules, the latter in the cells of the fat body. Mello and Cuocolo (1943) dealt with some aspects of the relations between *M. domestica* and *H. muscae* in the manure of horses, chiefly from the viewpoint of xenodiagnosis. From their experiments, they concluded that house fly eggs and up to 24-hour-old maggots are not infected; the main infection of maggots is on the second day, almost no infection occurring on the third day. Horse manure was infective for the house fly for 9 days. In maggots the larvae are found in the cytoplasm of the fat body cells, which undergo necrosis. In pupae, the parasitized cells are destroyed and what remains is a cyst-like wall containing the *Habronema* larva. In the adult fly, the infective larvae leave the cyst; on the first day they are almost exclusively in the abdomen; on the second and third days, they are distributed mainly in the head, but also in the thorax and abdomen; from the fourth day on the infective larvae are localized in the head of the fly. This has been confirmed by Waddell (1969). It was shown earlier by Van Saceghem (1918) that 70% of adult house flies were infested with *H. muscae* after they had been reared in the manure of a parasitized horse. The life cycle of *Habronema* species is illustrated in Figure 53.

Because flies are true biological hosts of *Habronema*, their role as vectors assumes a clarity and directness rarely encountered in fly-borne disease. Other nematodes, e.g. *Thelazia, Stephanofilaria*, and *Syngamus trachea*, use flies as true intermediate hosts, and their life cycles in relation to flies are discussed elsewhere in these last two chapters. In light of fly involvement, prophylactic measures include hygienic manure disposal to control flies and kill the worm larvae. Measures should also be taken against flies in stables and against fly contamination of food and water. Flies may aggravate the sores with their excreta and saliva, causing them to persist far beyond the time it takes for fly-protected sores to heal (Constantino, 1958; Iwanoff, 1934).

THELAZIOSIS (Eyeworm Infections)

Thelazia rhodesii occurs in cattle, sheep, goats, and buffalo in Europe, Asia, and Africa and has been found in California. *T. californiensis* occurs in sheep, deer, cats, and dogs in the United States. *T. skrjabini* occurs in calves. *T. gulosa* occurs in cattle in France and Sumatra. *T. callipaeda* is found primarily under the nictitating membrane of dogs in India, Burma, and China, but has been reported from man four times in China. These nematodes are commonly referred to as eyeworms and inhabit the conjunctival sac and lachrymal ducts of animals and occasionally man.

T. rhodesii is found in the tear ducts between the eyes and the lids of cattle. During migration, it may leave the ducts and be found on the surface of, or in, the eyeball, or under the nictitating membrane. The movements of the worm irritate the eye considerably, causing a free flow of tears and injection of blood vessels, and sometimes severe pain and nervous symptoms. At first the eye is not seriously affected, but Faust (1928) observed that in the course of time the repeated scratching of the surface of the eyeball by the serrated cuticle of the worm causes the formation of scar tissue, and the eye gradually develops a cloudiness, progressing outward from the worm nest, which ultimately reduces vision. Cattle are sometimes blinded by *Thelazia* in Africa and Asia.

Klesov (1949) concluded from his studies in the Ukraine that *T. rhodesii* larvae develop up to the invasive stage within *Musca larvipara* and *M. autumnalis*, which are both tenacious and persistent feeders at the eyelids. Other species of *Musca* and *Morellia* may be hosts, but *M. domestica* and *S. calcitrans* have no contact with the nematode and there-

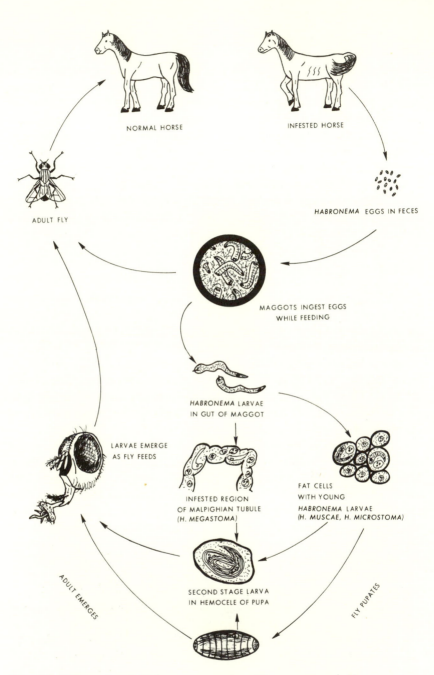

NORMAL HORSE

INFESTED HORSE

ADULT FLY

HABRONEMA EGGS IN FECES

MAGGOTS INGEST EGGS
WHILE FEEDING

HABRONEMA LARVAE
IN GUT OF MAGGOT

LARVAE EMERGE
AS FLY FEEDS

INFESTED REGION
OF MALPIGHIAN TUBULE
(*H. MEGASTOMA*)

FAT CELLS
WITH YOUNG
HABRONEMA LARVAE
(*H. MUSCAE, H. MICROSTOMA*)

ADULT EMERGES

SECOND STAGE LARVA
IN HEMOCELE OF PUPA

FLY PUPATES

FIG. 53. Life cycle of *Habronema* species. *H. muscae* and *H. megastoma* typically develop in *Musca domestica*, while *H. microstoma* is a parasite of *Stomoxys calcitrans*.

340   FLIES AND ANIMAL DISEASES

fore do not figure in its biology. Invasion of experimental animals by *Thelazia* occurred through direct contact by *Musca* flies with the eyes and not through the mouth or nose. Klesov found that *T. rhodesii* females give birth to larvae in the conjunctival region of the eyes and first-stage larvae are found free in the tear ducts. Larvae are produced at the beginning of April and May, and flies are infested at this time only if cattle are brought outdoors. The earlier the animals go outdoors, the earlier and more intensively, given favorable conditions, their infestation takes place. Contrariwise, even during the most dangerous period (June to September) cattle kept indoors remain free of *Thelazia* because the vector species are exophilic. Of 1290 flies, predominantly *M. larvipara* and *M. autumnalis*, caught near infected cattle in the summer of 1948, Klesov found 39 infested with *T. rhodesii*. In Tashkent, maximum activity of *M. larvipara* coincides with the appearance of mature worms in the eyes of cattle (Tukhmanyants et al., 1963).

Krastin (1950) showed that certain flies of the genus *Musca*, which cluster around the eyes of cattle, serve as intermediate hosts for *T. rhodesii*, which is harbored by over 90% of cattle in late summer in parts of eastern Siberia. Klesov (1950) and Krastin (1949, 1950, 1952) have shown that the true intermediate hosts of *T. rhodesii* are *Musca larvipara* and *M. convexifrons*, that those of *T. gulosa* are *M. larvipara* and *M. amica*, and that of *T. skjrabini* is possibly *M. amica*.

Klesov (1950, 1951) found that the first-stage larvae of *T. rhodesii* enter the gut of the fly via the proboscis from the eye secretions of the animal host and penetrate into the ovarian follicles of the fly (*M. larvipara*); only female symbovine flies attack the animals for food and become vectors. In the follicle, they develop into second-stage larvae, which grow and molt to become the third, infective larvae. The third-stage larvae leave the ovarian follicles and migrate to the proboscis of the fly, from which they are transferred to cattle (Fig. 54). The developmental time in the fly is 15 to 30 days. When infected flies were allowed access to calves, or when infective larvae were experimentally introduced into calves, adult *T. rhodesii* appeared in 20 to 25 days. When the infective larvae of *T. gulosa* were put on the conjunctiva of a calf, adults appeared in 7 days. Dissection of 5038 flies revealed 3.5% infected with *T. rhodesii*.

Krastin (1950) obtained infective larvae of *T. gulosa* from *M. amica* and placed them on the eye region of a calf; 6 weeks later he obtained one adult worm.

Thelaziosis of cattle is prevalent in eastern Slovakia, and Világiová (1962) found that, from investigations of over 3000 pasture flies collected mainly in association with animals, the larvae of *T. rhodesii* and

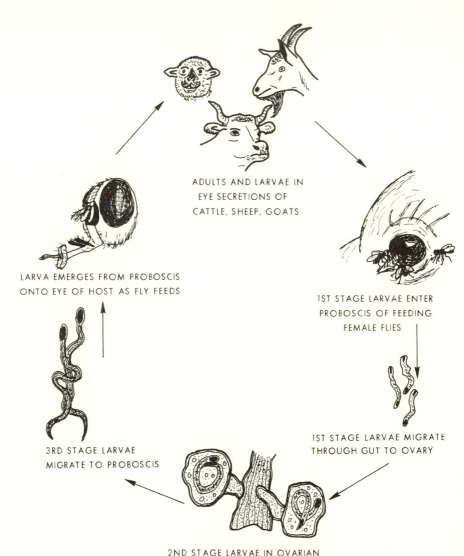

ADULTS AND LARVAE IN
EYE SECRETIONS OF
CATTLE, SHEEP, GOATS

LARVA EMERGES FROM PROBOSCIS
ONTO EYE OF HOST AS FLY FEEDS

1ST STAGE LARVAE ENTER
PROBOSCIS OF FEEDING
FEMALE FLIES

1ST STAGE LARVAE MIGRATE
THROUGH GUT TO OVARY

3RD STAGE LARVAE
MIGRATE TO PROBOSCIS

2ND STAGE LARVAE IN OVARIAN
FOLLICLES ATTACHED TO OVIDUCT

FIG. 54. Life cycle of *Thelazia rhodesii*.

T. gulosa were isolated from the gonads of *M. autumnalis* and *M. larvipara*. Világiová (1967) reports that these larvae developed to the infective stage in *M. autumnalis* in 28-32 days, whereas larvae of *T. skrjabini* took longer.

In California, *Fannia benjamini* females cluster around the eyes of deer and horses, feeding upon the eye secretions, and they also imbibe blood from sores and wounds made by biting flies. Because of their eye-frequenting behavior, they are suspected of being the intermediate host and vector of the California eyeworm, *T. californiensis*. Burnett et al. (1957) obtained development of eyeworm larvae in *F. canicularis*, and also reported finding similar developmental forms in wild-caught *F. benjamini* near San Bernardino.

The observations to date appear to indicate that *Thelazia* species are not host specific but may develop in a number of muscoid flies, though it is not clear whether there may not be host differences. Among other questions remaining to be answered is the possibility that *Thelazia* may overwinter in diapausing adult flies.

Bibliography

Ackert, J.
1920. On the life of *Davainea tetragona* (Molin), a fowl tapeworm. Jour. Parasit., *6*:28-34.
Adams, A.R.D., and Lionnet, F. E.
1933. An outbreak of surra among the wild deer (*Cervus unicolor var*) of Mauritius. Jour. Comp. Path. Ther., *46*:165-167.
Adelheim, R.
1919. Zur Epidemiologie der Ruhr. Hyg. Rundschau, *29*:1-7.
Adler, S., and Theodor, O.
1957. Transmission of disease agents by phlebotomine sand flies. Ann. Rev. Entom., *2*:203-226.
Aitken, T.H.G., Downs, W. G., and Anderson, C. R.
1958. Parasitic *Philornis* flies as possible sources of arbor virus infections (Diptera, Anthomyidae). Proc. Soc. Exp. Biol. Med., *99*:635-637.
Akatov, V. A.
1955. O dlitel'nosti sokhraneniĭa trikhomonad u domashnei mukhi. [On the length of survival of trichomonads in house flies.] Veternariia, *32*:84.
Alcivar, Z. C., and Campos, R. F.
1946. Las moscas, como agentes vectores de enfermedades entericas en Guayaquil. Rev. Ecuatoriana Hig. Med. Trop., *3*:3-14.
Aldrovandi, U.
1602. *Historiae de insectis. Liber tertius*. Bologna, see pp. 342-376.
Aleksander, L. A., and Dansker, V. N.
1935. Rol' mukh v rasprostranenii kishechnykh parazitov. [Role of flies in the dissemination of intestinal parasites.] Trudy Leningrad Inst. Epidem. Bacter. imeni Pastera, *2*:169-179. (Engl. summ. pp. 233-234.)
Aleksandrov, K. A.
1938. Transport of tuberculosis bacteria by flies. Gigiena i Sanitaria No. 9, pp. 50-52.
Alessandrini, G.
1912. Vitalità del vibrione colerigeno nelle mosche. Ann. Igiene (Sper.) New Series, *22*:634-650.
Alexander, R. A.
1948. The 1944 epizootic of horsesickness in the Middle East. Onderstepoort Jour. Veter. Sci. Anim. Indust., *23*:77-92.
Alibert, J. L.
1806. Descriptions des maladies de la peau. Paris, 232 pp., see p. 164.
Aly, M.
1940. Amoebic and bacillary dysentery in Egypt. Lab. Med. Progr., *1*:27-33.
An, S.
1933. Microorganisms found on *Musca domestica* in Keijo. Jour. Chosen Med. Assoc., *23*:1443-1452.

Anastas'ev, N. M.
  1952. Problems of general hygiene in dysentery prophylaxis [in Russian]. Zhurn. Mikrob. Epidem. Immun., *5*:39-41.
Anderson, J. F., and Frost, W. H.
  1912. Transmission of poliomyelitis by means of the stable fly (*Stomoxys calcitrans*). Publ. Health Repts., *27*:1733-1735.
Anderson, J. F., and Frost, W. H.
  1913. Poliomyelitis. Further attempts to transmit the disease through the agency of the stable fly (*Stomoxys calcitrans*). Publ. Health Repts., *28*:833-837.
Anderson, J. R., and Poorbaugh, J. H.
  1964. Observations on the ethology and ecology of various Diptera associated with Northern California poultry ranches. Jour. Med. Entom., *1*:131-147.
Anderson, J. R., and Poorbaugh, J. H.
  1968. The face fly, *Musca autumnalis*, a new livestock fly is now moving toward California. Calif. Agric., *22*:4-6.
André, C.
  1908. Les mouches comme agents de dissemination du bacilli de Koch. 6th Intl. Congr. Tuberculosis, Washington, D.C., pp. 162-166.
Anonymous
  1874. Jahresbericht über die Leistungen und Fortschritte in der gesammten Medicin. Virchow-Hirsch's Jahresber., *4*:692.
Anonymous
  1893. Bacteriology of the leprosy commission. Lancet, May 13, pp. 1153-1154.
Anonymous
  1909. Annual report of the Sanitary Commissioner with the government of India, see p. 50.
Anonymous
  1911. Third biennial report of the State Board of Health and Vital Statistics of Minnesota, 1909-1910, see pp. 205-209.
Anonymous
  1915. Thirty-second annual report of the Superintendent of Health of the city of Providence, for the year 1914, see pp. 34-35.
Anonymous
  1928. Cholera spread by flies. Jour. Amer. Med. Assoc., *90*:2048.
Anonymous
  1936. Blood parasites of cattle. Australian Council Scient. Indust. Res. Rept., *9*:31.
Anonymous
  1945. Diarrhoea and dysentery. Field Ser. Hyg. Notes, India, pp. 338-349.
Anonymous
  1956. *Shigella* finds ideal environment. Publ. Health Rept., *71*:1242-1243.
Anonymous
  1962. Expert committee on trypanosomiasis. Wld. Health Org. Tech. Rept. Ser., pp. 5-57.
Anonymous
  1964. Enteric infections. Report of a WHO expert committee. Wld. Health Org. Tech. Rept. Ser., No. 288, 35 pp.
Anonymous
  1968. Microbiological aspects of food hygiene. Report of WHO expert committee. Wld. Health Org. Tech. Rept. Ser. No. 399, 64 pp.
Anonymous
  1971. A waterborne epidemic of salmonellosis in Riverside, California, 1965. A collaborative report. Amer. Jour. Epidem., *93*:33-48.

Antonipulle, P.
1957. Some field observations on the breeding places of *Siphunculina funicola* (Diptera, Oscinidae) in Ceylon. Ceylon Jour. Sci. Biol. Sci., *1*:1-5.

Ara, F.
1933. Ulteriori ricerche sull' importanza della mosca nella diffusione della febbre tifoide. Ig. Moderna, *26*:327-335.

Ara, F., and Marengo, U.
1932. Sull' importanza della mosca nella diffusione della febbre tifoide. Bull. Soc. Ital. Biol., *7*:150-155.

Aradi, M. P.
1956. Die Bedeutung des Fliegenproblems vom Gesichtspunkt der Hygiene in Unseren Lagern und Neuere Untersuchungen in Zusammenhang mit der Rolle des Fliegen-Vektors [in Hungarian]. Különlenyomat a Katonaorvosi Szemle, *8*:406-413.

Aradi, M. P., and Mihályi, F.
1971. Seasonal investigations of flies visiting food markets in Budapest. Acta. Zool. Acad. Sci. Hung., *17*:1-10.

Araujo, E. de
1934. Nota sobre a veiculaçao da bouba. Bahia Med., *5*:384-385.

Arizumi, S.
1934. On the potential transmission of *B. leprae* by certain insects. Taiwan igakkai zasshi [abstract section], *33*:54-56.

Armstrong, D. B.
1914a. Flies and diarrhoeal diseases. N.Y. Assoc. for Improving the Conditions of the Poor. Bur. Publ. Health Hyg., Dept. Local Welfare, No. 79, 29 pp.

Armstrong, D. B.
1914b. An investigation into the relationship of the house fly to disease—the special importance of the fly in infant welfare. Amer. Jour. Publ. Health, *4*:185-196.

Armstrong, H. E.
1905. Remarks on the question of the aerial dissemination of smallpox infection round smallpox hospitals. Jour. Roy. Sanit. Inst., *26*:193-199.

Arskiĭ, V. G., Gadzhei, E. F., Zatsepin, N. I., and Yasinskiĭ, A. V.
1961. The role of flies in the seasonal character of dysentery [in Russian]. Zhurn. Mikrob. Epidem. Immun., *32*:27-32.

Asahina, S., Ogata, K., Noguchi, Y., Uchida, S., and Murata, M.
1963. Detection of polioviruses from flies and cockroaches captured during 1961 epidemics in Kumamoto Prefecture. Japanese Jour. Sanit. Zool., *14*:28-31.

Asami, S.
1934. Concerning the carrying of the leprosy bacillus by insects. Intl. Jour. Leprosy, *2*:465-469.

Aserkoff, B., Schroeder, S. A., and Brachman, P. S.
1970. Salmonellosis in the United States—a five-year review. Amer. Jour. Epidem., *92*:13-24.

Attimonelli, R.
1940. Di un metodo semplice per la colorazione degli organi di motilità dei protozoi. Pathologica, *32*, No. 579, p. 44.

Aubertot, M.
1923. Sur la dissémination et la transport de nématodes du genre *Rhabditis* par les diptères. Compt. Rend. Acad. Sci., *176*:1257-1260.

Auché, A.
1906. Transport des bacilles dysentériques par les mouches. Compt. Rend. Soc. Biol., *61*:450-452.

Awati, P. R.
 1920. Bionomics of houseflies IV. Some notes on the life-history of *Musca*. Indian Jour. Med. Res., *8*:80-88.
Aylett, W. R.
 1896. Do flies spread tuberculosis? The Virginia Med. Semi-Monthly, *1*:163-164.
Ayyar, T.V.R.
 1917. Notes on the life history and habits of the "eye fly," *Siphonella funicola*, de Meij. Madras. Agric. Dept. Yearbook, pp. 76-83.
Babudieri, B.
 1958. Animal reservoirs of leptospires. Ann. N.Y. Acad. Sci., *70*:393-413.
Bacot, A. W.
 1911a. On the persistence of bacilli in the gut of an insect during metamorphosis. Trans. Entom. Soc. London, Part II, pp. 497-500.
Bacot, A. W.
 1911b. The persistence of *Bacillus pyocyaneus* in pupae and imagines of *Musca domestica* raised from larvae experimentally infected with the bacillus. Parasit., *4*:68-74.
Bacot, A. W.
 1912. LXX. On the survival of bacteria in the alimentary canal of fleas during metamorphosis from larva to adult. Jour. Hyg., Plague Suppl. 1, pp. 655-664.
Baer, W. S.
 1931. The treatment of chronic osteomyelitis with the maggot (larva of the blow fly). Jour. Bone Joint Surg., *29*:438-475.
Bahr, L.
 1952. Nogle undersøgelser vedrørende "sommermastitis." Dansk MaanedsKr. Dyrlaeg., *62*:367-394.
Bahr, L.
 1953. "Summer-Mastitis" (Preliminary Rept.; summary). Proc. XV Intl. Veter. Congr. Stockholm, Vol. 2, Part 1, pp. 849-852; Part 2, p. 348.
Bahr, L.
 1955. *Hydrothaea irritans* Fallén as a carrier of infection. Acta Path. Microb. Scand. Suppl., *108*:107-108.
Bahr, P. H.
 1912. Report to the London School of Tropical Medicine on investigations on dysentery in Fiji during the year 1910. London School Hyg. Trop. Med. Res. Memoir Ser., see pp. 21-25, 28, 53-56.
Bahr, P. H.
 1914. A study of epidemic dysentery in the Fiji Islands, with special reference to its epidemiology and treatment. British Med. Jour., *1*:294-296.
Bajon, B.
 1781. Ubhandlungen von Krankheiten auf der Insel Cayenne und dem Franzosischen Guiane. Keyser Publ., Vol. 2, 135 pp., see pp. 52, 53, 68, 69. Transl. from French ed., Mémoires pour servir à l'histoire de la Guyane et de Cayenne. Paris, 1777.
Bakry, G.
 1955. Preliminary investigations of the seasonal incidence of *Musca sorbens* Wied. in relation to acute ophthalmias in Egypt. Jour. Egyptian Med. Assoc., *38*:507-511.
Baldrey, F.S.H.
 1911. The evolution of *Trypanosoma evansi* through the fly: *Tabanus* and *Stomoxys*. Jour. Trop. Vet. Sci., *6*:271-282.
Balzam, N.
 1937. Destin de la flore bacteriénne pendant la métamorphose de la mouche à viande (*Calliphora eryt[h]rocephala*). Ann. Inst. Pasteur, *58*:181-211.

348 BIBLIOGRAPHY

Bancroft, E.
1769. An essay on the natural history of Guiana, in South America. London, see pp. 385-387.

Bang, F. B., and Glaser, R. W.
1943. The persistence of poliomyelitis virus in flies. Amer. Jour. Hyg., *37*:320-324.

Barinskiĭ, F. G.
1952. A combined system of measures for dysentery prophylaxis. Zhurn. Mikrob. Epidem. Immun., *11*:38-43.

Barlow, J. S., and House, H. L.
1956. Ethylene oxide for sterilizing diets. Science, *123*:229.

Barnard, C. C.
1952a. Yaws and flies. Past and present opinions on the role of flies in the transmission of framboesia tropica. Part I. Jour. Trop. Med. Hyg., *55*: 100-114.

Barnard, C. C.
1952b. Past and present opinions on the role of flies in the transmission of framboesia tropica. Jour. Trop. Med. Hyg., *55*:135-141.

Barnett, H. S., Parneke, W. E., Lee, R. D., and Wagner, E. D.
1957. Observations on the life cycle of *Thelazia californiensis* Price, 1930. Jour. Parasit., *43*:433-439.

Barnhart, C. S., and Chadwick, L. E.
1953. A "fly factor" in attractant studies. Science, *117*:104-105.

Barrett, W. L. Jr.
1937. Natural dispersion of *Cochlyomyia americana*. Jour. Econ. Entom., *30*: 873-876.

Barros, R. Donoso, and Urzúa, L. Ferrada
1947. Hallazgo de moscas naturalmente infectadas por bacilos acido-alcohol-resistentes (probablemente grupo Hansen) en leprosarias de Isla de Pascua. Hoja Tisiol., *7*:133-134.

Bartlett, D. E.
1966. (Personal communication).

Bassett, D.C.J.
1967. (Personal communication).

Bassett, D.C.J.
1970. *Hippelates* flies and streptococcal skin infection in Trinidad. Trans. Roy. Soc. Trop. Med. Hyg., *64*:138-147.

Bay, D. E., Pitts, C. W., and Ward, G.
1968. Oviposition and development of the face fly in feces of six species of animals. Jour. Econ. Entom., *61*:1733-1735.

Bay, E. C., and Legner, E. F.
1963. The prospect for the biological control of *Hippelates collusor* (Townsend) in southern California. Proc. & Papers 31st Conf. California Mosquito Contr. Assoc., pp. 76-79.

Beck, M.
1910. Experimentelle Beiträge zur Infektion mit *Trypanosoma gambiense* und zur Heilung der menschlichen Trypanosomiasis. Arb. Kaiser. Gesundheitsamte, *34*:318-376.

Beck, M. D., Muñoz, J. A., and Scrimshaw, N. S.
1957. Studies on diarrheal diseases in Central America. I. Preliminary findings on cultural surveys of normal population groups in Guatemala. Amer. Jour. Trop. Med. Hyg., *6*:62-71.

Beesley, W. N.
1968. Observations on the biology of the ox warble-fly (*Hypoderma*: Diptera, Oestridae). II. Bacteriostatic properties of larval extracts. Ann. Trop. Med. Parasit., *62*:8-12.

Begg, M., and Sang, J. H.
   1950.   A method for collecting and sterilizing large numbers of *Drosophila* eggs. Science, *112*:11-12.
Beliaeva, M.
   1960.   In: Trudy Krasnoiarskogo Sel'skokhoziaistrennogo Inst., *6*:95 and on.
Bell, J. F., and Moore, G. J.
   1971.   Susceptibility of carnivores to rabies virus administered orally. Amer. Jour. Epidem., *93*:176-182.
Bengston, I. A.
   1922.   Preliminary note on a toxin-producing anaerobe isolated from the larvae of *Lucilia caesar*. U.S. Publ. Health Repts., *37*:164-170.
Bengston, I. A.
   1933.   Seasonal acute conjunctivitis occurring in the Southern States. U.S. Publ. Health Repts., *48*:917-926.
Bennett, G. F.
   1955.   Studies on *Cuterebra emasculator* Fitch 1856 (Diptera: Cuterebridae) and a discussion of the status of the genus *Cephenemyia* Ltr. 1818. Canad. Jour. Zool., *33*:75-98.
Benson, O. L., and Wingo, C. W.
   1963.   Investigation of the face fly in Missouri. Jour. Econ. Entom., *56*:251-258.
Berberian, D. A.
   1938.   Successful transmission of cutaneous leishmaniasis by the bites of *Stomoxys calcitrans*. Proc. Soc. Exper. Biol. Med., *38*:254-256.
Berberian, D. A.
   1948.   The use of DDT residual spray in malaria control and its effect on general sanitation in rural districts. Jour. Palestine Arab Med. Assoc., *3*: 49-61.
Beresoff, W. F.
   1914.   Die schlafenden Fliegen als Infektionsträger. Centralbl. Bakt. I. Abt. Orig., *74*:244-250.
Berkaloff, A., Berreur, P., and Cals, P.
   1967.   Multiplication du virus Sindbis et métamorphose de *Calliphora erythrocephala* (Meig.). C. R. Acad. Sci. Paris, *264* (Série D): 1545-1548.
Berndt, K. P.
   1969.   Calliphoriden -Zucht ohne Geruchsbelaestigung. [Breeding *Calliphora* without unpleasant odors.] Angew. Parasit., *10*:233-236.
Berrier, H.
   1940.   Contribution à l'étude des substances du type auxinique dans le règne animal. Bull. Biol. France et Belgique, Suppl. 27, pp. 112-115, 151-159, 181-183, 196, 198-200.
Bertarelli, E.
   1909.   Diffusione del tifo, colle mosche e mosche portatrici di bacilli specifici nelle case dei tifosi. (*Richerche sperimentali.*) Bull. Soc. Med. Parma, *2*:262-272.
Bertarelli, E.
   1910.   Verbreitung des Typhus durch die Fliegen. Fliegen als Trägerinnen spezifischer Bacillen in den Häusern von Typhuskranken. Centralbl. Bakt. I. Abt., Orig., *53*:486-495.
Berti, A. L.
   1958.   Importancia del control de moscas en la incidencia de diarreas y enteritis en los tropicos. Sextos Congresos Internacionales de Medicina Tropical y de Paludismo, 23 pp.

350   BIBLIOGRAPHY

Bettini, S.
  1965.   Acquired immune response of the house fly, *Musca domestica* (Linnaeus), to injected venom of the spider *Latrodectus mactans tredecimguttatus* (Rossi). Jour. Invert. Path., 7:378-383.
Bevan, L.E.W.
  1917.   Report of the government Veterinary Bacteriologist for the year 1916, Southern Rhodesia. 17 pp.
Beyer, G. E.
  1920.   Supplementary report on disease-carrying flies in public markets. Louisiana St. Bd. Health Quart. Bull., *11*:102-107.
Beyer, G. E.
  1925.   The bacteriology of market flies of New Orleans. Louisiana St. Bd. Health Quart. Bull., *16*:110-116.
Bhatia, H. L.
  1935.   The role of *Tabanus orientis* Wlk. and *Stomoxys calcitrans* Linn. in the mechanical transmission of rinderpest. Indian Jour. Veter. Science and Anim. Husb., *5*:2-22.
Bigham, J. T.
  1941.   *Hippelates* (eye gnat) investigations in the southeastern states. Jour. Econ. Entom., *34*:439-444.
Birk, W.
  1932.   Die Uebertragung des Paratyphus Breslau durch die *Stomoxys calcitrans*. Zentralbl. Bakt. Parasit. Infekt. Hyg., I Abt., *124*:280-300.
Bishopp, F. C.
  1915.   Flies which cause myiasis in man and animals. Some aspects of the problem. Jour. Econ. Entom., *8*:317-329.
Bishopp, F. C., Dove, W. E., and Parman, D. C.
  1915.   Notes on certain points of economic importance in the biology of the house fly. Jour. Econ. Entom., *8*:466-474.
Bishopp, F. C.
  1917.   Some problems in insect control about abattoirs and packing houses. Jour. Econ. Entom., *10*:269-277.
Bishopp, F. C., and Laake, E. W.
  1921.   Dispersion of flies by flight. Jour. Agric. Res., *21*:729-766.
Bishopp, F. C.
  1922-3.   Limberneck of fowls produced by fly larvae. Jour. Parasit., *9*:170-173.
Bishopp, F. C.
  1939.   The stable fly: how to prevent its annoyance and its losses to live stock. Farmer's Bull., U. S. Dept. Agric., No. 1097, 18 pp.
Blanc, G., Bruneau, J., and Chabaud, A.
  1950.   Quelques essais de transmission de la toxoplasmose par arthropodes piqueurs. Ann. Inst. Pasteur, *78*:277-280.
Bocharto, S.
  1712.   *Hierozoicon, sive bipartitum opus de animalibus. S. scripturae. Pars prior libris IV. De avibus, serpentibus, insectis, aquaticis, & fabulosis animalibus agit.* Holland. See pp. 496-501.
Bodenheimer, F. S.
  1928.   Geschichte der Entomologie bis Linné. Bd. I. W. Junk, Berlin.
Bodenheimer, F. S.
  1960.   Animal and man in bible lands. E. J. Brill, Leiden. 232 pp. See pp. 73-74.
Bodkin, G. E.
  1917.   Report of the economic biologist. In: British Guiana Dept. of Sci. Agric. Rept. for the Year 1916. 14 pp.

Bogdanow, E. A.
1906. Über das Züchten der Larven der gewöhnlichen Fleischfliege (*Calliphora vomitoria*) in sterilisierten Nehrmitteln. Pfluger's Arch., *113*: 97-105.

Bogoiavlenskiĭ, N. A., and Demidova, A. Ĩa.
1928. Rol' mukh v perenesnii ĩaits paraziticheskikh cherveĭ. [Role of flies in dissemination of parasitic worm eggs.] Vrach. Gaz. no. 16, pp. 1101-1103.

Bohart, G. E., and Gressitt, J. L.
1951. Filth-inhabiting flies of Guam. Bernice P. Bishop Museum Bull. 204, 152 pp.

Boikov, B. V.
1932. Rol' mukh v rasprostranenii briushnogo tifa i drugikh zheludochno-kishechnykh zabolevaniĭ. [Role of flies in dissemination of typhoid and other stomach and intestinal diseases.] Zhurn. Mikrob. Epidem. Immun., no. 7-8, pp. 26-39.

Bolaños, R.
1959. Frecuencia de *Salmonella* y *Shigella* en moscas domésticas colectadas en la ciudad de San José. Rev. Biol. Trop., 7:207-210.

Bollinger, O.
1874. Ueber die Milzbrandseuche in den baierischen Alpen. Deutsches Arch. Klinische Med., *14*:269-290.

Bollinger, O.
1875. Anthrax in man. In: Ziemssen's Cyclopedia of the Practice of Medicine, *3*:409.

Bolton, H. T., and Hansens, E. J.
1970. Ability of the house fly, *Musca domestica*, to ingest and transmit viable spores of selected fungi. Ann. Entom. Soc. Amer., *63*:98-100.

Bolwig, N.
1946. Senses and sense organs of the house fly larvae. Vid. Medd. dansk nat.-hist. Foren, *109*:81-217.

Bos, A.
1932. Overbrengingsproeven van hoenderpokken door *Anopheles maculipennis* Mg., *Theobaldia annulata* Schr. en *Stomoxys calcitrans*. Tijdschr. Diergeneesk., *59*: 191-194.

Boudreau, F. G., Brain, C. K., and McCampbell, E. F.
1914. Acute poliomyelitis, with special reference to the disease in Ohio, and certain transmission experiments. Ohio State Bd. Health Monthly Bull., *4*:23-62.

Bouet, G., and Roubaud, E.
1912. Expériences de transmission des trypanosomiases animales del'Afrique Occidentale française, par les *Stomoxes*. Bull. Soc. Path. Exot., Paris, *5*: 544-550.

Bouffard, G.
1907a. Sur l'étiologie de la souma, trypanosomiase de Soudan français. Compt. Rend. Soc. Biol., Paris, *62*:71-73.

Bouffard, G.
1907b. La souma, trypanosomiase du Soudan français. Ann. Inst. Pasteur, *21*: 587-592.

Bouillant, A. M., and Hanson, R. P.
1965a. Epizootiology on mink enteritis: I, Stability of mink enteritis virus in feces exposed to natural environmental factors. Canadian Jour. Comp. Med. Veter. Science, *29*:125-128.

Bouillant, A. M., Lee, V. H., and Hanson, R. P.
1965b. Epizootiology on mink enteritis: II, *Musca domestica* L. as a possible vector of virus. Canadian Jour. Comp. Med. Veter. Science, *29*:148-152.

Bourke, A.T.C.
1964. An evaluation of the role of *Phormia regina* in transmission of eastern encephalitis. Publ. Health Rept., *79*:522-524.

Bowmer, E. J.
1964. The challenge of salmonellosis major public health problem. Amer. Jour. Med. Sci., *247*:467-501.

Boyd, J.S.K.
1957. Dysentery: some personal experiences and observations. Trans. Roy. Soc. Trop. Med. Hyg., *51*:471-487.

Bozhenko, V. P., and Kniazevskiĭ, A. N.
1948. Osenniaia mukha-zhigalka *Stomoxys calcitrans* L. kak perenoschik tula-remii. [The autumn biting fly *Stomoxys calcitrans* L. as a carrier of tularemia.] Izvest. Akad. Nauk. Kazakh. SSR, Seriia Parasit., *6*:62-66.

Bradley, R.
1724. New improvements of planting and gardening, both philosophical and practical. 4th ed., London. [Cited by Hewitt, Canad. Entom., *4*:396-399, 1915.]

Brain, C. K.
1913. *Stomoxys calcitrans* Linn., Part II. Ann. Entom. Soc. Amer., *6*:197-201.

Brandariz, C.
1958. Posible papel de la mosca domestica en la etiologia de la ulceras esti-vales. Gac. Veter. (Buenos Aires), *20*:84-87.

Branford, R.
1919. Note on an outbreak of surra at the government cattle farm, hissar, and on cases treated. Agri. Jour. India, *14*:762-773.

Breasted, J. H.
1912. A history of Egypt, 2nd ed., Charles Scribner's Sons, I-XXIX, 634 pp. See p. 301.

Brefeld, O.
1870. Entwicklungsgeschichte der *Empusa muscae* und *Empusa radicans*. Bot. Zeit., *28*:161-166, 177-186.

Brefeld, O.
1873. Untersuchungen über die Entwicklung der *Empusa muscae* und *Empusa radicans* und die durch sie verursachten Epidemien der Stubenfliegen und Raupen. Abhandl. Nature Gesellsch. Halle., *12*:1-50.

Briggs, J. D.
1958. Humoral immunity in lepidopterous larvae. Jour. Exp. Zool., *138*:155-188.

Brock, R.R.A.
1957. Typhoid fever in Darwin, 1955. Med. Jour. Australia, *2*:78-82.

Brookes, V. J., and Fraenkel, G.
1958. The nutrition of the larva of the housefly, *Musca domestica* (L.). Physiol. Zool., *31*:208-223.

Brooks, M. A., and Kurtti, T. J.
1971. Insect cell and tissue culture. Ann. Rev. Entom., *16*:27-52.

Brown, A.W.A.
1936. Excretion of ammonia and uric acid in muscid larvae. Jour. Exp. Biol., *13*:131-139.

Brown, G. C.
1955. Virus excretion and antibody response in clinical and subclinical cases of poliomyelitis. Ann. New York Acad. Sci., *61*:989-997.

Brown, J. F.
1965. Preliminary investigations on transmission of *Moraxella bovis* by the face fly, *Musca autumnalis*. Clemson University; Thesis, August, 1965.

Bruce, D., Hamerton, A. E., Bateman, H. R., and Mackie, F. P.
  1910.   Trypanosome diseases of domestic animals in Uganda. I. *Trypanosoma pecorum*. Proc. Roy. Soc. Lond., *82*:468-479.
Bruch, H. A., Ascoli, W., Scrimshaw, N. S., and Gordon, J. E.
  1963.   Studies of diarrheal disease in Central America. Jour. Trop. Med. Hyg., *12*:567-579.
Brues, C. T., and Sheppard, P.A.E.
  1912.   The possible etiological relation of certain biting insects to the spread of infantile paralysis. Jour. Econ. Entom., *5*:305-324.
Brust, M., and Fraenkel, G.
  1955.   The nutritional requirements of the larvae of a blow fly, *Phormia regina* (Meig.). Physiol. Zool., *28*:186-204.
Brygoo, E. R., and LeNoc, P.
  1962.   Les salmonelles des cameleons malagaches. II. Etude expérimentale. Extrait Arch. Inst. Pasteur Madagascar, *30*:105-115.
Brygoo, E. R., Sureau, P., and LeNoc, P.
  1962.   Virus et germes fécaux des mouches de l'agglomération urbaine de Tananarive. Bull. Soc. Path., Exot., *55*:866-881.
Buchanan, M. B.
  1897.   Cholera diffusion by flies. Ind. Med. Gaz. *3*:86-87.
Buchanan, R. M.
  1907.   The carriage of infection by flies. Lancet. *2*:216-218.
Buchanan, R. M.
  1913.   *Empusa muscae* as a carrier of bacterial infection from the house-fly. British Med. Jour., *2*:1369-1374.
Budd, W.
  1862.   Observations on the occurrence (hitherto unnoticed) of malignant pustule in England. Lancet, Aug. 9, pp. 164-165.
Bull, L. B.
  1919.   A contribution to the study of habronemiasis: a clinical, pathological, and experimental investigation of a granulomatous condition of the horse—habronemic granuloma. Proc. Roy. Soc. South Australia, *43*:85-141.
Bulling, E., Bakri, G., and Kirchberg, E.
  1959.   Necrophage Fliegen als Salmonellenverbreiter. Intl. Sympos. über Schädliche Fliegen. Zeitsch. Angew. Zool., *46*:331-332.
Burgdorfer, W.
  1956.   The possible role of ticks as vectors of leptospirae. I. Transmission of *Leptospira pomona* by the argasid tick, *Ornithodoros turicata*, and the persistence of this organism in its tissues. Exp. Parasit., *5*:571-579.
Burgess, R. W.
  1951.   The life history and breeding habits of the eye gnat, *Hippelates pusio* Loew, in the Coachella Valley, Riverside County, California. Amer. Jour. Hyg., *53*:164-177.
Burke, J. P., Ingall, D., Klein, J. O., Gezon, H. M., and Finland, M.
  1971.   *Proteus mirabilis* infections in a hospital nursery traced to a human carrier. New England Jour. Med. *284*:115-121.
Burnett, H. S., Parneke, W. E., Lee, R. D., and Wagner, E. D.
  1957.   Observations on the life cycle of *Thelazia californiensis* Price 1930. Jour. Parasit., *43*:433.
Butler, T. H.
  1915.   Some aspects of ophthalmology in Palestine. The Ophthalmoscope, pp. 60-69.
Buxton, P. A. and Hopkins, G.H.E.
  1927.   Researches in Polynesia and Melanesia, Part III, Medical Entomology, London School Hyg. and Trop. Med., Mem., *1*:51-85.

Buxton, P. A.
1920.   The importance of the housefly as a carrier of *E. histolytica*. British Med. Jour., *1*:142-144.

Buxton, P. A.
1955.   Natural history of tsetse-flies. London Soc. Hyg. Trop. Med., Mem. 10, 766 pp.

Bychkov, V. A.
1932.   O dlitel'nosti khraneniĭa mukhami *Bacterium prodigiosum*. [The duration of the persistence of *Bacterium prodigiosum* in flies.] Parasit. Sborn. Zool. Inst. Akad. Nauk SSSR, *3*:149-159.

Cabasso, V.
1942.   Préparation d'une tuberculine très active à partir d'un bacille acido-résistant saprophyte. Bull. Hyg., *17*:660.

Calandruccio, S.
1906.   Ulteriori richerche sulla *Taenia nana*. Boll. Accad. Gioenia, Catania. Fasc. 89, pp. 15-19.

Campbell, A. W., Cleland, J. B., and Bradley, B.
1918.   A contribution to the experimental pathology of acute poliomyelitis (infantile paralysis). Med. Jour. Australia, *1*:123-128.

Cantwell, G. E., and Franklin, B. A.
1966.   Inactivation by irradiation of spores of *Bacillus thuringiensis* var. *thuringiensis*. Jour. Invert. Path., *8*:256-258.

Cao, G.
1898.   Sul passaggio dei microorganismi attraverso l'intestino di alcuni insetti. Ufficiale Sanit., *11*:337-348.

Cao, G.
1906a. Nuove osservazioni sul passaggio dei microorganismi a traverso l'intestino di alcun insetti. Ann. Ig. Sper., *16*:339-368.

Cao, G.
1906b. Sul passaggio dei germi a traverso le larve di alcuni insetti. Ann. Ig. Sper., *16*:645-664.

Cardamatis, J. P., and Melissidis, A.
1911.   Du rôle probable de la mouche domestique dans la transmission des "Leishmania." Bull. Soc. Path. Exot., *4*:459-461.

Carment, A. G.
1916.   Public Health Division's Report for 1916. 2. Trypanosomiases. Zanzibar Prot. Rept. on Med. Div. for Year 1915. pp. 101-104.

Carmichael, J.
1933.   Annual Report, 1933, of the Veterinary Pathol., Entebbe. Research. (3) Trypanosomiasis. Uganda Prot., Ann. Rept. Vet. Dept. Year ended 1933, pp. 29-45.

Carmichael, J.
1948.   Discussion. The epidemiology of trypanosomiasis in man and animals. Proc. Roy. Soc. Med., *41*:551-558.

Carter, H. J.
1861.   On a bisexual nematoid worm which infests the common house-fly (*Musca domestica*) in Bombay. Trans. Med. Phys. Soc. Bombay. n.s. *6*:12-16.

Castellani, A.
1907.   Experimental investigations on framboesia tropica (yaws). Jour. Hyg., *7*:558-569.

Castiglioni, A.
1927.   Storia dell medicina. Milano, Societá Editrice "Unitas." 959 pp. See pp. 72-73.

Castiglioni, M. C., and Raimondi, G. R.
1961.   First results of tissue culture in *Drosophila*. Experientia, *17*:88-90.

Cattani, G.
  1886.   Studi sul colera. Gazz. Ospit., *7*:611-612.
Causey, O. R.
  1932.   Sterilization and growth of the eggs and larvae of the blowfly. Amer. Jour. Hyg., *15*:276-286.
Cavaceppi, L.
  1951.   Le cognizioni degli antichi sulle mosche nel "*De animalibus insectis*" di Ulisse Aldrovandi. Riv. Parassit., *12*:60-64.
Cazalbou, L.
  1907.   A propos de l'étiologie de la souma. Compt. Rend. Hebdom. Soc. Biol., *62*:1104-1106.
Čeledovà, V., Prokešová, N., Havlík, B., and Müller, O.
  1963.   Demonstration of migration of *Musca domestica* from pig sheds into human dwellings. Jour. Hyg. Epidem. Microb. Immun., *7*:360-370.
Celli, A., and Alessi, G.
  1888.   Transmissibilitá dei germi patogeni mediante le dejezioni delle mosche. Bull. Soc. Lancisiana Osped. Roma. *1*:5-8.
Chaloner, R.
  1961.   Bible flies and fleas. Entom. Monthly Mag., *97*:26.
Chambers, H. D.
  1938a.   Warmth and humidity as predisposing factors in the incidence of yaws. Trans. Roy. Soc. Trop. Med. Hyg., *31*:451-456.
Chambers, H. D.
  1938b.   *Yaws (Framboesia tropica)*. Churchill, London. pp. 24-25, 49, 65-69, 72-76.
Champlain, R. A., and Fisk, F. W.
  1956.   The digestive enzymes of the stable fly, *Stomoxys calcitrans* (L.). Ohio Jour. Sci., *56*:52-62.
Chandler, A. C., and Read, C. P.
  1961.   Introduction to parasitology, with special reference to the parasites of man. John Wiley and Sons, Inc. New York, 822 pp., see p. 63.
Chang, J. T., and Wang, M. Y.
  1958.   Nutritional requirements of the common housefly, *Musca domestica vicina* Macq. Nature, *181*:566.
Chang, K.
  1940.   Domestic flies as mechanical carriers of certain human intestinal parasites in Chengtu. West China Border Res. Soc. Jour., Ser. B, *12*:92-98.
Chantemesse, A., and Borel, F.
  1905.   Mouches et choléra. Bull. Acad. Med. (Paris). *54*:252-259.
Chatelain, P.
  1917.   Traitement de la lymphangite épizootique. Rev. Gén. Méd. Vet., *26*: 289 and on.
Chatton, E.
  1912.   *Leptomonas* de deux Borborinae (Muscides). Evolution de *Leptomonas legerorum* n. sp. Compt. Rend. Soc. Biol., *73*:286-289.
Chatton, E., and Krempf, A.
  1911.   Sur le cycle évolutif et la position systématique des protistes du genre *Octosporea* Fl., parasites des muscides. Bull. Soc. Zool. France, *36*:172-179.
Chatton, E., and Leger, M.
  1912.   Trypanosomides et membrane péritrophique chez les Drosophiles. Culture et évolution. Compt. Rend. Soc. Biol., *72*:453-456.
Chaudhuri, R. P., Kumar, P., and Khan, M. H.
  1966.   Role of stable-fly, *Stomoxys calcitrans*, in the transmission of surra in India. Indian Jour. Vet. Sci., *36*:18-28.

Chaverri, P.
1970.   La estomatitis vesicular. Bol. Ofic. Sanit. Panamer., March.
Cheatum, E. L., Reilly, J. R., and Fordham, S. C.
1957.   Botulism in game farm pheasants. Trans. No. Amer. Wildlife Conf., 22:170-179.
Chebotarevich, N. D.
1937.   O biologii komnatoĭ mukh *Musca vicina* i bor'ba s neĩu v srednei Azii; *in* Problemy parazit. Fauna Turkmenistana. Izd. Ak. Nauk SSSR.
Cheng, T-H.
1967.   Frequency of pinkeye incidence in cattle in relation to face fly abundance. Jour. Econ. Entom., 60:598-599.
Child, F. S., and Roberts, E. F.
1931.   The treatment of chronic osteomyelitis with live maggots. New York State Jour. Med., 31:937-943.
Chillcott, J. G.
1960.   A revision of the nearctic species of Fanniinae (Diptera: Muscidae). Canad. Entom., 92: Suppl 14, 295 pp.
Chinery, W. A.
1965.   Mechanical transmission of some disease pathogens by insects—a review. Ghana Jour. Sci., 5:249-263.
Chmelicek, J. F.
1899.   My observations on the typhoid-fever epidemic in southern camps, and its treatment. New York Med. Jour., 70:193-198.
Chow, C. Y.
1940.   The common blue-bottle fly, *Chrysomyia megacephala*, as a carrier of pathogenic bacteria in Peiping, China. Chinese Med. Jour., 57:145-153.
Chu, I-Wu, and Axtell, R. C.
1971.   Fine structure of the dorsal organ of the house fly larva, *Musca domestica* L. Zeits. Zellforsch., 117:17-34; also 127:287-305, 1972; 130:489-495, 1972.
Clapham, P.
1939.   On flies as intermediate hosts of *Syngamus trachea*. Jour. Helmin., 17: 61-64.
Clark, P. F., Fraser, F. R., and Amoss, H. L.
1914.   The relation to the blood of the virus of epidemic poliomyelitis. Jour. Exp. Med., 19:223-233.
Cleland, J. B.
1912.   The relationship of insects to disease in man in Australia. Second Rept. Govt. Bur. Microb. for 1910-11, pp. 141-158.
Cleland, J. B.
1913.   Insects and their relationship to disease in man in Australia. Trans. Australasian Med. Congr. 1:548-570.
Cobb, J. O.
1905.   Is the common house fly a factor in the spread of tuberculosis? Amer. Med., 9:475-477.
Cochrane, E.W.W.
1912.   A small epidemic of typhoid fever in connection with specifically infected flies. Jour. Roy. Army Med. Corps, 18:271-276.
Coffey, J. H., and Schoof, H. F.
1949.   The control of domestic flies. Comm. Disease Center, U.S. Publ. Health Service, Brochure, Atlanta, Ga.
Coffey, M. D.
1966.   Studies on the association of flies (Diptera) with dung in southeastern Washington. Ann. Entom. Soc. Amer., 59:207-218.

Cohn, F.
  1855. *Empusa muscae* und die Krankheit der Stubenfliegen. Nov. Act. Acad.
         Caes. Leopold.-Carol. Nat. Cur., *25*: (Part I), see pp. 301, 309, 316-
         327, 342-345, 352-356.
Cole, F. R., and Schlinger, E. I.
  1969. The flies of western North America, Univ. of Calif. Press, Berkeley and
         Los Angeles, 693 pp.
Coleman, P. J., and Maier, P. P.
  1956. Investigation of diarrhea in a migrant labor camp. Publ. Health Rept.,
         71:1242.
Copeman, S. M., Howlett, F. M., and Merriman, G.
  1911. An experimental investigation on the range of flight of flies. Repts. Local
         Govt. Bd. Public Hlth. Med. Subjects, n.s. No. 53. Further Repts. on
         Flies as carriers of infection, 9 pp.
Corbo, S.
  1949. La mortalitá infantile per malattie intestinali in rapporto con l'irrora-
         zione di DDT e Octa-Klor. Arch. Ital. Pediat. Puericoltura, *13*:261-272.
Corbo, S.
  1951a. Sull'azione profilattica delle irrorazioni di DDT e Octa-Klor verso le
         malattie gastroenteriche nell'infanzia. Arch. Ital. Pediat. Puericoltura,
         14:324-330.
Corbo, S.
  1951b. La mosca domestica principale responsabile della mortalitá infantile per
         malattie gastroenteriche. Riv. Parassit., *12*:37-45.
Corbo, S.
  1953. La mosca domestica principale responsabile della mortalitá infantile per
         malattie gastroenteriche. (7 anni di osservazioni.) Riv. Parassit., *14*:55-59.
Coulanges, P., and Mayoux, A.
  1970. Isolement de salmonelles dans des eaux d'egout de la ville de Tananarive.
         Arch. Inst. Past. Madagascar, *39*:35-39.
Coutinho, J. O., Taunay, A. de E., and Lima, L. P. deC.
  1957. Importância da *Musca domestica* como vector de agentes patogênicos
         para o homen. Rev. Inst. Adolpho Lutz. São Paulo, *17*:5-23.
Cova, García, P.
  1956. Las moscas problema de salud pública y organización del servicio de
         aseo urbano y domiciliario en la ciudad de Valencia. Conf. Sociedad
         Venezolana de Salud Publ., 43 pp.
Cox, G. L., Lewis, F. C., and Glynn, E. E.
  1912. The number and varieties of bacteria carried by the common house-fly
         in sanitary and unsanitary city areas. Jour. Hyg., *12*:290-319.
Cragg, J. B., and Hobart, J.
  1955. A study of a field population of the blowflies *Lucilia caesar* (L.) and
         L. sericata (Mg.). Ann. Appl. Biol., *43*:645-663.
Craig, T. C.
  1894. The transmission of the cholera spirillum by the alimentary contents and
         intestinal dejecta of the common house-fly. Med. Rec. *46*:38-39.
Crawford, M.
  1926. Development of *Habronema* larvae in drosophilid flies. Jour. Comp.
         Path. Therap., *39*:321-323.
Crawford, W.
  1933. Rinderpest-transmission of infection by contact [letter to the editor].
         Indian Jour. Veter. Science and Anim. Husb., *3*:399-401.
Cross, H. E.
  1923. A further note on surra transmission experiments with *Tabanus albi-
         medius* and ticks. Roy. Soc. Trop. Med. Hyg., London Trans., *26*:469-
         474.

358   BIBLIOGRAPHY

Curbelo, A., and Arango, M. C.
  1945. *Eberthella typhi* en la *Musca domestica*. Inst. Finlay (Inst. Nat. Hyg.), Habana, Sección de publicaciones cientificas biblioteca y museo, 43 pp.
Currie, D. H.
  1910. Studies on leprosy. IX. Mosquitoes in relation to the transmission of leprosy. X. Flies in relation to the transmission of leprosy. Public Hlth. Bull. no. 39: pp. 3-50.
Curry, J. J.
  1902. A report of an acute, fatal, epidemic disease affecting horses and other animals; with studies on the mode of transmission, etc. Amer. Med., 4:95-99.
Dale, J.
  1922. Flies on a sanitary site and typhoid in a boys' home. Med. Jour. Australia, June 24, pp. 694-695.
Dansauer
  1907. Erfahrungen und Beobachtungen über Ruhr in Südwestafrika. Arch. Schiffs Trop. Hyg., *11*:45-94.
Darling, S. T.
  1911a. Murrina, a trypanosomal disease of equines in Panama. Jour. Infect. Dis., *8*:467-485.
Darling, S. T.
  1911b. The probable mode of infection and the methods used in controlling an outbreak of equine trypanosomiasis (*Murrina*) in the Panama Canal Zone. Parasit., *4*:83-86.
Darling, S. T.
  1912. Experimental infection of the mule with *Trypanosoma hippicum* by means of *Musca domestica*. Jour. Exp. Med., *15*:365-369.
Darling, S. T.
  1913. The part played by flies and other insects in the spread of infectious diseases in the tropics with special reference to ants and to the transmission of *Tr. hippicum* by *Musca domestica*. Trans. 15th Intl. Cong. Hyg. Demogr., Washington, pp. 182-185.
Davaine, C.
  1868. Expérience relatives à la durée de l'incubation des maladies charbonneuses et à la quantité de virus necessaire à la transmission de la maladie. Bull. Acad. Méd., *33*:816-821.
Davaine, C.
  1870. Études sur la contagion du charbon chez les animaux domestiques. Bull. Acad. Méd., *35*:215-235.
Davé, K. H., and Wallis, R. C.
  1965. Survival of type 1 and type 3 poliovaccine virus in blowflies (*Phaenicia sericata*) at 40°C. Proc. Soc. Exp. Biol. Med., *119*:121-124.
Davies, W. M.
  1934. The sheep blowfly problem in North Wales. Ann. Appl. Biol., *21*:267-282.
DeCapito, T.
  1963. Isolation of *Salmonella* from flies. Amer. Jour. Trop. Med. Hyg., *12*:892.
Déclat, G.
  1865. Nouvelles applications de l'acide phénique en médicine et en chirurgie aux affections occasionneés par les microphytes—les microzoaires—les virus, les ferments. Paris. [Relevant passages quoted verbatim in: Kelly, H. A. 1901. A historical note upon Diptera as carriers of diseases— Paré—Déclat. Johns Hopkins Hosp. Bull., *12*:240-242.]

DeCoursey, J. D., McGuire, C. D., Otto, J. S., and Durant, R. C.
    1956.   The role of flies in the transmission of non-enteric diseases in the Middle
            East. Research Rept., NAMRU. 3, Cairo, Egypt.
Deegener, P.
    1904.   Die Entwicklung des Darmkanals der Insekten wahrend der Metamor-
            phose. Teil I. *Cybistes roeseli* Curtis. Zool. Jahrb. Anat. Von Spengel,
            20:499-676.
DeGeer, C.
    1782.   Abhandlungen zur Geschichte der Insecten. Transl. into German by
            von Götze. Nurnberg; Vol. 4, see p. 38.
De Kock, G., Robinson, E. M., and Keppel, J. J.
    1940.   Swine fever in South Africa. Onderstepoort Jour. Vet. Sci., *14*:31-93.
de la Paz, G. C.
    1938a. The bacterial flora of flies caught in foodstores in the city of Manila.
            Monthly Bull. Bureau Health, *18*:1-20.
de la Paz, G. C.
    1938b. The breeding of flies in garbage and their control. Monthly Bull. Bureau
            Health, *18*:515-519.
Delcourt, A., and Guyenot, E.
    1912.   Nécessité de la détermination des conditions, sa possibilité à chez les
            drosophiles-technique. Bull. Sci. France Belg., *45*:249-332.
de Mello, F., and Cabral, J.
    1926.   Les insectes sont-ils susceptibles de transmettre la lèpre? Bull. Soc. Path.
            Exot., *19*:774-778.
de Mello, M. J., and Cuocolo, R.
    1943a. Alguns aspetos das relaçôes do *Habronema muscae* (Carter, 1861) com
            a mosca doméstica. Arquiv. Inst. Biol. (São Paulo), *14*:227-234.
de Mello, M. J., and Cuocolo, R.
    1943b. Técnica para o xenodiagnóstico da habronemose gastrica dos equideos.
            Arquiv. Inst. Biol., *14*:217-226.
de Mello, M. J., and Pereira, C.
    1946.   Determinismo da evasão das larvas de *Habronema* sp. da tromba da
            mosca doméstica. Arquiv. Inst. Biol. (São Paulo), *17*:259-266 (English
            Summary).
Demetriadès, M.
    1894.   L'ophtalmie purulente d'Egypte et ses rapports avec le trachome; son
            étiologie. Centralbl. prakt. Augenheilk. *18*:412-413.
Deoras, P. J., and Jandw, A. S.
    1943.   The housefly and its control. Indian Fmg., *4*:565-568.
Depner, K. R.
    1969.   Distribution of the face fly, *Musca autumnalis* (Diptera: Muscidae), in
            western Canada and the relation between its environment and popula-
            tion density. Canad. Entomol., *101*:97-100.
Derbeneva-Ukhova, B. L.
    1947.   Kharakter raspredeleniĭa mukh po pomeshcheniĭam naselennogo punkta.
            Med. Parasit. i Parasit. Bolenzni, *16*:72-79.
Derbeneva-Ukhova, V. P.
    1952.   Mukhi i ikh epidemiologicheskoe znachenie. [Flies and their epidemio-
            logical significance.] Moscow, Medgiz, 271 pp.
Derbeneva-Ukhova, V. P.
    1967.   Effect of climatic factors on significance of flies as potential vectors.
            Wiadomosci Parazytol., *13*:591-594.
Derbeneva-Ukhova, V. P., Bragin, E. A., and Skvortsov, A. A.
    1943.   Materialy k nauchnoĭ tematike po mukham perenoschikam infektsii.
            [Materials on the scientific question of the fly as carrier of infections.]
            Med. Parasit., *12*:39-44.

de Sauvages, F. B.
1768. *Nosologia methodica sistens morbum classes.* Edit. ult. Amstelodami, Vol. 2, see p. 556.

Descazeaux, J., and Morel
1933. Diagnostic biologique (xénodiagnostic) des Habronémoses gastriques du cheval. Bull. Soc. Path. Exot., *26*:1010-1014.

Dethier, V. G., and Rhoades, M. V.
1954. Sugar preference-aversion functions for the blowfly. Jour. Exp. Zool., *126*:177-203.

Dhennin, L., de Balsac, H. H., Verge, J., and Dhennin, Louis
1961. Du rôle des parasites dans la transmission naturelle et expérimentale du virus de la fievre aphteuse. Rec. Med. Veter., *137*:95-104.

Dick, G. F.
1911. A study of the dysentery occurring at the Cook County institutions at Dunning, Ill., in 1910. Jour. Infec. Dis., *8*:386-398.

Dick, R.
1925. The epidemiology and administrative control of anterior poliomyelitis. Med. Jour. Australia, *1*:536-541.

Dieben, C.P.A.
1928. Enkele Surraoverbrengingsproevn met *Stomoxys calcitrans* en *Ctenocephalus canis.* Nederl., *40*:57-83.

Diesing, K. M.
1861. Kleine helminthologische Mitteilungen. Sitzungsk. Math.-Naturwiss. Class Akad. Wissenschaft., *43*:269-282.

Diffloth, P.
1921. La mouche des étables. Vie agricole et rurale, *10*:101-104.

Dikmans, G.
1933. Anaplasmosis IV. The carrier problem. Jour. Amer. Vet. Med. Assoc., *35*:862-870.

Dinulescu, G.
1930. Le *Stomoxys calcitrans*, ses attaques sur les chevaux et le bétail en Roumanie. Ann. Parasit., *8*:71-74.

Dishon, T.
1956. Dissert. Hebrew Univ. Jerusalem. [Not seen.]

Dobzhansky, T., and Wright, S.
1943. Genetics of natural populations. X. Dispersion rates in *Drosophila pseudoobscura.* Genetics, *28*:304-340.

Donatien, A., and Lestoquard, F.
1923. Le *Debab* naturel du chien. Transmission par les *Stomoxes.* Bull. Soc. Path. Exot., Paris, *16*:168-170.

Donatien, A., and Parrot, L.
1922. Trypanosomiase naturelle du chien au Sahara. Bull. Soc. Path. Exot., Paris, *15*:549-551.

Dow, R. P.
1959. A dispersal of adult *Hippelates pusio*, the eye gnat. Ann. Entom. Soc. Amer., *52*:372-381.

Dow, R. P., Bigham, J. T., and Sabrosky, C. W.
1951. Sequel to "*Hippelates* (eye gnat) investigations in the southeastern states" by John T. Bigham. Proc. Entom. Soc. Wash., *53*:263-271.

Dow, R. P., and Hines, V. D.
1957. Conjunctivitis in Southwest Georgia. Publ. Health Repts., *72*:441-448.

Dow, R. P., and Hutson, G. A.
1958. The measurement of adult populations of the eye gnat, *Hippelates pusio.* Ann. Entom. Soc. Amer., *51*:351-360.

Downey, T. W.
1963. Polioviruses and flies: Studies on the epidemiology of enteroviruses in an urban area. Yale Jour. Biol. Med., *35*:341-352.

Dreguss, M. N., and Lombard, L. S.
1954. Experimental studies in equine infectious anaemia. Univ. of Pennsylvania Press, p. 203.

Duca, M., Duca, E., Tomescu, E., and Oana, C.
1958. Cercetări asupra rolului mustei de casă in transmiterea infectiei cu virus Coxsackie. [Investigations on the role of the house fly in the transmission of Coxsackie virus infection.] Studii Cercetări Inframicrobiol. *9*:31-39.

Dudfield, R.
1912. Diarrhoea in 1911. Proc. Roy. Soc. Med., *5*:99-148.

Dudgeon, L. S.
1919. The dysenteries: bacillary and amoebic. British Med. Jour., *1*:448-451.

Duke, H. L., Mettam, R.W.M., and Wallace, J. M.
1934. Observations on the direct passage from vertebrate to vertebrate of recently isolated strains of *Trypanosoma brucei* and *Trypanosoma rhodesiense*. Trans. Roy. Soc. Med. Hyg., *28*:77-84.

Duméril, A.
1835. Rapport sur une observation de M. Vallot, relative à une sorte de teigne. Compt. Rend. Acad. Sci., Paris, *1*:101-103.

Duncan, J. T.
1926. On a bactericidal principle present in the alimentary canal of insects and arachnids. Parasit., *18*:238-252.

Dunne, A. B.
1902. Typhoid fever in South Africa: its cause and prevention. British Med. Jour., *1*:622.

DuToit, R. M.
1944. The transmission of blue-tongue and horse-sickness by *Culicoides*. Onderstepoort Jour. Vet. Sci. Animal Indust., *19*:7-16.

Dykhno, M. A., and Sukhova, M. N.
1952. Duration of survival of dysentery microbes in synanthropic flies of different species under experimental conditions. [In Russian.] *In*: Problems of prophylaxis and treatment of dysentery. (Izd. AMN, 1952), pp. 72-77 [not seen].

Eaves, G. N., and Mundt, J. O.
1960. Distribution and characterization of streptococci from insects. Jour. Insect. Path., *2*:289-298.

Echalier, G., and Ohanessian, A.
1970. In vitro culture of *Drosophila melanogaster* embryonic cells. In Vitro, *6*:162-172.

Echalier, G., Ohanessian, A., and Brun, G.
1965. Cultures "primaires" de cellules embryonnaires de *Drosophila melanogaster* (Insecte Diptère). Compt. Rend. Acad. Sci. Paris, *261*:3211-3213.

Eddy, G. W., Roth, A. R., and Plapp, F. W.
1962. Studies on the flight habits of some marked insects. Jour. Econ. Entom., *55*:603-607.

Edington, A., and Coutts, J. M.
1907. A note on a recent epidemic of trypanosomiasis at Mauritius. Lancet, Oct. 5, pp. 952-955.

Edwards, D. K.
1960. Effects of experimentally altered unipolar air-ion density upon the amount of activity of the blowfly, *Calliphora vicina* R.D. Canad. Jour. Zool., *38*:1079-1091.

Eide, P. E., and Chang, T.-H.
  1969. Cell cultures from dispersed embryonic house fly tissues. Exp. Cell Res.,
      54:302-308.

Elkin, I. I., Editor, translated from the Russian by Pringle, C. R.
  1961. A course in epidemiology. Pergamon Press, New York, pp. 200-207;
      233-241; 250-253; 261-263.

Elliott, L. P.
  1965. Staphylococci of bovine mastitis: their ecology in the dairy herd and
      its environment. Characterization of the isolates. Doct. Dissert., Univ.
      Wisconsin.

Elmassian, M., and Migone, E.
  1903. Sur le mal de caderas flagellose parésiante des équidés sud-américains.
      Ann. Inst. Pasteur, *17*:264-267.

Emmel, L.
  1949. Die Rolle der Fliegen als Krankheits Übertrager. Untersuchungen zur
      Frage der Bedeutung der *Musca domestica* bei der Übertragung der Bak-
      terienruhr. Zeitsch. Hyg., *129*:288-302.

Esten, W. M., and Mason, C. J.
  1908. Sources of bacteria in milk. Bull. 51, Storrs Agric. Exp. Station, pp.
      94-98.

Evtodienko, V. G.
  1968. Survival rate of *Shigella* bacteria on the body surface and in the intes-
      tines of flies. [In Russian.] Kishechnye Infek. Resp. Mezhvedomsb, *2*:
      80-81.

Ewing, H. E.
  1942. The relation of flies (*Musca domestica* Linn.) to the transmission of
      bovine mastitis. Amer. Jour. Veter. Res., *3*:295-299.

Ewing, W. H.
  1962. Sources of *Escherichia coli* cultures that belonged to O antigen groups
      associated with infantile diarrheal disease. Jour. Infect. Dis., *110*:114-
      120.

Eyre, J.W.H., McNaught, J. G., Kennedy, J. C., and Zammit, T.
  1907. Report upon the bacteriological and experimental investigations during
      the summer of 1906. Rpts. Comm. Medit. Fever Roy. Soc. London, Part
      VI: 96-99.

Faichnie, N.
  1909a. *Bacillus typhosus* in flies. Jour. Roy. Army Med. Corps, *13*:672-675.

Faichnie, N.
  1909b. Fly-borne enteric fever: the source of infection. Jour. Roy. Army Med.
      Corps, *13*:580-584.

Faichnie, N.
  1921. Means of infection in fly-borne disease. So. African Med. Rec., *19*:
      438-441.

Faichnie, N.
  1929. The etiology of enteric fever: personal views and experiences. Jour. Med.
      Assoc. So. Africa, *3*:669-675.

Fales, J. H., Bodenstein, O. F., Mills, G. D. Jr., and Wessel, L. H.
  1964. Preliminary studies on face fly dispersion. Ann. Entom. Soc. Amer., *57*:
      135-137.

Faust, E. C.
  1928. Studies on *Thelazia callipaeda*. Jour. Parasit., *15*:75-86.

Fedorov, V. G.
  1962. O roli komnatnoi mukhi (*Musca domestica* L.) v rasprostranenii zimoi yaits gel'mintov i tsist kishechynky prosteishikh. [The role of the house fly (*Musca domestica* L.) in the dissemination of helminth ova and the cysts of intestinal Protozoa during the winter.] Med. Parazit. Parazit. Bolezni, *31*:618-620.

Fermi, C.
  1912. Fliegenlarven und Tollwutvirus. Lyssizide Wirkung und Virusübertragung. Centralbl. Bakter., *61*:93-97.

Ferris, D. H., Hanson, R. P., Dicke, R. J., and Roberts, R. H.
  1955. Experimental transmission of vesicular stomatitis virus by Diptera. Jour. Infect. Dis., *96*:184-192.

Ficker, M.
  1903. Typhus und Fliegen. Vorläufige Mitteilung. Archiv. Hyg., *46*:274-283.

Field, J. W.
  1951. Yaws. The Institute for Medical Research 1900-1950 (by various authors). Jubilee Volume No. 25, Government Press (Malaya), Kuala Lumpur.

Firth, R. H., and Horrocks, W. H.
  1902. An inquiry into the influence of soil fabrics, and flies in the dissemination of enteric infection. British Med. Jour., *2*:936-943.

Fish, N. A., Mitchell, W. R., and Barnum, D. A.
  1967. A report of a natural outbreak of botulism in pheasants. Can. Veter. Jour., *8*:10-16.

Fletcher, F., and Haub, J. G.
  1933. Digestion in blowfly larvae, *Phormia regina* (Meigen), used in the treatment of osteomyelitis. Ohio Jour. Sci., *33*:101-109.

Fletcher, T. B.
  1916. Report of the imperial pathological entomologist. Rept. Agri. Res. Inst., Pusa, pp. 78-84.

Flexner, S., and Clark, P. F.
  1911. Contamination of the fly with poliomyelitis virus. Jour. Amer. Med. Assoc., *56*:1717-1718.

Floyd, T. M., and Cook, B. H.
  1953. The housefly as a carrier of pathogenic human enteric bacteria in Cairo. Jour. Egyptian Publ. Health Assoc., *28*:75-85.

Flu, P. C.
  1915. Epidemiologische studiën over de cholera te Batavia 1909-1915. Gen. Tijdsch., *55*:863-924.

Fontana, P., and Severino-Brea, R.
  1961. Centro estudio profilaxis hidatidosis. Comm. VII Intl. Cong. Hidatid. Rome, pp. 283-286.

Ford, J.
  1964. The geographical distribution of trypanosome infections in African cattle populations. Bull. Epiz. Dis. Afr., *12*:307-320.

Foster, N. M., Jones, R. H., and McCrory, B. R.
  1963. Preliminary investigations on insect transmission of blue-tongue virus in sheep. Amer. Jour. Veter. Res., *24*:1195-1200.

Fox, E.C.R.
  1920-1. Naga sore. Indian Jour. Med. Res., *8*:694-699.

Francis, E.
  1914. An attempt to transmit poliomyelitis by the bite of *Lyperosia irritans*. Jour. Infect. Dis., *15*:1-5.

Francis, E.
  1919. Deer fly fever or Pahvant Valley plague. Publ. Health Rept., *34*:2061-2062.

364 BIBLIOGRAPHY

Francis, E., and Mayne, B.
1921. Experimental transmission of tularemia by flies of the species *Chrysops discalis*. Publ. Health Rept., *36*:1738-1746.

Francis, E.
1937. Cecil's Text Book of Medicine, p. 350. Summary of present knowledge of tularemia. 1928. Med., *7*:411.

Francis, T. Jr., Brown, G. C., and Penner, L. R.
1948. Search for extrahuman sources of poliomyelitis virus. Jour. Amer. Med. Assoc., *136*:1088-1092.

Francis, T. Jr., Brown, G. C., and Ainslie, J. D.
1953. Poliomyelitis in Hidalgo County, Texas, 1948. Poliomyelitis and Coxsackie viruses in privy specimens. Amer. Jour. Hyg., *58*:310-318.

Fraser, A., Ring, R. A., and Stewart, R. K.
1961. Intestinal proteinases in an insect, *Calliphora vomitoria* L. Nature, *192*: 999-1000.

Fraser, H.
1909. Surra in the federated Malay states. Jour. Trop. Vet. Sci., *4*:345-389.

Fraser, H., and Symonds, S. L.
1908. Surra in the federated Malay states. Studies Inst. Med. Res., *9*:1-35.

Frenkel, J., Dubey, J., and Miller, N.
1969. *Toxoplasma gondii*: fecal forms separated from eggs of the nematode *Toxocara cati*. Science, *164*:432-433.

Frenkel, J. K., Dubey, J. P., and Miller, N. L.
1970. *Toxoplasma gondii* in cats: Fecal stages identified as coccidian oocysts. Science, *167*:893-896.

Friend, W. G.
1955. Problems in nutritional studies on phytophagous insects. Symposium Rept. Entom. Soc. Ontario, *86*:13-17.

Frings, M. R., and Frings, H.
1946. A potometer for rapid measurements of ingestion by haustellate insects. Science, *103*:22-23.

Froggatt, W. W.
1914. Sheep maggot flies. Agric. Gaz. N.S.W., Sydney, *25*:756-758.

Frye, W. W., and Meleney, H. E.
1932. Investigations of *Endamoeba histolytica* and other intestinal Protozoa in Tennessee: IV. A study of flies, rats, mice and some domestic animals as possible carriers of the intestinal Protozoa of man in a rural community. Amer. Jour. Hyg., *16*:729-749.

Fuller, C.
1913. An unusual outbreak of *Stomoxys calcitrans* following floods. Agric. Jour. Un. So. Afric., *5*:922-924.

Fuller, H. B.
1913. Myths of American history. Munsey's Magazine, May.

Gabaldón, A., Berti, A. L., and Jove, J. A.
1956. El saneamiento en la lucha contra la gastroenteritis y colitis. I. Congresso Venezolano de Salud Publica y III Conferencia de Unidades Sanitarias, 55 pp.

Galli-Valerio, B.
1905a. Die Verbreitung und Verhütung der Helminthen des Menschen. Therap. Monatsh., *19*:339-347.

Galli-Valerio, B.
1905b. Notes parasitologie technique parasitologique. Centralbl. Bakter., *39*: 233-242.

Galli-Valerio, B.
1905c. Notes de parasitologie et de technique parasitologique. Centralbl. Bakt. Abt. I., Orig., *39*:230-247.

Galli-Valerio, B.
  1908. Recherches experimentales sur une sarcine pathogène. Centralbl. Bakt.,
          Parasit. Infekt., I. Abt., Orig., *47*:177-186.
Galloway, I. A.
  1937. Fifth progress report of the Foot and Mouth Disease Research Commit-
          tee, H. N. Stationery Office, London, pp. 345-349.
Galun, R., and Fraenkel, G.
  1957. Physiological effects of carbohydrates in the nutrition of a mosquito,
          Aedes aegypti and two flies, *Sarcophaga bullata* and *Musca domestica.*
          Jour. Cell. Comp., Physiol., *50*:1-23.
Galuzo, I. G., and Rementzova, M. M.
  1956. Transmitters and reservoirs of the brucellosis infection in nature. [In
          Russian.] Entomol. Obozr., *35*:560-569.
Galuzo, I. G., and Krivkova, A. M.
  1970. Novye kanaly tsirkulyatsii toksoplasm v prirode. [New paths of circula-
          tion of *Toxoplasma* in nature.] Izvest. Akad. Nauk Kazakh. SSR, Seriya
          Biol., *5*:38-44.
Gandel'sman, B. I., Zvorykin, N. A., and Sukhacheva, K. A.
  1947. K voprosu o mikroflore mukh v ochagakh kishechnykh infektsiĭ. [On the
          microflora of flies in foci of intestinal infections.] Zhurn. Mikrob. Epidem.
          Immun., *8*:56-60.
Ganon, J.
  1908. Cholera en vliegen. Geneesk. Tijdschr. Nederl. Indië. *48*:227-233.
Garcia, R., and Radovsky, F. J.
  1962. Haematophagy by two non-biting muscid flies and its relationship to
          tabanid feeding. Canad. Entom., *94*:1110-1116.
Garrison, F. H.
  1917. An introduction to the history of medicine. 2nd Ed. Philadelphia, Lon-
          don; W. B. Saunders Co. 905 pp. See pp. 55, 57.
Gaud, J., and Faure, P.
  1951. Effet de la lutte antimouches sur l'incidence des maladies oculaires dans
          le Sud Marocain. Bull. Soc. Path. Exot., *44*:446-448.
Gaud, J., Laurent, J., and Faure, P.
  1954. Biologie de *Musca sorbens* et role vecteur probable de cette espèce en
          pathologie humaine au Maroc. Bull. Soc. Path. Exot., *47*:97-101.
Gaud, J., Maurice, A., Faure, P., and Lalu, P.
  1950. Expériences de lutte contre les mouches au Maroc. Bull. Inst. Hyg.
          Maroc, *10*:55-71.
Gear, H. S.
  1944. Hygiene aspects of the El Alamein victory, 1942. British Med. Jour.,
          1:383-387.
Gear, J., Cuthbertson, E., and Ryan, J.
  1962. A study of South African strains of trachoma virus in experimental ani-
          mals. Ann. New York Acad. Sciences, 98 (Art. 1): 197-200.
Geiger, W.
  1937. Virusschweinepest und Afrikanische Virusseuche der Schweine. Thesis
          Dr. Med. Vet. Habil. Vet. High School, Hanover.
Generali, G.
  1886? Una larva di nematode della mosca commune. Atti Soc. Modena, Ser. 3,
          2:88-89.
Georgiou, G. P.
  1966. Distribution of insecticide-resistant house flies on neighboring farms.
          Jour. Econ. Entom., *59*:341-346.
Gerberich, J. B.
  1948. Rearing house-flies on common bacteriological media. Jour. Econ.
          Entom., *41*:125-126.

366 BIBLIOGRAPHY

Gerberich, J. B.
1951a. Passing of *Staphylococcus aureus* Rosenbach through the metamorphosis of the house-fly. Proc. Minn. Acad. Sci., *16*:35.

Gerberich, J. B.
1951b. Transmission of bacteria through the metamorphosis of the house fly and the longevity of such an association. Ph.D. Dissertation. Ohio State Univ., Columbus, Ohio, 82 pp.

Gerberich, J. B.
1952. The housefly (*Musca domestica* Linn.), as a vector of *Salmonella pullorum* (Retteger) Bergy, the agent of white diarrhea of chickens. Ohio Jour. Sci., *52*:287-290.

Germann, T.
1896. Augenärztliche Beobachtungen in Syrien und Palästina; speciell über das Trachom in diesen Ländern. Centralbl. prakt. Augenheilk., *20*:386-400.

Ghai, D. P., Kalra, S. L., and Jaiswal, V. N.
1969. Epidemiology of diarrhoea in infants and preschool children in a rural community near Delhi. Indian Pediat., *6*:263-271.

Gibbons, R. J.
1937. An epidemic of bacillary dysentery. Canad. Publ. Health Jour., *28*:278-281.

Gilbert
1795. Recherches sur les causes des maladies charbonneuses. [Not seen.]

Giles, G. M.
1906. Anatomy of biting flies of the genera *Stomoxys* and *Glossina*. Jour. Trop. Med., *9*:99-102, 153-156, 169-173, 182-185, 198-202, 217-219, 235-236.

Gill, C. A., and Lal, R. B.
1930. The epidemiology of cholera with special reference to transmission. A preliminary report. Indian Jour. Med. Res., *18*:1255-1297.

Gill, G. D.
1955. Filth flies of Central Alaska. Jour. Econ. Entom., *48*:648-653.

Gilmour, D., Waterhouse, D. F., and McIntyre, G. A.
1946. An account of experiments undertaken to determine the natural population density of the sheep blowfly, *Lucilia cuprina* Wied. Australian Council Sci. Indust. Res., Bull. No. 195, 39 pp.

Gingrich, R. E.
1960. Development of a synthetic medium for aseptic rearing of larvae of *Stomoxys calcitrans* (L.). Jour. Econ. Entom., *53*:408-411.

Gip, L., and Svensson, S. A.
1968. Can flies cause the spread of dermatophytosis? Acta Derm.-Venereol., *48*:26-29.

Glaser, R. W.
1923. The survival of bacteria in the pupal and adult stages of flies. Amer. Jour. Hyg., *3*:469-480.

Glaser, R. W.
1924. The relation of microorganisms to the development and longevity of flies. Amer. Jour. Trop. Med., *4*:85-107.

Glick, P. A.
1939. The distribution of insects, spiders, and mites in the air. U.S.D.A. Tech. Bull. No. 673, 150 pp.

von Goethe, J. W.
1842. Goethe's Werke, vollständige Ausgabe letzter Hand. Bd. 58, p. 175.

González Herrejón, S., and Ortiz Lombardini, C.
1938. Es *Simulium haematopotum* un vector del mal del pinto? Medicina, *18*: 631-638.

Gómez, J.
1879. Du carathés ou tachés endémique des Cordillères. Thèse Faculté de Médecine de Paris.

Gorbacheva, E. A., Semenikhina, A. D., and Sadykova, V. R.
1967. Materialy o mekhanicheskom perenose yaits gel'mintov mukhami. [Information on the mechanical transport of helminth eggs by flies.] Med. Zh. Uzb., 2:70 (not seen).

Gorbatow, O.
1951. Om vatten-, mjölk- och flughygien och deras förhållande till sommardiarrefall i en landsortskommun. Nordisk Hyg. Tidskrift, pp. 225-239.

Gordon, F. B.
1943. Studies on the survival of poliomyelitis virus in insects. Jour. Bact., *45*:77.

Gorodetskiĭ, A. S., and Kuznetsov, B. P.
1937. Dal'nost'polëta komnatnoĭ mukhi, *in book*: Mukhi, Kiev, see pp. 88-92.

Gosio, B., Alessandrini, G., and Russo, C.
1924. Contribution à l'étude des propagations infectieuses par les insectes, en ce qui concerne plus particulièrement la peste bubonique. Seen in Rev. Appl. Entom., *13*:181.

Gosio, B.
1925. Über die Verbreitung der Bubonenpesterreger durch Insektenlarven. Archiv. Schiffs. Trop. Hyg., *29*: (Beiheft 1): 134-139.

Gourlay, R. J., and Marsh, M.
1965. An outbreak of yaws in a suburban community in Jamaica. Amer. Jour. Trop. Med. Hyg., *14*:777-779.

Graham-Smith, G. S.
1910. Observations on the ways in which artificially infected flies (*Musca domestica*) carry and distribute pathogenic and other bacteria. Repts. Loc. Govt. Bd. Public Health Med. Subjs. London, n.s., *40* (3): 1-40.

Graham-Smith, G. S.
1911a. Some observations on the anatomy and function of the oral sucker of the blow-fly (*Calliphora erythrocephala*). Jour. Hyg. *11*:390-408.

Graham-Smith, G. S.
1911b. Further observations on the ways in which artificially infected flies (*Musca domestica* and *Calliphora erythrocephala*) carry and distribute pathogenic and other bacteria. Repts. Loc. Govt. Bd. Public Health Med. Subjs. London, n.s., *53*:31-48, and *16*:31-48.

Graham-Smith, G. S.
1912a. An investigation into the possibility of pathogenic microorganisms being taken up by the larva and subsequently distributed by the fly. 41st Ann. Rept. Loc. Govt. Bd., Supp. Rept. Med. Officer, 1911-1912, pp. 330-335.

Graham-Smith, G. S.
1912b. An investigation of the incidence of the micro-organisms known as non-lactose-fermenters in flies in normal surroundings and in surroundings associated with epidemic diarrhoea. 41st Ann. Rept. Loc. Govt. Bd. Supp. Rept. Med. Officer, 1911-1912, pp. 304-329.

Graham-Smith, G. S.
1913. Further observations on non-lactose fermenting bacilli in flies, and the sources from which they are derived, with special reference to Morgan's bacillus. Rept. Loc. Govt. Bd. Publ. Health and Med. Subj., n.s., No. 85, pp. 43-46.

Graham-Smith, G. S.
1914. Flies in relation to disease. Non-bloodsucking flies. University Press, Cambridge, 389 pp.

Graham-Smith, G. S.
1930a. The Oscinidae (Diptera) as vectors of conjunctivitis, and the anatomy of their mouth parts. Parasit., *22*:457-467.

Graham-Smith, G. S.
1930b. Further observations on the anatomy and function of the proboscis of the blow-fly, *Calliphora erythrocephala* L. Parasit., *22*:47-115.
Graham-Smith, G. S.
1934. The alimentary canal of *Calliphora erythrocephala* L., with special reference to its musculature and to the proventriculus, rectal valve and rectal papillae. Parasit., *26*:176-248.
Graham, R., and Boughton, I. B.
1924. *Clostridium botulinum* Type C associated with limberneck-like disease in chickens and ducks. Jour. South African Vet. Med. Assoc., *17*:723.
Grassi, B.
1883. Malefizi delle mosche. Gazz. Ospedali, *59*:309-310.
Grassi, B., and Rovelli, G.
1889. Embryologische Forschungen an Cestoden. Centralbl. Bakt. Parasit. Infekt. I. Abt., Orig., *5*:370-377.
Gray, D. P., and Bannister, G. L.
1961. Studies on blue tongue. I. Infectivity of the virus in the sheep ked, *Melophagus ovinus* (L.). Canad. Jour. Comp. Med. Vet. Sci., *25*:230-232.
Green, A. A.
1951. The control of blowflies infesting slaughter-houses. I. Field observations of the habits of blowflies. Ann. Appl. Biol., *38*:475-494.
Greenberg, B.
1954. A method for the sterile culture of housefly larvae, *Musca domestica* L. Canad. Entom., *86*:527-528.
Greenberg, B.
1959a. House fly nutrition. I. Quantitative study of the protein and sugar requirements of males and females. Jour. Cell. Comp. Physiol., *53*:169-178.
Greenberg, B.
1959b. Persistence of bacteria in the developmental stages of the housefly. I. Survival of enteric pathogens in the normal and aseptically reared host. Amer. Jour. Trop. Med. Hyg., *8*:405-411.
Greenberg, B.
1959c. Persistence of bacteria in the developmental stages of the housefly. II. Quantitative study of the host-contaminant relationship in flies breeding under natural conditions. Amer. Jour. Trop. Med. Hyg., *8*:412-416.
Greenberg, B.
1959d. Persistence of bacteria in the developmental stages of the housefly. III. Quantitative distribution in prepupae and pupae. Amer. Jour. Trop. Med. Hyg., *8*:613-617.
Greenberg, B.
1959e. Persistence of bacteria in the developmental stages of the housefly. IV. Infectivity of the newly emerged adult. Amer. Jour. Trop. Med. Hyg., *8*:618-622.
Greenberg, B.
1960. Host-contaminant biology of muscoid flies. I. Bacterial survival in the pre-adult stages and adults of four species of blow flies. Jour. Insect. Path., *2*:44-54.
Greenberg, B.
1961. Mite orientation and survival on flies. Nature, *190*:107-108.
Greenberg, B.
1962a. Host-contaminant biology of muscoid flies. II. Bacterial survival in the stable fly, false stable fly, and the little house fly. Jour. Insect. Path., *4*:216-223.

Greenberg, B.
1962b. Host-contaminant biology of muscoid flies. III. Effect of hibernation, diapause, and larval bactericides on normal flora of blow-fly prepupae. Jour. Insect Path., *4*:415-428.

Greenberg, B.
1964. Experimental transmission of *Salmonella typhimurium* by houseflies to man. Amer. Jour. Hyg., *80*:149-156.

Greenberg, B.
1965. Flies and disease. Sci. Amer., *213*:92-99.

Greenberg, B.
1966. Bacterial interactions in gnotobiotic flies. Symposium on Gnotobiology. IX Intl. Congr. Microb. Moscow, 1966, pp. 371-380.

Greenberg, B.
1968a. Micro-potentiometric pH determinations of muscoid maggot digestive tracts. Ann. Entom. Soc. Amer., *61*:365-368.

Greenberg, B.
1968b. Model for destruction of bacteria in the midgut of blow fly maggots. Jour. Med. Entom., *5*:31-38.

Greenberg, B.
1968c. Gnotobiotic insects in biomedical research. *In*: Advances in germfree research and gnotobiology. Ed. Myakawa, M., and Luckey, T. D. CRC Press, Cleveland, Ohio, see pp. 410-416.

Greenberg, B.
1969. Sterile culture of *Musca sorbens*. Ann. Entom. Soc. Amer., *62*:450.

Greenberg, B.
1970. Sterilizing procedures and agents, antibiotics and inhibitors in mass rearing of insects. Bull. Entom. Soc. Amer., *16*:31-36.

Greenberg, B.
1970. Species distribution of new structures on fly antennae. Nature, *28*:1338-1339.

Greenberg, B., and Archetti, I.
1969. In vitro cultivation of *Musca domestica* L. and *Musca sorbens* Wiedemann tissues. Exp. Cell. Res., *54*:284-287.

Greenberg, B., and Bornstein, A. A.
1964. Fly dispersion from a rural Mexican slaughterhouse. Amer. Jour. Trop. Med. Hyg., *13*:881-886.

Greenberg, B., and Burkman, A.
1963. Effect of B-vitamins and a mixed flora on the longevity germ-free adult houseflies, *Musca domestica* L. Jour. Cell. Comp. Physiol., *62*:17-22.

Greenberg, B., Kowalski, J. A., and Klowden, M. J.
1970. Factors affecting the transmission of *Salmonella* by flies: Natural resistance to colonization and bacterial interference. Infec. Immun., *2*:800-809.

Greenberg, B., and Miggiano, V.
1963. Host-contaminant biology of muscoid flies. IV. Microbial competition in a blowfly. Jour. Infect. Dis., *112*:37-46.

Greenberg, B., and Paretsky, D.
1955. Proteolytic enzymes in the house fly, *Musca domestica* (L.). Ann. Entom. Soc. Amer., *48*:46-50.

Greenberg, B., Varela, G., Bornstein, A., and Hernandez, H.
1963. Salmonellae from flies in a Mexican slaughterhouse. Amer. Jour. Hyg., *77*:177-183.

Grey, G.
1838. Journal of two expeditions of discovery in Northwest and Western Australia. Vol. 1, see p. 81.

Griffith, B. T.
1952. A study of antibiosis between wind-borne molds and insect larvae from wind-borne eggs. Jour. Allergy, *23*:375-382.

Gross, H., and Preuss, U.
1951. Infektionsversuche an Fliegen mit darmpathogenen Keimen. I. Mitteilung. Versuche mit Typhus und Paratyphus-bakterien. Zentralbl. Bakt., Parasit. Infekt. Hyg., I. Abt., Orig., *156*:371-377.

Gross, H., and Preuss, U.
1953. Infektionsversuche an Fliegen mit darmpathogenen Keimen. II. Mitteilung. Zentralbl. Bakt., Parasit. Infekt. Hyg., I. Abt. Orig., *160*:526-529.

Gudnadóttir, M. G.
1960. Studies of the fate of type 1 polioviruses in flies. Jour. Exp. Med., *113*:159-176.

Guinée, P.A.M.
1963. Experimental studies on the origin and significance of antibiotic-resistant *Escherichia coli* in animals and man. Drukkerij G. van Dijk N.V., Breukelen, 75 pp.

Gurney, W. B., and Woodhill, A. R.
1926. Investigations on sheep blowflies. Part I. Range of flight and longevity. Agric. Dept. New South Wales, Science Bull., No. 27, pp. 4-28.

Gutberlet, J. E.
1916. Studies on the transmission and prevention of cestode infection in chickens. Jour. Amer. Veter. Med. Assoc., *49*:218-237.

Gutberlet, J. E.
1920. On the life history of the chicken cestode, *Hymenolepis carioca* (Magalhaes). Jour. Parasit., *6*:35-38.

Gwatkin, R., and Dzenis, L.
1952. Studies in pullorum disease. XXIX. Bacteriological examination of blowflies (*Lucilia* sp.) and houseflies (*Musca domestica*) which during their larval stage had fed on chicks infected with *Salmonella pullorum*. Canad. Jour. Comp. Med. Veter. Sci., *16*:148-150.

Gwatkin, R., and Fallis, A. M.
1938. Bactericidal and antigenic qualities of the washings of blowfly maggots. Canad. Jour. Res., *16*:343-352.

Gwatkin, R., and Mitchell, C. A.
1944. Transmission of *Salmonella pullorum* by flies. Canad. Jour. Publ. Health, *35*:281-285.

Haddow, A. J., and Lumsden, W.H.R.
1935. *Fannia canicularis* L. and *F. scalaris* Fab. as agents of human myiasis, with an abstract of recorded cases. Surgo, *1*:1 and on.

Haddow, A. J., and Thomson, R.C.M.
1937. Sheep myiasis in Southwest Scotland with special reference to the species involved. Parasit., *29*:96.

Hadorn, E.
1968. Transdetermination in cells. Sci. Amer., *219*:110-120.

Hadorn, E., and Garcia-Bellida, A.
1964. Zur Proliferation von *Drosophila*-Zellkulturen im Adultmilieu. Rev. Suisse Zool., *71*:576-582.

Haeser, H.
1882. Lehrbuch der Geschichte der Medicin und der epidemischen Krankheiten. G. Fisher, Jena., see p. 728.

Hafez, M.
1941. Investigations into the problem of fly control in Egypt. Bull. Soc. Fouad Entom., *25*:99-144.

Hafez, M., and Attia, M. A.
1958a. Studies on the ecology of *Musca sorbens* Wied. in Egypt; on the developmental stages of *Musca sorbens* Wied. with special reference to larval behavior. Bull. Soc. Entom. Egypt, *42*:83-161.

Hafez, M., and Attia, M. A.
1958b. The relation of *Musca sorbens* Wied. to eye diseases in Egypt. Bull. Soc. Entom. Egypt, *42*:275-283.

Hafez, M., and Attia, M. A.
1958c. Studies on the ecology of *Musca sorbens* Wied. in Egypt. Bull. Soc. Entom. Egypt, *42*:83-121.

Hafez, M., and Gamal-Eddin, F. M.
1959a. Ecological studies on *Stomoxys calcitrans* L. and *sitiens* Rond. in Egypt, with suggestions on their control. Bull. Soc. Entom. Egypt, *43*:245-254.

Hafez, M., and Gamal-Eddin, F. M.
1959b. On the feeding habits of *Stomoxys calcitrans* L. and *sitiens* Rond., with special reference to their biting cycle in nature. Bull. Soc. Entom. Egypt, *43*:291-301.

Haines, T. W.
1953. Breeding media of some common flies. I. Urban areas. Amer. Jour. Trop. Med. Hyg., *2*:933-940.

Hale, J. H., Davies, T.A.L., and Hin, W.K.N.C.
1960. Flies in aeroplanes as vectors of faecal-borne disease. Tran. Roy. Soc. Trop. Med. Hyg., *54*:261-262.

Hall, D. G.
1932. Some studies on the breeding media, development, and stages of the eye gnat, *Hippelates pusio* Loew (*Diptera*: Chloropidae). Amer. Jour. Hyg., *16*:854-864.

Hall, D. G.
1948. The blowflies of North America, Thomas Say Foundation, 477 pp.

Hall, G. N.
1927. Res. Div. Ann. Rept. Veter. Dept. Uganda year ended 1926, pp. 12-16. (Transmission experiments.)

Hallock, H. C.
1940. II. The Sarcophaginae and their allies in New York. Jour. New York Entom. Soc., *48*:201-231.

Hamer, W. H.
1908. Nuisance from flies. Rept. Publ. Health Comm. London County Council, No. 1207, pp. 1-6.

Hamilton, A.
1903. The fly as a carrier of typhoid. An inquiry into the part played by the common house fly in the recent epidemic of typhoid fever in Chicago. Jour. Amer. Med. Assoc., *40*:576-583.

Hamilton, C.S.P.
1939. Investigation into the natural breeding places of the *Siphunculina funicola* fly in Assam. Indian Med. Gaz., *74*:210-215.

Hammer, O.
1942. Biological and ecological investigations on flies associated with pasturing cattle and their excrement. Videnskabelige Meddelelser fra Dansk naturhistorisk forening i København, *105*:141-393 (appeared as separate on 15 Nov. 1941).

Hanec, W.
1956. A study of the environmental factors affecting the dispersion of house flies (*Musca domestica* L.) in a dairy community near Fort Whyte, Manitoba. Canad. Entom., *88*:270-272.

Hansen, H. J.
1903. The mouth-parts of *Glossina* and *Stomoxys*. *In*: Austen, E.E., Monogr. Tsetse-Flies, pp. 105, 110-117.

Hansens, E. J.
1963. Fly populations in dairy barns. Jour. Econ. Entom., *56*:842-844.

Hansens, E. J., and Valiela, I.
1967. Activity of the face fly in New Jersey. Jour. Econ. Entom., *60*:26-28.

Hanson, R. P.
1952. The natural history of vesicular stomatitis. Bact. Rev., *16*:179-204.

Harada, F.
1954. Investigations of hookworm larvae. IV. On the fly as a carrier of infective larvae. Yokohama Med. Bull., *5*:282-286.

Harber, A. F.
1913. Epizootic lymphangitis and its treatment. Vet. Jour., *69*:408-411.

Hardy, A. V.
1959. Diarrhoeal diseases of infants and children. Mortality and epidemiology. Bull. Wld. Health Org., *21*:309-319.

Hardy, A. V., Watt, J., Peterson, J., and Schlosser, E.
1942. Studies of the acute diarrheal diseases. VIII. Sulfaguanidine in the control of *Shigella dysenteriae* infections. Publ. Health Rept., *57*:529-535.

Hardy, A. V., and Watt, J.
1948. Studies of the acute diarrheal diseases. XVIII. Epidemiology. Publ. Health Rept., *63*:363-378.

Harris, A. H., and Down, H. A.
1946. Studies of the dissemination of cysts and ova of human intestinal parasites by flies in various localities on Guam. Amer. Jour. Trop. Med., *26*:789-800.

Harris, R. L., Frazar, E. D., Grossman, P. D., and Graham, O. H.
1966. Mating habits of the stable fly. Jour. Econ. Entom., *59*:634-636.

Harvey, R.W.S., and Price, T. H.
1962. *Salmonella* serotypes and *Arizona* paracolons isolated from Indian crushed bone. Monthly Bull. Min. Health Publ. Health Lab. Ser., *21*: 54-57.

Hastings, J.
1911. A dictionary of the Bible. Charles Scribner's Sons, see p. 25.

Havlík, B., and Baľová, B.
1961. A study of the most abundant synanthropic flies occurring in Prague. Acta. Soc. Entom. Czech., *58*:1-11.

Havlík, B.
1964. Sanitarny problem synantropijnych much (Diptera) wielkiej Pragi. Wiad. Parazytol., *10*:588-589.

Hawley, J. E., Penner, L. R., Wedberg, S. E., and Kulp, W. L.
1951. The role of the house fly, *Musca domestica*, in the multiplication of certain enteric bacteria. Amer. Jour. Trop. Med. *31*:572-582.

Hayward, E. H.
1904. The fly as a carrier of tuberculous infection. New York Phil. Med. Jour., *80*:643-644.

Hecker
1899. Untersuchungen zur Bekämpfung der Maul-und Klauenseuche. Berl. Thierärztl. Wochenschr., *34*:407-411.

Hecker, A., and Otto, R.
1911. Beiträge zur Lehre von der sogenannten Weilschen Krankheit. Veröffentl. Geb. Militär-Sanitätsw.: 46.

Hegner, R.
1928.   Experimental studies on the viability and transmission of *Trichomonas hominis*. Amer. Jour. Hyg., *8*:16-34.

Heinz, H. J.
1949.   Methodik zur Untersuchung des Darminhaltes von Fliegen auf pathogene Darmprotozoen des Menschen. Zentrabl. Bakt. (Abt. 1), *153*:106-108.

Heinz, H. J., and Brauns, W.
1955.   The ability of flies to transmit ova of *Echinococcus granulosus* to human foods. So. African Jour. Med. Sci., *20*:131-132.

Hemphill, F. M.
1948.   Trends of diarrheal disease mortality in the United States, 1941-1946, inclusive. Publ. Health Rept., *63*:1699-1711.

Hennig, W.
1968.   Die Larvenformen der Dipteren. Eine Übersicht über die bisher bekannten Jugendstadien der zweiflugeligen Insekten. Akademie-Verlag, Berlin, in three parts, Part III, 628 pp.

Henning, M. W.
1956.   Animal diseases in South Africa. 3rd ed., Central News Agency Ltd., South Africa.

Henry, S. M., and Cotty, V. F.
1957.   The rearing of aseptic adult houseflies for physiological studies. Contrib. Boyce Thompson Inst., *19*:227-229.

Hermann, P.
1965.   The house fly as a disease carrier. Pest Control, *33*:16-20.

Hermans, E. H.
1931.   Framboesia tropica. Acta Leidensia, Batavia, *6*:1-168, see pp. 36-38.

Herms, W. B.
1926.   Hippelates flies and certain other pests of the Coachella Valley, California. Jour. Econ. Entom., *19*:692-695.

Herms, W. B.
1928.   The effect of different quantities of food during the larval period on the sex ratio and size of *Lucilia sericata* Meigen and *Theobaldia incidens* (Thom.). Jour. Econ. Entom., *21*:720-729.

Herms, W. B.
1953.   Medical entomology, MacMillan, New York, pp. 484.

Herodotus
ca. 520 B.C.   The histories, Book IV, Trans. by Aubrey de Selincourt, 602 pp. Harmondsworth, Mddx.: Penguin Books: 602 pp., 1954. See p. 310.

Herreng, F.
1967.   Étude de la multiplication de l'arbovirus "Sindbis" chez la drosophile. Compt. Rend. Acad. Sci. Paris, *264*:2854-2857.

Herzog, M.
1905.   Zur Frage der Pestverbreitung durch Insecten. Zeitschr. Hyg., *49*:268-282.

Hewitt, C. G.
1912a.   An account of the bionomics and the larvae of the flies *Fannia (Homalomyia) canicularis* L., and *F. scalaris* Fab. and their relation to myiasis of the intestinal and urinary tracts. Repts. Loc. Govt. Bd. Publ. Health Med. Subjs. London, n.s., *66*:15-21.

Hewitt, C. G.
1912b.   Observations on the range of flight of flies. Rept. Loc. Govt. Bd. Publ. Health Med. Subjs., n.s., *66*. Further Repts. on flies as carriers of infection; Great Britain, 5 pp.

Hewitt, C. G.
1914.   The house-fly. *Musca domestica* Linn. Its structure, habits, development, relation to disease, and control. Cambridge Univ. Press, London, 382 pp.

Higgins, A. R., Floyd, T. M., and Kader, M. A.
  1955. Studies in shigellosis. II. Observations on incidence and etiology of diar-
      rheal disease in Egyptian village children. Amer. Jour. Trop. Med. Hyg.,
      4:271-280.
Hightower, B. G., and Adams, A. L.
  1969. Dispersal and local distribution of laboratory-reared sterile screw-worm
      flies released in winter. Jour. Econ. Entom., 62:259-261.
Hightower, B. G., Adams, A. L., and Alley, D. A.
  1965. Dispersal of released irradiated laboratory-reared screw-worm flies. Jour.
      Econ. Entom., 58:373-374.
Hightower, B. G.
  1963. Nocturnal resting places of the screw-worm fly. Jour. Econ. Entom., 56:
      498-500.
Hightower, B. G., and Alley, D. A.
  1963. Local distribution of released laboratory-reared screw-worm flies in rela-
      tion to water sources. Jour. Econ. Entom., 56:798-802.
Hill, G. F.
  1919. Relationship of insects to parasitic diseases in stock. Part I. The life his-
      tory of *Habronema muscae*, *H. microstoma*, and *H. megastoma*. Proc.
      Roy. Soc. Victoria, 31:11-66.
Hindle, E.
  1914. Flies in relation to disease: blood-sucking flies. Cambridge Univ. Press,
      398 pp.
Hindle, E., and Merriman, G.
  1914. The range of flight of *Musca domestica*. Report upon experiments. Jour.
      Hyg., 14:23-45; *also in* Repts. Loc. Govt. Bd. Publ. Health Med. Subjs.,
      1913, n.s., No. 85, pp. 20-41.
Hinton, H. E.
  1960. The chorionic plastron and its role in the eggs of the Muscinae (Dip-
      tera). Quart. Jour. Microsc. Sci., 101:313-332.
Hinton, H. E.
  1970. Insect eggshells. Sci. Amer., 223:84-91.
Hirschfelder, H., and Wolf, J.
  1938. Die Bedeutung von Insekten und Zecken für die Epidemiologie der Maul-
      und Klauenseuche. Zeitsch. Hyg. Zool. Schädlingsbekämpfg., 20:142-147.
Hirschhorn, N., and Greenough III, W. B.
  1971. Cholera. Sci. Amer., 225:15-21.
Hobbs, B. C.
  1961. Public health significance of *Salmonella* carriers in livestock and birds.
      In: Symposium on bacteria of the intestine. Jour. Appl. Bact., 24:340-
      352.
Hobson, R. P.
  1931. On an enzyme from blow-fly larvae (*Lucilia sericata*) which digests col-
      lagen in alkaline solution. Biochem. Jour., 25:1458-1463.
Hobson, R. P.
  1932. Studies on the nutrition of blowfly larvae. 4. The normal rôle of micro-
      organisms in larval growth. Jour. Exp. Biol., 9:366-377.
Hobson, R. P.
  1938. Sheep blow-fly investigations. VII. Observations on the development of
      eggs and oviposition in the sheep blow-fly, *Lucilia sericata*, Mg. Ann.
      App. Biol., 25:573-582.
Hocking, B.
  1953. The intrinsic range and speed of insect flight. Trans. Roy. Entom. Soc.
      London, 104:223-345.

Hodge, C. F.
  1913. The distance house flies, blue bottles, and stable flies may travel over water. Science, *38*:512-513.
Hoelscher, C. E., Combs, R. L. Jr., and Brazzel, J. R.
  1968. Horn fly dispersal. Jour. Econ. Entom., *61*:370-373.
Hoffmann, S.
  1950. Die hygienische Bedeutung der Fliegen. Travaux Chimie Alimentaire Hyg. Berne, *41*:189-222.
Hofmann, E.
  1888. Ueber die Gefahr der Verbreitung der Tuberculose durch unsere Stubenfliege. Correspondenzbl. Ärtz. Kreis-Bezirk. Königr. Sachs., *44*:130-133.
Holmboe, F. V.
  1941. Flies and the mastitis problem. Norsk Veter.-Tidssk., Oslo, *53*:314-317 (noted by title only).
Holz, J.
  1953. Die Bedeutung von *Musca domestica* als Überträger von *Trichomonas foetus*. Tiarärztl. Umschau, *8*:396-397.
Honeij, J. A., and Parker, R. R.
  1914. Leprosy: flies in relation to the transmission of the disease. Jour. Med. Res., *30*:127-130.
Horikawa, M.
  1958. Developmental genetic studies of tissue cultured eye discs of *Drosophila melanogaster*. I. Growth, differentiation, and tryptophan metabolism. Cytologia, *23*:468-477.
Horikawa, M., and Fox, A. S.
  1964. Culture of embryonic cells of *Drosophila melanogaster* in vitro. Science, *145*:1437-1439.
Horikawa, M., and Kuroda, Y.
  1959. *In vitro* cultivation of blood cells of *Drosophila melanogaster* in a synthetic medium. Nature, *184*:2017-2018.
Horikawa, M., and Sugahara, T.
  1960. Studies on the effects of radiation on living cells in tissue culture. I. Radiosensitivity of various imaginal discs and organs in larvae of *Drosophila melanogaster*. Radiation Res., *12*:266-275.
Hormaeche, E., Peluffo, C. A., and Ricaud de Pereyra, V.
  1944. A new *Salmonella* type, *Salmonella carrau*, with special reference to the 1, 7 . . . phases of the Kaufmann-White classification. Jour. Bact., *47*: 323-326.
Hormaeche, E., Surraco, N. L., Peluffo, C. A., and Aleppo, P. L.
  1943. Causes of infantile summer diarrhea. Amer. Jour. Diseases Children, *66*: 539-551.
Horn, A., and Huber, E.
  1911. Zur Frage der Verbreitung paratyphus B-ähnlicher Bakterien durch Fliegen. Zschr. Inf-Krakh. der Haustiere, *10*:443-453.
Hornby, H. E.
  1926. Studies on rinderpest immunity: (2) Methods of infection. Veter. Jour., *82*:348-351, 354-355.
Horstmann, D. M., Niederman, J. C., Riordan, J. T., and Paul, J. R.
  1959. The trial use of Sabin's attenuated type I poliovirus vaccine in a village in southern Arizona. Amer. Jour. Hyg., *70*:169-184.
Hoskins, M.
  1933. An attempt to transmit yellow fever virus by dog fleas. (*Ctenocephalides canis* Curt) and flies (*Stomoxys calcitrans* Linn.). Jour. Parasit., *19*: 299-303.

Howard, C. W.
 1917. Insect transmission of infectious anemia of horses. Jour. Parasit., *4*: 70-79.
Howard, C. W., and Clark, P. F.
 1912. Experiments on insect transmission of the virus of poliomyelitis. Jour. Exp. Med., *16*:850-859.
Howard, L. O.
 1900. A contribution to the study of the insect fauna of human excrement. Proc. Wash. Acad. Sci., *2*:541-604.
Howe, L.
 1888. On the influence of flies in the spread of Egyptian ophthalmia. 7th Intl. Congr. Ophth., Wiesbaden: Becker u Hess, pp. 323-328.
Howell, D. E., Sanborn, C. E., Rozenboom, L. E., Stiles, G. W., and Moe, L. H.
 1941a. The transmission of anaplasmosis by horseflies (Tabanidae). Okla. Agric. Exp. Sta. Tech. Bull. T-11.
Howell, D. E., Stiles, G. W., and Moe, L. H.
 1941b. The transmission of anaplasmosis by mosquitoes (Culicidae). Jour. Amer. Veter. Med. Assoc., *99*:107-110.
Hudson, H.
 1915. How dangerous is the house fly? Modern Hospital, *4*:359-360.
Hughes, R. D.
 1970. The seasonal distribution of bushfly (*Musca vetustissima* Walker) in southeast Australia. Jour. Anim. Ecol., *39*:691-706.
Hurlbut, H. S.
 1950. The recovery of poliomyelitis virus after parenteral introduction into cockroaches and houseflies. Jour. Infect. Dis., *86*:103-104.
Hummadi, M. K., and Maki, B.
 1970. Studies on the breeding of *Stomoxys calcitrans* stable fly. Bull. Iraq Nat. Hist. Mus. Univ. Baghdad, *4*:21-26.
Hunt, D., and Johnson, A. L.
 1923. Yaws, a study based on over 2,000 cases treated in American Samoa. U.S. Nav. Med. Bull., *18*:599-607.
Hunter, W.
 1905. The spread of plague infection by insects. Centralbl. Bakt., *40*:43-55.
Hunziker, H., and Reese, H.
 1922. Die Basler Pockenepidemie von 1921 unter besonderer Berücksichtigung der Verbreitung der Pocken durch Fliegen. Schweiz. Med. Wochenschr., *3*:469-476.
Huq, M.
 1961. African horse sickness. Veter. Record, *73*:123.
Idina, M. S.
 1959. The period of survival in faeces of dysentery bacilli. Jour. Microb. Epidem. Immun., *30*:103-107.
Ikeme, M. M.
 1967. Kerato-conjunctivitis in cattle in the plateau area of Northern Nigeria. A study of *Thelazia rhodesi* as a possible aetiological agent. Bull. Epizoot. Dis. Afr., *15*:363-367. (Cited here but not in text. *Thelazia* larvae were isolated from a minute fly of the genus *Musca* found near animal quarters.)
Illingworth, J. F.
 1923. Insect fauna of hen manure. Proc. Hawaiian Entom. Soc., *5*:270-273.
Illingworth, J. F.
 1926a. The common muscoid flies, occurring about sweet-shops in Yokohama, Japan. Proc. Hawaiian Entom. Soc., *6*:260-261.

Illingworth, J. F.
  1926b.   Observations on *Chrysomyia megacephala* (Fabr.), our common blow-fly in the Orient. Proc. Hawaiian Entom. Soc., *6*:253-255.
Illingworth, J. F.
  1926c.   Notes on *Chrysomyia megacephala* (Fabr.) (Diptera). Proc. Hawaiian Entom. Soc., *6*:266.
Illingworth, J. F.
  1926d.   Notes on *Sarcophaga fuscicauda* Böttcher (Diptera). Proc. Hawaiian Entom. Soc., *6*:262-265.
Il'Yashenko, L. Ya
  1964.   Migrations of flies (*Musca domestica vicina*) in rural areas. Med. Parazitol. Parazitarn. Bolezni, *33*:9-13.
Imray, J.
  1873.   On framboesia or yaws, as the disease has existed in the island of Dominica, West Indies. (Appendix D in Milroy, G., see pp. 72-83.)
Ingram, R. L., Larsen, J. R. Jr., and Pippen, W. F.
  1956.   Biological and bacteriological studies on *Escherichia coli* in the housefly, *Musca domestica*. Amer. Jour. Trop. Med. Hyg., *5*:820-830.
Ishijima, H.
  1967.   Revision of the third stage larvae of synanthropic flies of Japan (Diptera: Anthomyiidae, Muscidae, Calliphoridae and Sarcophagidae). Japanese Jour. San. Zool., *18*:47-100.
Ivashkin, V. M.
  1959.   Epizootologiya parabronematoza zhvachnykh. [The epizootiology of *Parabronema* of ruminants.] Tr. Gel'mintol. Lab. Akad. Nauk SSSR, *9*:97-105.
Ivashkin, V. M., Khromova, L. A., and Shmitova, G. Y.
  1963.   Stephanofilariasis in cattle. Veterinariya, *40*:36-39.
Iwanoff, X.
  1934.   Über Sommerwunden beim Rinde. Arch. Tierheilk, *67*:261-270.
Jack, R. W.
  1917.   Natural transmission of trypanosomiasis (*T. pecorum* group) in the absence of tsetse-fly. Bull. Entom. Res., *8*:35-41.
Jack, R. W., and Williams, W. L.
  1937.   The effect of temperature on the reaction of *Glossina morsitans* Westw. to light. A preliminary note. Bull. Entom. Res., *28*:499-503.
Jackson, D. D.
  1909.   The house-fly at the bar. Indictment. Guilty or not guilty? Merchants' Assoc. of New York, April, 48 pp.
Jackson, G. J., Herman, R., and Singer, I., Editors
  1969.   Immunity to parasitic animals. Vol. I. Appleton-Century-Crofts, New York, 292 pp.
Jacob, P., and Klopstock, M.
  1910.   Die Uebertragung der Tuberkulose durch Fliegen. Tuberculosis, *9*:496-510.
Jacobs, J., Guinée, P.A.M., Kampelmacher, E. H., and van Keulen, A.
  1963.   Studies on the incidence of *Salmonella* in imported fish meal. Zentralbl. Veter., *6*:542-550.
James, M. T.
  1947.   The flies that cause myiasis in man. U.S.D.A. Misc. Publ. No. 631, 175 pp.
James, M. T., and Harwood, R. F.
  1969.   Herm's Medical Entomology, 6th edition. Macmillan, 484 pp.

Japanese Commission
1914. Report on the results obtained by the special committee for investigation of infectious anemia of the horse. Horse Administration Bureau, Tokyo. Veter. Jour., *70*:604-627.

Jarvis, E. M.
1918. Report on ixodic lymphangitis. Veter. Jour., *74*:44-53.

Jausion, H., and Dekester, M.
1923. Sur la transmission comparée des kystes d'*Entamoeba dysenteriae* et de *Giardia intestinalis* par les mouches. Arch. Inst. Past. Algérie, *1*:154-155; also Arch. Inst. Past. l'Afrique du Nord, *3*:154-155.

Jellison, W. L.
1950. Geographical distribution of "deer-fly fever" and the biting fly *Chrysops discalis* Williston. Publ. Health Rept., *65*:1321-1329.

Jenkins, D. W., Casida, J., Ingram, R. L., and Larsen, J. R.
1954. Unpublished data discussed in Exp. Parasit., *3*:474-490.

Jensen, J. A., and Fay, R. W.
1951. Tagging of adult house flies and flesh flies with radioactive phosphorus. Amer. Jour. Trop. Med., *31*:523-530.

Jepson, J. P.
1908. Notes on experiments in colouring flies, for purposes of identification. Repts. Loc. Govt. Bd. Publ. Health Med. Subjs., n.s., No. 16, Further Prel. Repts. on flies as carriers of infection, pp. 4-9.

Jettmar, H. M.
1940. Some experiments on the resistance of the larvae of latrine fly, *Chrysomyia megacephala*, against chemicals. Chinese Med. Jour., *57*:74-85.

Jettmar, H. M.
1953. Uber bakteriostatische Stoffe im Darm der Hypoderma larvae. Z. Jour. Hyg., *137*:61-66.

Jewell, C. H.
1904. Contagious ulcerative lymphangitis. Amer. Veter. Rev., *28*:34-37.

Johan, B.
1933. Jelentés a M. kir. országos közegészségügyi intézet 1932. Évben végzett munkájáról. [Report of Work of Roy. Hung. Publ. Health Inst. for 1932.] Pp. 90-92; Engl. Summ. p. 190.

Johnston, T. H.
1912. Notes on some Entozoa. Proc. Roy. Soc. Queensland, *24*:68-91.

Johnston, T. H., and Bancroft, M. J.
1920. The life history of *Habronema* in relation to *Musca domestica* and native flies in Queensland. Proc. Roy. Soc. Queensland, *32*:61-88.

Johnston, T. H., and Tiegs, O. W.
1923. Notes on the biology of some of the more common Queensland muscoid flies. Proc. Roy. Soc. Queensland, *34*:77-104.

Joly
1901. Mission hydrographique de l'aviso-transport la Rance à Madagascar (1899-1900). Arch. Méd. Nav., *75*:401-463.

Jones, C. L.
1969. Biology of the face fly: Migration of larvae. Jour. Econ. Entom., *62*:255-256.

Jones, C. M.
1963. Research on the face fly during 1962. Proc. North Central Branch, Entom. Soc. Amer., *18*:53.

Jones, C. M.
1966. Stable flies. *In*: Insect colonization and mass production. Ed. C. N. Smith, Academic Press, see pp. 145-152.

Jones, E. B.
  1941. A fly-borne epidemic of enteric fever. Med. Officer, 65:65-67.
Jones, F. S., and Little, R. B.
  1923. An infectious ophthalmia of cattle. Jour. Exp. Med., 38:139-148.
Jones, F. S., and Little, R. B.
  1924. The transmission and treatment of infectious ophthalmia of cattle. Jour. Exp. Med., 39:803-810.
Jones, F.W.C.
  1907. Notes on enteric fever prevention in India. Jour. Roy. Army Med. Corps, 8:22-34.
Jones, G. W.
  1962. The significance of the plague fly. An historical note. Virginia Med. Monthly, 89:87-89.
Jones, H. L.
  1915. The treatment of trypanosomiasis in cattle caused by the *Trypanosoma pecorum*. Jour. Comp. Path. Ther., 28:154-156.
Joós, I.
  1936. Ueber die Gefahr der Infektion mit Typhusbakterien durch Fliegen als Bakterienüberträger. Zentralbl. Bakt., Parasit. Infekt., 137:223-226.
Josefson, A.
  1912. II. Experimental investigations with the object of determining the possibility of transmission of infantile paralysis by means of dead objects and by flies. Rept. State Med. Inst. Sweden, pp. 170-178.
Jowett, W.
  1911. Further note on a cattle trypanosomiasis of Portuguese East Africa. Jour. Comp. Path., 24:21-40.
Joyeux, C.
  1920. Cycle évolutif de quelques cestodes. Supp. Bull. Biol. France et Belgique, pp. 146-147.
Kadner, C. G., and LaFleur, F. M.
  1951. The vitamin requirements of *Phaenicia sericata* (Meigen) larvae (Diptera: Calliphoridae). Wasmann Jour. Biol., 9:129-136.
Kamal, A. S.
  1958. Comparative study of thirteen species of sarcosaprophagous Calliphoridae and Sarcophagidae (Diptera). I. Bionomics. Ann. Entom. Soc. Amer., 51:261-271.
Kamyszek, F.
  1965a. Wpływ niektórych czynników na mozliwość biernego przenoszenia patogennych grzybów przez muchy domowe (*Musca domestica*). [Effect of certain factors on the passive transmission of pathogenic fungi by domestic flies.] Wiademości Parazyt., 11:567-572.
Kamyszek, F.
  1965b. Mucha domowa (*Musca domestica*) jako przenosiciel grzybic. [The house fly [*Musca domestica*] as a carrier of fungi.] Med. Weter., 21:622-624.
Kamyszek, F.
  1967. Z badań nad rolą wpleszcza owczego (*Melophagus ovinus*) w przenoszeniu grzybów chorobotwórczych. Med. Weter., 23:139-141.
Kaneko, K., and Kano, R.
  1960. Experimental transmission of *Salmonella pullorum* with larvae of *Sarcophaga peregrina*. Jap. Jour. Sanit. Zool., 11:66-71.
Karmann, P.
  1928. Die Weideeuterentzündung der Kühe und Färsen in Ostfriesland. Zeitsch. Infektionskrank. Parasit. Krank. Hyg. der Haustiere, 34:122-151.

Kåss, E.
1954. Undersøkelser over *Toxoplasma* og toxoplasmose. Bakter. Inst. Rikshospitalet, Oslo. 101 pp.

Keller, O.
1913. Die Antike Tierwelt. Zweiter Bd. Verlag von Wilhelm Engelmann, Leipzig, see pp. 447-454.

Kennedy, J. C.
1906. VIII. Experiments on mosquitoes and flies. Repts. Comm. Medit. Fever Roy. Soc. London, Part IV: 83-84.

Khan, S. M.
1922. Rept. Veter. Sec. for the year 1920. Zanzibar Prot. Rept. Med. Div., pp. 44-48.

Khera, S. S., and Sharma, G. L.
1967. Important exotic diseases of livestock including poultry. Indian Coun. Agric. Res., New Delhi, 179 pp.

Killough, R. A., Hartsock, J. G., Wolf, W. W., and Smith, J. W.
1965. Face fly dispersal, nocturnal resting places, and activity during sunset as observed in 1963. Jour. Econ. Entom., *58*:711-715.

Killough, R. A., and McClellan, E. S.
1965. Face fly oviposition studies. Jour. Econ. Entom., *58*:716-719.

Killough, R. A., and McKinstry, D. M.
1965. Mating and oviposition studies of the stable fly. Jour. Econ. Entom., *58*: 489-591.

Kinghorn, A., and Yorke, W.
1912. Trypanosomes infecting game and domestic stock in the Luangwa Valley, North Eastern Rhodesia. Ann. Trop. Med. Parasit., *6*:301-315.

Kinsely, A. T.
1909. Equine infectious anemia. Amer. Veter. Rev., *36*:45-46.

Kirchberg, E.
1950. Müllkasten und Müllplatze—wichtige Fliegenbrutstätten der Grossstadt. Bln. Gesundheitsbl., *1*:376-378.

Kircheri, A.
1668. *Scrutinium physico—medicum contagiosae luis, quae pestis dicitur.* Rome, see pp. 144, 145.

Kirk, R.
1945. Spread of infective hepatitis. Lancet, Jan. 20, pp. 80-81.

Klesov, M. D.
1949. Biological studies on the nematode *Thelazia rhodesi* Desm. Zool. Zhurn., *28*:515-522.

Klesov, M. D.
1950. The biology of two nematodes of the genus *Thelazia* Bosc, 1819, parasites of the eye of cattle. Dokl. Akad. Nauk SSSR, *70*:549-551.

Klesov, M. D.
1951. Question of the biology of nematode of the genus *Thelazia* Bosc, 1819. Veterinar., *28*:22-25.

Klock, J. W., Pimentel, D., and Stenburg, R. L.
1953. A mechanical fly-tagging device. Science, *118*:48-49.

Knight, Sister M. R.
1962. Rhythmic activities of the alimentary canal of the black blow fly, *Phormia regina* (Diptera: Calliphoridae). Ann. Entom. Soc. Amer., *55*:380-382.

Knipling, E. F., and Rainwater, H. T.
1937. Species and incidence of dipterous larvae concerned in wound myiasis. Jour. Parasit., *23*:451-455.

Knopf, S. A.
1899.   Prophylaxis and treatment of pulmonary tuberculosis. Blakiston's Son and Co., Phila., p. 40.
Knuckles, J. L.
1959.   Studies on the role of *Phormia regina* (Meigen) as a vector of certain enteric bacteria. Dissert. Abst., 20(4).
Knuckles, J. L.
1963.   An apparatus for studying the transfer of pathogenic bacteria from fly to fly. Turtox News, *41*:284-285.
Knuckles, J. L.
1964.   A fly rack assembly useful in fecal collection studies. Turtox News, *42*: 82-83.
Knuckles, J. L.
1967.   An apparatus for rearing bacteria-free blowfly larvae and pupae. Turtox News, *45*:76-77.
Kobayashi, H.
1919.   Flies in Korea. Report 1 [in Korean]. Jour. Chosen Med. Soc. No. 24 [Rev. Appl. Entom. (B), 7:142].
Koch, H. A.
1963.   Zur Ökologie von *Trichophyton verrucosum* Bodin 1902. Int. Symp. Med. Mycol., 1963, pp. 75-79.
Koch, H. A.
1964.   Fliegen als Überträger von Dermatophyten. Zeit. Derm. Vener. Verw. Geb., *7*:365-366.
Koch, W.
1886.   Milzbrand und Rauschbrand. Verlag von Ferdinand Enke., pp. 23-25.
Koster, H.
1817.   Travels in Brazil in the years from 1809 to 1815. Philadelphia, Vol. 2, 323 pp., see pp. 235, 236.
Kowalewsky, A.
1887.   Beiträge zur Kenntniss der nachembryonalen Entwicklung der Musciden. Zeit. Wiss. Zool., *45*:542-594.
Kraepelin, K.
1883.   Zur Anatomie und Physiologie des Rüssels von *Musca.* Zeitsch. Wiss. Zool., *39*:683-716.
Krastin, N. I.
1949.   The decipherment of the cycle of development of the nematode *Thelazia rhodesi* (Desmarest, 1927), parasitizing the eyes of cattle. Dokl. Akad. Nauk SSSR, *64*:885-887.
Krastin, N. I.
1950.   Determination of the cycle of development of *Thelazia gulosa*, parasite of the eye of cattle. Dokl. Akad. Nauk SSSR, *70*:549-551.
Krastin, N. I.
1952.   Determination of the cycle of development of *Thelazia skrjabini* (Erschow, 1928), a parasite of the eye of cattle. Dokl. Akad. Nauk SSSR, n.s., *82*:829-831.
Krontowski, A.
1913.   K voprosu o rasprostranenie tifa i dizenterii mukhami. Kiev: Tipografiia Korchak-Novitskogo: 1913 (i.e. as separate pamphlet). Referat: Vrachebnaia Gazeta, 1913, No. 11, p. 417. [Zur Frage über die Typhus und Dysenterieverbreitung durch Fliegen.] Centralbl. Bakt., Parasit. Infekt., *68*:586-590.
Kubo, K.
1920.   Common species of flies in houses of Manchuria. Tokyo Med. News., No. 2180.

Kuenen, W. A., and Swellengrebel, W. H.
1913. Die Entamöben des Menschen und ihre practische Bedeutung. Zentrabl. Bakt., *71*:378 and on (see pp. 401-403).
Kuhns, D. M., and Anderson, T. G.
1944. A fly-borne bacillary dysentery epidemic in a large military organization. Amer. Jour. Publ. Health, *34*:750-755.
Kumar, P., Ranbir, S., and Sehgal, B. S.
1970. Fly control and diarrheal morbidity in a rural community in U.P. Indian Jour. Med. Sci., *24*:285-291.
Kumm, H. W.
1935a. The natural infection of *Hippelates pallipes* Loew with the spirochaetes of yaws. Trans. Roy. Soc. Trop. Med. Hyg., *29*:265-272.
Kumm, H. W.
1935b. The digestive mechanism of one of the West Indian "eye gnats," *Hippelates pallipes* Loew. Ann. Trop. Med. Parasit., *29*:283-298.
Kumm, H. W., and Turner, T. B.
1936. The transmission of yaws from man to rabbits by an insect vector, *Hippelates pallipes* Loew. Amer. Jour. Trop. Med., *16*:245-262.
Kumm, H. W., Turner, T. B., and Peat, A. A.
1935. The duration of motility of the spirochaete of yaws in a small West Indian fly, *Hippelates pallipes* Loew. Amer. Jour. Trop. Med., *15*:209-223.
Kunert, H., and Schmidtke, L.
1952. Die Bedeutung der nichtstechenden Fliegen für die Verschleppung von Leptospiren. Zeit. Trop. Med. Parasit., *3*:475-486.
Kunert, H., and Schmidtke, L.
1953. Zur Übertragung von *Toxoplasma gondii* durch Fliegen. Zeit. Hyg., *136*:163-173.
Kunike, G.
1927. Untersuchungen über die Rolle der Fliegen als Überträger der Maul-und Klauenseuche auf Meerschweinchen. Zeitsch. Desinfekt. Gsndhtw., *19*:115-117.
Kupka, K., Nižetič, B., and Reinhards, J.
1968. Sampling studies on the epidemiology and control of trachoma in southern Morocco. Bull. Wld. Health Org., *39*:547-566.
Kural, H., and Ayberk, N.
1936. Trachoma and anti-trachoma work in Turkey. Quart. Bull. Health Org. League of Nations, *5*:353-366.
Kuroda, Y.
1954. The culture of the eye-discs of *Drosophila melanogaster*. Jap. Jour. Genet., *29*:163.
Kvasnikova, P. A.
1931. Flies observed in human dwellings and outhouses in the town of Tomsk [in Russian]. Wiss. Ber. Biol. Fak. Tomsk St. Univ. i, No. 1, pp. 9-47.
Laarman, J. J.
1956. Transmission of experimental toxoplasmosis by *Stomoxys calcitrans*. Doc. Med. Geog. Trop., *8*:293-298.
Laboulbène, A., and Follin
1848. Note sur la matière pulvérulente qui recouvre la surface du corps des *Lixus* et de quelques autres insectes. Ann. Soc. Entom. France, II série, *6*:301-306.
Lackany, A.
1963. Flies as carriers of typhoid bacilli in Alexandria. Alexandria Med. Jour., *9*:186-193.

Lainson, R., and Southgate, B. A.
    1965. Mechanical transmission of *Leishmania mexicana* by *Stomoxys calcitrans*. Trans. Roy. Soc. Trop. Med. Hyg., *59*:716.

Laird, M.
    1952. Insects collected from aircraft arriving in New Zealand during 1951. Jour. Aviation Med., *23*:280-285.

Lal, R. B., Ghosal, S. C., and Mukherji, B.
    1939. Investigations on the variation of vibrios in the house fly. Ind. Jour. Med. Res., *26*:597-609.

Lamborn, W. A.
    1933. Med. Entom. Rept. 1932. Ann. Med. Rept. for year ended 1932, Nyasaland Prot., pp. 54-59.

Lamborn, W. A.
    1934. The annual report of the Medical Entomologist for 1933. Pp. 61-63.

Lamborn, W. A.
    1935. Annual report of the Medical Entomologist for 1934. Ann. Med. Rept. Nyasaland, 1934, pp. 65-69.

Lamborn, W. A.
    1936a. The experimental transmission to man of *Treponema pertenue* by the fly *Musca sorbens* Wd. Jour. Trop. Med. Hyg., *39*:235-239.

Lamborn, W. A.
    1936b. Annual report of the Medical Entomologist for 1935. Nyasaland Prot., Ann. Med. Sanit. Rept. for 1935, Zomba, pp. 50-52.

Lamborn, W. A.
    1937. The haematophagous fly *Musca sorbens* Wied., in relation to the transmission of leprosy. Jour. Trop. Med. Hyg., *40*:37-42.

Lamborn, W. A.
    1937. Annual report of the Medical Entomologist for 1935. Nyasaland Prot., Ann. Med. Sanit. Rept. for 1936, Zomba, pp. 43-44.

Lamborn, W. A.
    1938. Annual report of the Medical Entomologist for 1938. II. Transmission of tuberculosis by *Musca sorbens*. Nyasaland Ann. Med. Sanit. Rept. 1938, pp. 40-48.

Lamborn, W. A.
    1939. Annual report of the Medical Entomologist for 1939. IV. *Musca sorbens* in relation to *Mycobacterium tuberculosis*. Nyasaland Ann. Med. Sanit. Rept. 1939, p. 31.

Lamborn, W. A.
    1955. The haematophagous fly as a possible vector of leishmania. Bull. Endem. Dis., *1*:239-249.

Lamborn, W. A., and Howat, C. H.
    1936. A possible reservoir host of *Trypanosoma rhodesiense*. Brit. Med. Jour., *1*:1153-1155.

Lambremont, E. N., Fisk, F. W., and Ashrafi, S.
    1959. Pepsin-like enzyme in larvae of stable flies. Science, *129*:1484-1485.

Landi, S.
    1960. Bacteriostatic effect of haemolymph of larvae of various botflies. Canad. Jour. Microb., *6*:115-119.

Landsberger, B.
    1934. Die Fauna des alten Mesopotamien nach der 14. Tafel der Serie Har-ra-Hubulla. Verlag Von S. Hirzel, Leipzig, 144 pp., see p. 25.

Lang, N. N.
    1940. On the question of the preservation of *Bacillus pestis* in the developing fly larvae. Vestnik Mikr. Epid. Parazitol., *19*:96-97.

Lapage, G.
1968. Veterinary parasitology. Charles C. Thomas, Springfield, Illinois, 1160 pp.

Larrey, D. J.
1829. Clinique chirurgicale, exercée particulièrement dans les camps et les hôpitaux militaires, depuis 1792 jusqu'en 1829. Chez Gabon, Paris, pp. 1-491, see pp. 51, 52, 449.

Lauber, I.
1920. Bakteriologische Untersuchungsergebnisse der Mannheimer Ruhrepidemie Juli bis November 1917. Centralbl. Bakt., Parasit. Infekt. I. Abt., Orig., 84:201-213.

LeBailly, C.
1924. Les mouches ne jouent pas de rôle dans la dissémination de la fièvre aphteuse. Compt. Rend. Acad. Sci., 179:1225-1227.

Lebert
1856. Die Pilzkrankheit der Fliegen. Verhand. Zürcherischen Natur. Gesell. vom 29, October.

Leboeuf, A.
1912. Dissémination du bacille de Hansen par la mouche domestique. Bull. Soc. Path. Exot., 5:860-868.

Leboeuf, A.
1913. Notes sur l'épidémiologie de la lèpre dans l'archipel calèdonien. Bull. Soc. Path. Exot., 6:551-556.

Leckie, V. C.
1925. Some notes on surra in the camel, its treatment and prevention. Vet. Jour., 81:281-292, 346-352, 398-404, 494-499, 546-553.

Ledingham, J.C.G.
1911. On the survival of specific microorganisms in pupae and imagines of *Musca domestica* raised from experimentally infected larvae. Experiments with *B. typhosus*. Jour. Hyg., 11:333-340.

Ledingham, J.C.G.
1920. Dysentery and enteric disease in Mesopotamia from the laboratory standpoint. An analysis of laboratory data during the eighteen months ending December 31, 1918. Jour. Roy. Army Med. Corps, 34:189-320.

Lee, D. J.
1968. Human myiasis in Australia. Med. Jour. Australia, 1:170-173.

Lee, V. H., Vadlamudi, S., and Hanson, R. P.
1962. Blow fly larvae as a source of botulinum toxin for game farm pheasants. Jour. Wildlife Manag., 26:411-413.

Leese, A. S.
1909. Experiments regarding the natural transmission of surra carried out at Mohand in 1908. Jour. Trop. Vet. Sci., 4:107-132.

Leidy, J.
1874. On a parasitic worm of the house fly. Proc. Acad. Nat. Sci. Phil., 26:139-140.

Legner, E. F., and Bay, E. C.
1963. The prospect for the biological control of *Hippelates collusor* (Townsend) in southern California. Proc. Calif. Mosquito Control Assoc., 31:76-79.

Legner, E. F., and Brydon, H. W.
1966. Suppression of dung-inhabiting fly populations by pupal parasites. Ann. Entom. Soc. Amer., 59:638-651.

Legner, E. F., and McCoy, C. W.
1966. The house fly, *Musca domestica* Linnaeus, as an exotic species in the western hemisphere incites biological control studies. Canad. Entom., *98*:243-248.

Legroux, R., and Second, L.
1945. La spore botulique dans la mouche *Piophila casei* L. Ann. Inst. Pasteur, *71*:464-466.

Lelikov, V. L.
1964. Role of flies in the distribution of typhoid fever in Leninabad. Tsentral'nyi Inst. Usovershenstvovaniia Vrachei, Trudy, Moscow, *68*:31-34 (cited here only).

Leloup, A. M., and Gianfelici, E.
1966. The persistence of neurosecretion in organotypic culture by a dipterous insect, *Calliphora erythrocephala*. Ann. Endocr. (Paris), *27*:506-508.

Lemery, N.
1751. Dizionario overo trattato universale delle droghe semplici. G. Bertella, Venice, see p. 230.

Lempke, B. J.
1962. Insecten gevangen op het lichtschip "Noord Hinder." Entom. Bericht. Nederland. Entom. Vereen., *22*:100-111.

Léon
1908. Bull. Méd. Natur. (Jassy). Nos. 9 and 10. (Quoted from Galli-Valerio.)

León, J. J.
1860. La tiña endémica de Tabasco, Chiapas y el sur de México. Soc. Mex. Geogr. Estad: Primera Bol., *8*:503-521.

León y Blanco, F.
1940. La noción del contagio y la idea del vector en el mal del pinto. Medicina, *20*:162-169.

León y Blanco, F., and Soberón y Parra, G.
1941. Nota sobre la trasmisión experimental del mal del pinto por medio de una mosca del genero *Hippelates*. Gac. Med. Mexico, *71*:534-539.

Leroux, P. H.
1944. Mechanical transmission of trypanosomiasis. Rept. Northern Rhodesia Dept. of Veterinary Services [not seen].

Lesage, A.
1921. Le choléra infantile et les mouches. Son isolement en pavillon spécial. Bull. Acad. Med., *86*:267-268.

Levaditi, C., and Kling, C.
1914. Le rôle des *Stomoxys calcitrans* dans la transmission de la poliomyélite aiguë épidémique. Zeitsch. Immun. Exp. Therapie, *22*:260-268.

Levinson, Z. H.
1960. Food of housefly larvae. Nature, *188*:427-428.

Levinson, Z. H., and Bergmann, E. D.
1959. Vitamin deficiencies in the housefly produced by anti-vitamins. Jour. Insect Physiol., *3*:293-305.

Levkovich, E. N., and Sukhova, M. N.
1957. Duration of retention and excretion of poliomyelitis virus by synanthropic flies and its relation to dissemination and prevention of poliomyelitis. Med. Parazit., Moskva, *26*:343-347.

Lewallen, L. L.
1954. Biological and toxicological studies of the little house fly. Jour. Econ. Entom., *47*:1137-1141.

Lewis, E. A.
1933. Observations on some Diptera and myiasis in Kenya Colony. Bull. Entom. Res., *24*:263-269.

386   BIBLIOGRAPHY

Lindquist, A. W., Yates, W. W., Hoffman, R. A., and Butts, J. S.
 1951. Studies of the flight habits of three species of flies tagged with radioactive phosphorus. Jour. Econ. Entom., *44*:397-400.
Lindsay, D. R., and Scudder, H. I.
 1956. Nonbiting flies and disease. Ann. Rev. Entom., *1*:323-346.
Lindsay, D. R., Stewart, W. H., and Watt, J.
 1953. Effects of fly control on diarrheal diseases in an area of moderate morbidity. Publ. Health Rept., *68*:361-367.
Liu, S., Chen, H., and Lien, J.
 1957. A brief study of the bionomics of fly breeding in Keelung City, Taiwan. Jour. Formosan Med. Assoc., *56*:417-424.
Livingston, S. K., and Prince, L. H.
 1932. The treatment of chronic osteomyelitis with special reference to the use of the maggot active principle. Jour. Amer. Med. Assoc., *98*:1143-1149.
Lobanov, A. M.
 1960. The role of flies in the epidemiology of intestinal infections. [In Russian.] Zhurn. Mikrob. Epidem. Immun., *31*:116-121.
Loeb, J., and Northrop, J. H.
 1916. Nutrition and evolution. Jour. Biol. Chem., *27*:309-312.
Logan, D. C.
 1944. Increase in diarrhoeal diseases. Brit. Med. Jour., *10*:795.
Lokshina, S. S., and Gorodetskiĭ, A. S.
 1956. Presence of microbes of the intestinal group in hibernating flies. Dizenterii, Kiev, Gosmedizdat Ukr., SSR, pp. 242-244 (not seen).
Lord, F. T.
 1904. Flies and tuberculosis. Boston Med. Surg. Jour., *151*:651-654.
Lörincz, F., and Makara, G.
 1936. Investigations into the fly density in Hungary in the years 1934 and 1935. League of Nations Quart. Bull. Health Org., *5*:219-227.
Love, J. A., and Gill, G. D.
 1965. Incidence of coliforms and enterococci in field populations of *Stomoxys calcitrans* (Linnaeus). Jour. Invert. Path., *7*:430-436.
Lowne, B. T.
 1893-5. The anatomy, physiology, morphology, and development of the blow-fly (*Calliphora erythrocephala*). R. H. Porter, London, Vol. II, 778 pp.
Lucas, J.M.S.
 1955. Transmission of *Trypanosoma congolense* in cattle under field conditions in the absence of tsetse flies. Veter. Record, *67*:403-406.
Luedke, A. J., Jochim, M. M., and Bowne, J. G.
 1965. Preliminary bluetongue transmission with the sheep ked *Melophagus ovinus* (L.). Canad. Jour. Comp. Med. Vet. Sci., *29*:229-231.
Lührs
 1919. Die ansteckende Blutarmut der Pferde. Zeitschr. Vet., *31*:369-457.
Luxmoore, E.J.H.
 1907. Report on outbreak of enteric among 17th Lancers. Jour. Roy. Army Med. Corps, *8*:494-495.
Lyon, G. M., and Price, A. M.
 1941. Spot maps of bacillary dysentery and of poliomyelitis. Jour. Pediatrics, *19*:628-631.
Lyons, F. M.
 1953. The control of trachoma and acute seasonal conjunctivitis. Bull. Ophthal. Soc. Egypt, *46*:137-150.
Lyons, F. M., and Abdine, G. E.
 1952. The effect of fly control on the epidemic spread of acute ophthalmia. Bull. Ophthal. Soc. Egypt, *45*:81-88.

Lyons, F. M., and Amies, C. R.
 1949. The epidemiology and prevention of the acute ophthalmias of Egypt. Bull. Ophthal. Soc. Egypt, *42*:116-139.
Lysenko, O.
 1957. Možnost přenosu mikroorganismů metamorfosou pakomára *Chironomus plumosus*. [Possibility of bacterial carryover in the metamorphosis of the midge *Chironomus plumosus*.] Československá Mikrob., 2:248-250.
Lysenko, O.
 1958. Mikroflora některých našich much. [Microflora of some flies of Czechoslovakia.] Československá Mikrob., *3*:51-53.
Lysenko, O., and Povolný, D.
 1961. The microflora of synanthropic flies in Czechoslovakia. Folia Microb., 6:27-32.
Macadam, I.
 1964. Observations on the effects of flies and humidity on the natural lesions of streptothricosis. Veter. Rec., *76*:194-198.
MacDougall, R. S.
 1909. Sheep maggot and related flies: their clasification, life-history, and habits. Trans. Highl. Agric. Soc. Scotland, *21*:135-174.
MacKaig, A.
 1902. Insects and cholera. Edin. Med. Jour., *12*:137-140.
Mackerras, M. J.
 1933. Observations on the life-histories, nutritional requirements and fecundity of blow-flies. Bull. Entom. Res., *24*:353-362.
Mackerras, M. J., and Freney, M. R.
 1933. Observations on the nutrition of maggots of Australian blowflies. Jour. Exp. Biol., *10*:237-246.
Mackerras, I. M., and Fuller, M. E.
 1937. A survey of the Australian sheep blowflies. Jour. Coun. Sci. Indust. Res. Austral., *10*:261-270.
MacLeod, J.
 1937. The species of Diptera concerned in cutaneous myiasis of sheep in Britain. Proc. Roy. Entom. Soc. London (A), *12*:127-133.
Macleod, J.
 1961. Arthropod transmission of micro-organisms. Nature, *191*:885-888.
MacLeod, J., and Donnelly, J.
 1957a. Some ecological relationships of natural populations of calliphorine blowflies. Jour. Anim. Ecol., *26*:135-170.
MacLeod, J., and Donnelly, J.
 1957b. Individual and group marking methods for fly-population studies. Bull. Entom. Res., *48*:585-592.
MacLeod, J., and Donnelly, J.
 1958. Local distribution and dispersal paths of blowflies in hill country. Jour. Anim. Ecol., *27*:349-374.
MacLeod, J., and Donnelly, J.
 1960. Natural features and blowfly movement. Jour. Anim. Ecol., *29*:85-93.
MaCrae, R.
 1894. Flies and cholera diffusion. Ind. Med. Gaz., *29*:407-412.
Maddox, R. L.
 1885a. Experiments on feeding some insects with the curved or "comma" bacillus, and also with another bacillus (*B. subtilis*?). Jour. Roy. Micr. Soc., Ser. II. *5*:602-607.
Maddox, R. L.
 1885b. Further experiments on feeding insects with the curved or "comma" *Bacillus*. Jour. Roy. Micr. Soc., Ser. II. *5*:941-952.

388 BIBLIOGRAPHY

Maier, P. O., Baher, W. C., Bogue, M. D., Kilpatrick, J. W., and Quarterman, K. D.
    1952.   Field studies on the resting habits of flies in relation to chemical con-
           trol. Part 1. In urban areas. Amer. Jour. Trop. Med. Hyg., *1*:1020-1025.
Maille
    1908.   Une épidémie de dysenterie à Cherbourg. Arch. Méd. Nav., *2*:91-109.
[de] Magarinos Torres, C.
    1925.   L'hémotropisme des larves mûres d'*Habronema muscae* (Carter 1861).
           Compt. Rend. Soc. Biol., *93*:38-39.
Magy, H. I., and Black, R. J.
    1962.   An evaluation of the migration of fly larvae from garbage cans in Pasa-
           dena, California. Calif. Vector Views, *9*:55-59.
Malke, H.
    1965.   Über das Vorkommen von Lysozym in Insekten. Zeitsch. Allg. Mikrob.,
           5:42-47.
Manning, J.
    1902.   A preliminary report on the transmission of pathogenic germs by the
           housefly. Jour. Amer. Med. Assoc., *38*:1291-1294.
Mansfield-Aders, W.
    1923-4.   Trypanosomiasis of stock in Zanzibar. Trans. Roy. Soc. Trop. Med.
           Hyg., *17*:192-200.
Manson-Bahr, P. H.
    1919.   Bacillary dysentery. Trans. Roy. Soc. Trop. Med. Hyg., *13*:64-72.
Markar'iants, L. A.
    1962.   Voprosy Epidemiol., Bakteriol., Gigieny, Parazitol. i Virusol. Tezisy
           Dokl. Dushanbe, pp. 138-139. [The role of flies in the epidemiology of
           ascaridosis.] (Not seen.)
Maroja, R. C., Ferro, T. L., and De Frietas, E. W.
    1956.   Enterobactérias patogénicas mecanicamente transportadas por moscas
           nacidade de Palmares, Pernambuco (Nota previa). Revista Serviço Espe-
           cial Saude Publ., *10*:741-746.
Marten, B.
    1720.   A new theory of consumptions: more especially of a phthisis or con-
           sumption of the lungs. I-XII, 186 pp., London. See pp. 61-76.
Martin, G., Leboeuf, A., and Roubaud, E.
    1908.   Expériences de transmission du "Nagana" par les stomoxes et par les
           moustiques du genre *Mansonia*. Bull. Soc. Path. Exot., Paris, *1*:355-358.
Martini, E.
    1903.   Ueber die Entwickelung der Tsetseparasiten in Säugethieren. Zeitsch.
           Hyg. Infect., *42*:341-349.
Maseritz, I. H.
    1934.   Digestion of bone by larvae of *Phormia regina*. Its relationship to bac-
           teria. Arch. Surg., *28*:589-607.
Matignon, J.-J.
    1898.   La peste bubonique en Mongolie. Ann. Hyg. Publ. Miéd. Légale, *39*:
           227-256.
Matthew, D. L., Dobson, R. C., and Osmun, J. V.
    1960.   The face fly is becoming a household pest. Pest Control, *28*:16.
Maw, M. G.
    1965.   Effects of air ions on duration and rate of sustained flight on the blow-
           fly, *Phaenicia sericata* Meigen. Canad. Entom., *97*:552-556.
Maxwell, J.
    1839.   Observations on yaws and its influence in originating leprosy. Edinburgh,
           also in Selected Essays and Monographs. New Sydenham Society, *161*:
           187-237.
Mayer, H.
    1965.   Investigaciones sobre toxoplasmosis. Bol. Ofic. San. Panam., *68*:485-497.

McCabe, L. J., and Haines, T. W.
    1957. Diarrheal disease control by improved human excreta disposal. Publ. Health Rept., *72*:921-928.

McCall, F. J.
    1926. Ann. Rept. Dept. Vet. Sci. Anim. Husb. Tanganyika Territory, pp. 12-14.

McCall, F. J.
    1928. Ann. Rept. Dept. Vet. Anim. Husb. Tanganyika Territory, pp 21-28.

McGuire, C. D., and Durant, R. C.
    1957. The role of flies in the transmission of eye disease in Egypt. Amer. Jour. Trop. Med. Hyg., *6*:569-575.

McNeil, E., and Hinshaw, W. R.
    1944. Snakes, cats, and flies as carriers of *Salmonella typhimurium*. Poultry Sci., *23*:456-457.

Mégnin, J. P.
    1874. Du transport et de l'inoculation des virus, charbonneux et autres, par les mouches. Compt. Rend. Acad. Sci., *79*:1338-1340.

Mellor, J.E.M.
    1920. Observations on the habits of certain flies, especially of those breeding in manure. Ann. Appl. Biol., *6*:53-88.

Melnick, J. L.
    1949. Isolation of poliomyelitis virus from single species of flies collected during an urban epidemic. Amer. Jour. Hyg., *49*:8-16.

Melnick, J. L.
    1950. Studies on the Coxsackie viruses: properties, immunological aspects and distribution in nature. Bull. New York Acad. Med., *26*:342-356.

Melnick, J. L., and Dow, R. P.
    1953. Poliomyelitis in Hidalgo County, Texas, 1948. Poliomyelitis and Coxsackie viruses from flies. Amer. Jour. Hyg., *58*:288-309.

Melnick, J. L., Emmons, J., Coffey, J. H., and Schoof, H.
    1954. Seasonal distribution of Coxsackie viruses in urban sewage and flies. Amer. Jour. Hyg., *59*:164-184.

Melnick, J. L., and Penner, L. R.
    1947. Experimental infection of flies with human poliomyelitis virus. Proc. Soc. Exp. Biol. Med., *65*:342-346.

Melnick, J. L., and Penner, L. R.
    1952. The survival of poliomyelitis and Coxsackie viruses following their ingestion by flies. Jour. Exp. Med., *96*:255-271.

Melnick, J. L., Shaw, E. W., and Curnen, E. C.
    1949. A virus isolated from patients diagnosed as non-paralytic poliomyelitis or aseptic meningitis. Proc. Soc. Exp. Biol. Med., *71*:344-349.

Melnick, J. L., and Ward, R.
    1945. Susceptibility of vervet monkeys to poliomyelitis virus in flies collected at epidemics. Jour. Infect. Dis., *77*:249-252.

Melnick, J. L., Ward, R., Lindsay, D. R., and Lyman, F. E.
    1947. Fly-abatement studies in urban poliomyelitis epidemics during 1945. Publ. Health Repts., *62*:910-922.

Melvin, R.
    1934. Incubation period of eggs of certain muscoid flies at different constant temperatures. Ann. Entom. Soc. Amer., *27*:406-410.

Menees, J. H.
    1962. The skeletal elements of the gnathocephalon and its appendages in the larvae of higher Diptera. Ann. Entom. Soc. Amer., *55*:607-616.

Meng, C. H., and Winfield, G. F.
1938. Studies on the control of fecal-borne diseases in North China. V. A pre-liminary study of the density, species make up, and breeding habits of the house frequenting fly population of Tsinan, Shantung, China. Chinese Med. Jour. Suppl. *2*:463-486.

Meng, C. H., and Winfield, G. F.
1944. Natural enemies of the common West China flies. Chinese Med. Jour., *62A*:89-92.

Meng, C. H., and Winfield, G. F.
1950. Studies on the control of fecal-borne diseases in North China. XVI. An approach to the quantitative study of the house frequenting fly popula-tion. D. The breeding habits of the common North China flies. Phil. Jour. Sci., *79*:175-200.

Menzel, R.
1924. Über die Verbreitung von Rhabditis-larven durch Dipteren. Zool. Anz., *58*:345-349.

Mercurialis
1577. *De pestes in universum, praesertim vero de Veneta et Patavina. Item de morbis cutaneis, et omnibus humani corporis excrementis* (not seen).

Metelkin, A. I.
1935. Significance of flies in the spread of coccidioses among animals and man. Med. Parazit., *4*:75-82 [in Russian].

Meyerhof, M.
1914. Beobachtungen über akute Konjunctivitis und Trachom der Säuglinge in Aegypten. Klin. Monatsbl. Augenheilk., *53*:334-378.

Meyerhof, M.
1928. The book of the ten treatises on the eye ascribed to Hunain Ibn Is-Haq (809-877 A.D.). Government Press, Cairo, 227 pp., see p. 56.

Michelbacher, A. E., Hoskins, W. M., and Herms, W. B.
1932. The nutrition of flesh fly larvae, *Lucilia sericata* (Meigen). I. The ade-quacy of sterile synthetic diets. Jour. Exp. Zool., *64*:109-128.

Mihályi, F.
1965. Rearing flies from faeces and meat infected under natural conditions. Acta Zool. Hung., *9*:153-164.

Mihályi, F.
1966. Flies visiting fruit and meat in an open-air market in Budapest. Acta Zool. Hung., *12*:331-337.

Mihályi, F.
1967a. Seasonal distribution of the synanthropic flies in Hungary. Ann. Hist.-Nat. Mus. Nat. Hung., *59*:327-344.

Mihályi, F.
1967b. Separating the rural and urban synanthropic fly faunas. Acta Zool. Hung., *13*:379-383.

Mihályi, F.
1967c. The danger-index of the synanthropic flies. Acta Zool. Acad. Sci. Hung., *13*:373-377.

Milam, D. F., and Meleney, H. E.
1931. Investigations of *Endamoeba histolytica* and other intestinal Protozoa in Tennessee: II. An epidemiological study of amoebiasis in a rural com-munity. Amer. Jour. Hyg., *14*:325-336.

Miller, T. A., and Treece, R. E.
1968. Some relationships of face fly feeding, ovarian development, and inci-dence on dairy cattle. Jour. Econ. Entom., *61*:250-257.

Millingen, J. G.
  1838. Curiosities of medical experience. Haswell, Barrington & Haswell, Phila-
         delphia, 372 pp., see p. 171.
Milroy, G.
  1873. Report on leprosy and yaws in the West Indies addressed to H. M. Sec-
         retary of State for the Colonies. London (C-729). Accounts and Papers,
         Session 6 Feb. 5, Vol. 50.
Minchin, E. A.
  1908. Investigations on the development of trypanosomes in tsetse-flies and
         other Diptera. Quart. Jour. Microscop. Sci., *52*:159-260 (see pp. 180,
         184).
Minett, E. P.
  1912. The question of flies as leprosy carriers. Jour. London School Trop. Med.,
         1:31-35.
Mirchamsy, H., Hazrati, A., Bahrami, S., and Shafyi, A.
  1970. Growth and persistent infection of African horse-sickness virus in a mos-
         quito cell line. Amer. Jour. Veter. Res., *31*:1755-1761.
Mitscherlich, E.
  1943. Die Übertragung der Kerato-Conjunctivitis infectiosa des Rindes durch
         Fliegen und die Tenazität von *Rickettsia conjunctivae* in der Aussenwelt.
         Deutsche Tropenmed. Zeit., *47*:57-64.
Mitzmain, M. B.
  1912a. Collected notes on the insect transmission of surra in carabaos. Philip-
          pine Agric. Rev., *5*:670-681.
Mitzmain, M. B.
  1912b. The role of *Stomoxys calcitrans* in the transmission of *Trypanosoma
          evansi*. Philippine Jour. Sci., *7*:475-519.
Mitzmain, M. B.
  1913. The bionomics of *Stomoxys calcitrans* Linnaeus; a preliminary account.
         Philippine Jour. Sci., *8*:29-48.
Mitzmain, M. B.
  1914a. Collected studies on the insect transmission of *Trypanosoma evansi*. I.
          The relation of *Tabanus striatus* to surra dissemination. Treas. Dept.,
          U. S. Publ. Health Serv., Hyg. Lab. Bull., *94*:7-39.
Mitzmain, M. B.
  1914b. II. Summary of experiments in the transmission of anthrax by biting
          flies. Treas. Dept., U. S. Publ. Health Serv., Hyg. Lab. Bull., *94*:41-48.
Mitzmain, M. B.
  1914c. An experiment with *Stomoxys calcitrans* in an attempt to transmit a fila-
          ria of horses in the Philippines. Amer. Jour. Trop. Dis. Prevent. Med.,
          2:759-763.
Mitzmain, M. B.
  1916. A digest of the insect transmission of disease in the orient with especial
         reference to the experimental conveyance of *Trypanosoma evansi*. New
         Orleans Med. Surg. Jour., *69*:416-424.
Miura, S., Sato, G., and Miyamae, T.
  1964. Occurrence and survival of *Salmonella* organisms in hatcher chick fluff
         from commercial hatcheries. Avian Diseases, *8*:546-554.
Mohammed
         Le recueil des traditions Mahome'tanes. Abou Abdallah Mohammed ibn
         Ismael el-Bokhari. 4 Vols., Edited by T.G.J. Juynboll, 1862-1908, Leiden.
         See Vol. 4, no. 76, chapt. 58, pp. 71-72.
Mohler
  1908. Three diseases of animals which have recently assumed importance to
         the State Sanitarian. Amer. Vet. Rev., *34*:198-199.

Mohler, J. R.
1920. Annual Reports of the Dept. of Agric. for the year 1919. Report of the Chief of the Bureau of Animal Industry. U. S. Dept. Agric., Washington, D. C. See pp. 125-126.

Mohler, J. R.
1926. Foot-and-mouth disease with special reference to the outbreak in California 1924 and Texas 1924-25. U. S. Dept. Agric. Circ., *400*:1-83.

Mohler, J. R.
1929. Report of the Chief of the Bureau of Animal Industry. U. S. Dept. Agric., Washington, D. C.

Mohrig, W., and Messner, B.
1968a. Immunreaktionen bei Insekten: I. Lysozym als grundlegender antibakterieller Faktor im humoralen Abwehrmechanismus der Insekten. [Immune reactions of insects: I. Lysozyme as fundamental antibacterial factor in the humoral defense mechanism of insects.] Biol. Zentralbl., *87*:439-470.

Mohrig, W., and Messner, B.
1968b. Immunreaktionen bei Insekten: II. Lysozym als antimikrobielles Agens im Darmtrakt von Insekten. [Immune reactions of insects: II. Lysozyme as an antibacterial agent in the intestinal tract of insects.] Biol. Zentralbl., *87*:705-718.

Monroe, R. E.
1962. A method for rearing housefly larvae aseptically on a synthetic medium. Ann. Entom. Soc. Amer., *55*:140.

Montgomery, R. E.
1921. On a form of swine fever occurring in British East Africa (Kenya Colony). Jour. Comp. Path., *34*:159-191 and 242-262.

Moore, H. A., de la Cruz, E., and Vargas-Mendez, O.
1965. Diarrheal disease studies in Costa Rica. IV. The influence of sanitation upon the prevalence of intestinal infection and diarrheal disease. Amer. Jour. Epidem., *82*:162-184.

Moore, W.
1893. Flies and disease. Med. Mag., *2*:1-8.

Moorehead, S., and Weiser, H. H.
1946. The survival of staphylococci food poisoning strain in the gut and excreta of the housefly. Jour. Milk Food Tech., *9*:253-259.

Morellini, M.
1952. Microbatteri tubercolari e paratubercolari nella mosca domestica adulta. Nuovi Ann. Igiene Microbiol., *3*:305-320.

Morellini, M.
1956. Rapporti fra mosca domestica e infezione tubercolare. Federaz. Ital. Tuber., pp. 1-10.

Morellini, M., and Saccà, G.
1953. Alcuni aspetti del meccanismo di diffusione del b. tubercolare da parte di *M. domestica* (studio sperimentale). Rend. Ist. Sup. Sanità, *16*:267-285.

Morgan, B. B.
1942. The viability of *Trichomonas foetus* (Protozoa) in the house fly (*Musca domestica*). Proc. Helminth. Soc. Washington, *9*:17-20.

Morhardt, P.-E.
1936. Les mouches et les maladies contagieuses. Presse Méd., *44*:1397-1398.

Morison, J.
1915. The causes of monsoon diarrhoea and dysentery in Poona. Indian Jour. Med. Res., *2*:950-976.

Morison, G. D.
1937. Some results of trapping the sheep blow-fly (*Lucilia sericata* Meigen). Scot. Jour. Agric., *20*:123-134.

Morison, J., and Keyworth, W. D.
1916. Flies and their relation to epidemic diarrhoea and dysentery in Poona. Indian Jour. Med. Res., *3*:619-627.

Morris, A. P., and Hansens, E. J.
1966. Dispersion of insecticide-resistant populations of the house fly, *Musca domestica* L. Jour. Econ. Entom., *59*:45-50.

Morris, H.
1920. Some carriers of anthrax infection. Jour. Amer. Vet. Med. Assoc., *56*: 606-608.

Morris, J.
1918. Blood-sucking insects as transmitters of anthrax or charbon. Louisiana State Univ. Agric. Exp. Sta. Baton Rouge, Bull. No. 163, pp. 3-15.

Morris, R. F.
1954. Note on the blow fly *Phaenicia sericata* (Mg.) (Diptera: Calliphoridae) in Newfoundland. Canad. Entom., *86*:356.

Moser, J. G.
1966. The purification and characterization of a proteinase from the larva of *Calliphora erythrocephala* Meigen. Biochem. Zhurn., *344*:337-352.

Moutia, A.
1928. Surra in Mauritius and its principal vector, *Stomoxys nigra*. Bull. Entom. Res., *19*:211-216.

Moutia, A.
1929. Flies injurious to domestic animals in Mauritius. Dept. Agric. Mauritius, Scientific Series Bull. No. 15, pp. 5-8.

Muirhead-Thomson, R. C.
1968. Ecology of insect vector populations. Academic Press, New York, 174 pp.

Mulla, M. S.
1959. Some important aspects of *Hippelates* gnats, with a brief presentation of current research findings. Proc. Papers Calif. Mosq. Control Assoc., *27*:48-52.

Mulla, M. S.
1962. The breeding niches of *Hippelates* gnats. Ann. Entom. Soc. Amer., *55*: 389-393.

Mulla, M. S.
1963. An ecological basis for the suppression of *Hippelates* eye gnats. Jour. Econ. Entom., *56*:768-770.

Mulla, M. S.
1966. Oviposition and emergence period of the eye gnat *Hippelates collusor*. Jour. Econ. Entom., *59*:93-96.

Mulla, M. S., and March, R. B.
1959. Flight range, dispersal patterns and population density of the eye gnat *Hippelates collusor* (Townsend). Ann. Entom. Soc. Amer., *52*:641-646.

Munch-Petersen, E.
1938. Bovine mastitis: survey of the literature to the end of 1935. Imperial Bureau Anim. Health Review, *1*:55-56.

Mundt, J. O., Johnson, A. H., and Khatchikian, R.
1958. Incidence and nature of enterococci on plant materials. Food. Res., *23*: 186-193.

Murvosh, C. M., and Thaggard, C. W.
1966. Ecological studies of the house fly. Ann. Entom. Soc. Amer., *59*:533-547.

Musgrave, W. E., and Clegg, M. T.
1903. *Trypanosoma* and trypanosomiasis, with special reference to surra in the Philippine Islands. Publ. Bur. Gov. Labs., Manila, *5*:1-248.

394   BIBLIOGRAPHY

Mussgay, M., and Suarez, O.
1962. Multiplication of vesicular stomatitis virus in *Aedes aegypti* (L.) mosquitoes. Virol., *17*:202-204.

Muzzarelli, E.
1925. Sul passaggio di alcuni microrganismi patogeni attraverso lo intestino delle mosche carnarie. Ann. Ig., *35*:219-228.

Nabarro, D., and Greig, E.D.W.
1905. Further observations on the trypanosomiases (human and animal) in Uganda. Rept. Sleeping Sickness Comm. Roy. Soc. London, pp. 8-47.

Nadzhafov, I. G.
1967. O roli razlichnykh vidov sinantropnykh mukh v rasprostranenii onkosfer *Taeniarhynchus saginatus*. [Role of different species of synanthropic flies in dissemination of oncospheres of *Taeniarhynchus saginatus*.] Med. Parazit. Parazitarn. Bolezni, *36*:144-149.

Nappi, A. J., and Stoffolano, Jr., J. G.
1971. *Heterotylenchus autumnalis*. Hemocytic reactions and capsule formation in the host, *Musca domestica*. Exp. Parasit., *29*:116-125.

Nash, J.T.C.
1903. The etiology of summer diarrhoea. Lancet, *1*:330.

Nash, J.T.C.
1905. The waste of infant life. Jour. Roy. Sanit. Inst., *26*:494-498.

Nash, J.T.C.
1913. Range of flight of *Musca domestica*. Lancet, 2:1585-1586.

Neal, J. C.
1897. *Hippelates* flies and sore eyes. U.S.D.A. Div. Entom. Bull. No. 7, new series.

Needham, J. G., Galtsoff, P. S., Lutz, F. E., and Welch, P. S.
1937. Culture methods for invertebrate animals. Dover edition reprint, 1959, 590 pp.

Nees von Esenbeck, C. G.
1831. On *Empusa muscae* in flies. Nova. Acta. Acad. Caes, Leop.-Car. Nat. Cur., *15* (part II).

Neitz, W. O.
1966. Bluetongue. Bull. Off. int. Epiz., *65*:1749-1758.

Newstead, R.
1906. On the life-history of *Stomoxys calcitrans*, Linn. Jour. Econ. Biol., *1*: 157-166.

Nicholls, L.
1911. The ulcer fly and its relation to yaws and other ulcerated conditions. Windward Islands (St. Lucia). Report of laboratory work, including sanitation and research, for the six months ending Sept. 30, 1910. Pp. 1-7.

Nicholls, L.
1912. The transmission of pathogenic microorganisms by flies in Saint Lucia. Bull. Entom. Res., *3*:81-88.

Nicoli, J., and Vattier, G.
1964. Culture de *Trypanosoma rhodesiense* sur tissus de pupes de glossines. Bull. Soc. Path. Exot., *57*:213-219.

Nicoll, W.
1911. Further reports (No. 4) on flies as carriers of infection. 3. On the part played by flies in the dispersal of the eggs of parasitic worms. Rept. Loc. Govt. Bd. Publ. Health and Med. Subj., London, n.s., No. 16, pp. 13-30.

Nicolle, C., and Cuenod, A.
1921. Étude expérimentale du trachome. Arch. Inst. Pasteur Afrique du Nord, Tunis, *1*:149-178.

Nielen, P. M.
    1780.   Verhandeling over de Indiaansche Pokken. Verhandelingen uitgegeven
            door de Hollandsche Mautschappije der Wetenschappen te Haarlem,
            XIX, Part 2, pp. 135-188.
Nielsen, B. O.
    1967.   On a migration of hover-flies (Dipt., Syrphidae) and sawflies (Hym.,
            Tenthredinidae) observed in Denmark, August 1967. Entom. Medd., *36*:
            215-224.
Nieschulz, O.
    1926.   Zoologische bijdragen tot het surraprobleem. III. Overbrengingsproeven
            met *Tabanus rubidus* Wied., *T. striatus* Fabr. en *Stomoxys calcitrans* L
            Ned. Ind. Blad. Diergeneesk., *38*:255-279.
Nieschulz, O.
    1927.   Zoologische bijdragen tot het surraprobleem. XIX. Overbrengingsproeven
            met *Stomoxys, Lyperosia, Musca* en *Stegomyia*. [Zoological contribution
            to the surra problem. XIX. Transmission experiments with *Stomoxys,
            Lyperosia, Musca*, and *Stegomyia*.] Ned. Ind. Blad. Diergeneesk., *39*:371-
            390.
Nieschulz, O.
    1928a. Enkele miltvuuroverbrengingsproeven met tabaniden, musciden, en mus-
            kieten. Ned. Ind. Blad. Diergeneesk., *40*:355-377.
Nieschulz, O.
    1928b. Zoologische Bieträge zum Surraproblem. XXII. Uebertragungsversuche
            mit *Anopheles fuliginosus* Gil. Centralbl. Bakt., Parasit. Infekt., 1. Abt.,
            Orig., *109*:327-330.
Nieschulz, O.
    1929.   Zoologische Beiträge zum Surraproblem. XXV. Ueber den Einfluss
            verschiedener Versuchtiere auf das Ergebnis von Surraübertragungsver-
            suchen mit *Stomoxys calcitrans*. Zentralbl. Bakt., Parasit. Infekt., *113*:
            80-89.
Nieschulz, O.
    1930.   Surraübertragungsversuche auf Java und Sumatra. Veeartsenijk, Meded.
            Ned.-Ind., No. 75, pp. 1-295. (Summary, see Archiv. Schiffs. Tropen-
            Hyg., *33*:257-266, 1929.)
Nieschulz, O.
    1933.   Über die Temperaturbegrenzung der Aktivitätsstufen von *Stomoxys cal-
            citrans*. Zeitschr. Parasit., *6*:220-242.
Nieschulz, O.
    1940.   Versuche über die unmittelbare Übertragung von *Trypanosoma congo-
            lense* durch *Stomoxys calcitrans*. Arch. Schiffs. Trop. Hyg., *44*:120-124.
Nieschulz, O.
    1941.   Uebertragungsversuche mit einem südeuropäischen Surrastamm und
            Stomoxys calcitrans. Zentralbl. Bakt., Parasit. Infekt., *146*:113-115.
Nieschulz, O., and Kraneveld, F. C.
    1929.   Experimentelle Untersuchungen über die Uebertragung der Büffelseuche
            durch Insekten. Zentralbl. Bakt., *113*:403-417.
Nishiyama, S.
    1958.   Studies on habronemiasis in horses. Bull. Fac. Agric. Kagoshima. Univ.,
            No. 7.
Niven, J.
    1905.   Summer diarrhoea. Manchester Ann. Health Repts., pp. 147-159.
Niven, J.
    1906.   Summer diarrhoea. Manchester Ann. Health Repts., pp. 82-96.
Niven, J.
    1910.   Summer diarrhoea and enteric fever. Roy. Soc. Med. Proc., *3*:131-216.

Noè, G.
1903. Studî sul ciclo evolutivo della *Filaria labiato-papillosa* Alessandrini. Atti Reale Accad. Lincei, Ser. 5, *12*:387-393.

Noguchi, H.
1925. Zentr. Hyg. Grenz., *11*:285.

Noguchi, H., and Kudo, R.
1917. The relation of mosquitoes and flies to the epidemiology of acute poliomyelitis. Jour. Exp. Med., *26*:49-57.

Norris, K. R.
1957. A method of marking Calliphoridae (Diptera) during emergence from the puparium. Nature, *180*:1002.

Norris, K. R.
1959. The ecology of sheep blowflies in Australia. *In*: Biogeography and ecology in Australia. *Monographiae Biologicae*, *8*:514-544.

Norris, K. R.
1966a. Notes on the ecology of the bushfly, *Musca vetustissima* Walk. (Diptera: Muscidae), in the Canberra district. Australian Jour. Zool., *14*:1139-1156.

Norris, K. R.
1966b. Daily patterns of flight activity of blowflies (Calliphoridae: Diptera) in the Canberra district as indicated by trap catches. Australian Jour. Zool., *14*:835-853.

Norton, O. M.
1911. A practitioner's view of swamp fever. Amer. Jour. Vet. Med., *6*:672-673.

Nuorteva, P.
1959. Studies on the significance of flies in the transmission of poliomyelitis. I. The occurrence of the *Lucilia* species (Dipt., Calliphoridae) in relation to the occurrence of poliomyelitis in Finland. Ann. Entom. Fenn., *25*: 1-24.

Nuorteva, P.
1963. Die Rolle der Fliegen in der Epidemiologie der Poliomyelitis. Sonderabdr. Anz. Schädl., *36*:149-155.

Nuorteva, P.
1965. The flying activity of blowflies (Diptera, Calliphoridae) in subarctic conditions. Ann. Entom. Fenn., *31*:242-245.

Nuorteva, P.
1966. Local distribution of blowflies in relation to human settlement in an area around the town of Forssa in South Finland. Ann. Entom. Fenn., *32*: 128-137.

Nuorteva, P., Kotimaa, T., Pohjolaineu, L., and Räsänen, T.
1964. Blowflies (Dipt., Calliphoridae) on the refuse depot of the city of Kuopio in Central Finland. Ann. Entom. Fenn., *30*:94-104.

Nuorteva, P., and Vesikari, T.
1966. The synanthropy of blowflies (Diptera, Calliphoridae) on the coast of the Arctic Ocean. Ann. Med. Exp. Fenn., *44*:544-548.

Nuttall, G.H.F.
1897. Zur Aufklärung der Rolle, welche die Insekten bei der Verbreitung der Pest spielen.—Ueber die Empfindlichkeit verschiedener Tiere für dieselbe. Centralbl. Bakt., *22*:87-97.

Nuttall, G.H.F.
1899. Die Rolle der Insekten, Arachniden (*Ixodes*) und Myriapoden als Träger bei der Verbreitung von durch Bakterien und Thierische Parasiten verursachten Krankheiten des Menschen und der Thiere. Hyg. Rundschau, *9*:209-220, 275-289, 393-409, 503-520, 606-620. Also Johns Hopkins Hosp. Repts., *8*:1-154.

Nuttall, G.H.F.
  1908. Insects as carriers of disease. Recent advances in our knowledge of the part played by blood-sucking arthropods (exclusive of mosquitoes and ticks) in the transmission of infective diseases. XIV. Internat. Cong. Hyg. Demogr., Berlin, *1*:195-206.
Nuttall, G.H.F., and Jepson, F. P.
  1909. The part played by *Musca domestica* and allied (non-biting) flies in the spread of infective diseases. Gr. Brit. Rept. Publ. Health, *16*:13-41.
Oda, T.
  1966. Studies on the dispersal of the house fly *Musca domestica vicina* by mark-and-release method. Endem. Dis. Bull. Nagasaki Univ., *8*:136-144 [in Japanese].
Ode, P. E., and Matthysse, J. G.
  1967. Bionomics of the face fly, *Musca autumnalis* De Geer. Cornell Univ. Agric. Exp. Stat. Memoir 402, 91 pp.
Odlum, W. H.
  1908. Are flies the cause of enteric fever? Jour. Roy. Army Med. Corps, *10*: 528-530.
Ogata, K., and Suzuki, T.
  1960. Release studies on the dispersion of the lesser house fly, *Fannia canicularis*, in the residential area of Bibai, Hokkaido. Boytu-Kagabu, *25*:51-57 (Kyoto) [in Japanese].
Ogawa, H., Nakamura, A., and Sakazaki, R.
  1968. Pathogenic properties of enteropathogenic *Escherichia coli* from diarrheal children and adults. Jap. Jour. Med. Sci. Biol., *21*:333-349.
Ogden, L. J., and Kilpatrick, J. W.
  1958. Control of *Fannia canicularis* (L.) in dairy barns. Jour. Econ. Entom., *51*:611-612.
Ohanessian, A., and Echalier, G.
  1967a. Multiplication du virus *Sindbis* chez *Drosophila melanogaster* (Insecte diptère), en conditions expérimentales. Compt. Rend. Acad. Sci. Paris, *264*:1356-1358.
Ohanessian, A., and Echalier, G.
  1967b. Multiplication of *Drosophila* hereditary virus (σ virus) in *Drosophila* embryonic cells cultivated *in vitro*. Nature, *213*:1049-1050.
Oho, D.
  1921. Über die Framboesie in Formosa. Proc. 4th Congr. Far East. Assn. Trop. Med., *2*:138-148.
Ojala, O., and Nuorteva, P.
  1966. Isolering av *Salmonella*—bakterier hos flugor; sodra Finland. Proc. 10th Congr. (1966) Nordic Veterinarians, Stockholm. [Offprint paged 1-6.]
Olin, G.
  1938. Off. Inter. d'Hyg. publ., *30*:10.
Olsufiev, N. G.
  1940. The role of *Stomoxys calcitrans* L. in the transmission and preservation of tularemia. Arkhiv. Biolog. Nauk, *58*:25-31 [in Russian].
Omodei-Zorini, A.
  1914. Lezioni di tisiologia. Idelson (not seen).
Orton, S. T., and Dodd, W. L.
  1910. Experiments on transmission of bacteria by flies with special relation to an epidemic of bacillary dysentery at the Worcester State Hospital, Massachusetts, 1910. Boston Med. Surg. Jour., *163*:863-868.
Osins'kii, S. O.
  1938. The role of insects that attack carcasses in the preservation and spread of the bacilli of anthrax in nature. Actes. Sci. Inst. Vet., *1*:50-58.

Osmond, A. E.
1909. The fly—an etiological factor in intestinal disease. Lancet, 7:394-403.
Ostashev, S. N.
1956. Flies as carriers of infectious diseases of domestic animals. Veter., 33: 75-76 [in Russian].
Ostashev, S. N.
1958. Sud'ba bakterii pri metamorfoze komnatnoi mukhi. Byul. Nauchn. -Tekh. Inform. Vses. Nauchn. -Issled. Inst. Vet. Sanit., 3:18-19 [not seen].
Ostrolenk, M., and Welch, H.
1942a. The common house fly (Musca domestica) as a source of pollution in food establishments. Food Res., 7:192-200.
Ostrolenk, M., and Welch, H.
1942b. The house fly as a vector of food poisoning organisms in food producing establishments. Amer. Jour. Publ. Health, 32:487-494.
Ozawa, Y., and Nakata, G.
1965. Experimental transmission of African horsesickness by means of mosquitoes. Amer. Jour. Veter. Res., 26:119-125.
Ozawa, Y., Nakata, G., Shad Del, F., and Nvai, S.
1966. Transmission of African horse-sickness by a species of mosquito, Aedes aegypti. Amer. Jour. Veter. Res., 27:695-697.
Pach, H.
1935. Übertragen die Hausfliegen den Bauchtyphus? Wien. Med. Woch., 85: 60-64.
Paffenbarger, R. S., Jr., and Watt, J.
1953. Poliomyelitis in Hidalgo County, Texas, 1948. Epidemiologic observations. Amer. Jour. Hyg., 58:269-287.
Paine, J. H.
1912. The house fly in its relation to city garbage. Psyche, 19:156-159.
Paramonov, S. J.
1933. Dipterenlarven als Mittelgegen die Gangräne, Osteomyelitis u. s. w. Zhurn. Bio-Zool. Tsiklu Kiev, 3 (7):73-83.
Paré, A.
1585. The apologie and treatise of Ambroise Paré containing the voyages made into divers places. Edited by G. Keynes. Univ. of Chicago Press, 1952. See pp. 68-70.
Paré, A.
1678. The works of that famous chirurgeon Ambrose Parey, trans. by Th. Johnson, Book X, p. 249, Book XI, p. 277. London.
Parish, H. E.
1937. Flight tests on screwworm flies. Jour. Econ. Entom., 30:740-743.
Parish, H. E., and Laake, E. W.
1935. Species of Calliphoridae concerned in the production of myiasis in domestic animals, Menard County, Texas. Jour. Parasit., 21:264-266.
Parisot, J., and Fernier, L.
1934. The best methods of treating manure-heaps to prevent the hatching of flies. League of Nations Quart. Bull. Health Organ., 3:1-31.
Parker, R. R.
1916. Dispersion of Musca domestica Linnaeus under city conditions in Montana. Jour. Econ. Entom., 9:325-354.
Parker, R. R.
1933. Certain phases of the problem of Rocky Mountain spotted fever. Arch. Path., 15:398-429.
Parman, D. C.
1920. Observations on the effect of storm phenomena on insect activity. Jour. Econ. Entom., 13:339-343.

Parodi, S. E.
1917. Acción patógena de los ixodideos. Anal. Soc. Rur. Argentina, *51*:111-128.

Parr, H.C.M.
1962. Studies on *Stomoxys calcitrans* (L.) in Uganda, East Africa. II. Notes on life-history and behavior. Bull. Entom. Res., *53*:437-443.

Parry, W. H.
1961. The survival of salmonellae in baked confectionery. Med. Officer, *105*: 197-202.

Patnaik, B.
1965. Personal communication.

Patnaik, B., and Roy, S. P.
1966. On the life cycle of the filariid *Stephanofilaria assamensis* Pande, 1936, in the arthropod vector *Musca conducens* Walker, 1859. Indian Jour. Anim. Health, *5*:91-101.

Patterson, R. A., and Fisk, F. W.
1958. A study of the trypsin-like protease of the adult stable fly, *Stomoxys calcitrans* L. Ohio Jour. Sci., *58*:299-310.

Patton, W. S.
1912. Preliminary report on an investigation into the etiology of oriental sore in Cambay. India Med. Dept., Sci. Mem. Off., No. 50:1-21.

Patton, W. S.
1921. Studies on the flagellates of the genera *Herpetomonas*, *Crithidia* and *Rhynchoidomonas*. No. 3. The morphology and life history of *Rhynchoidomonas siphunculinae* sp. nov., parasitic in the malpighian tubes of *Siphunculina funicola* de Meijere. Indian Jour. Med. Res., *8*:603-612.

Patton, W. S.
1930. Insects, ticks, mites and venomous animals of medical and veterinary importance. Part II. Public Health. Liverpool School of Trop. Med., 740 pp.

Patton, W. S.
1933. Studies on the higher Diptera of medical and veterinary importance. A revision on the genera of the tribe Muscini, subfamily Muscinae, based on a comparative study of the male terminalia. I. The genus *Musca* Linnaeus. II. A practical guide to the palaearctic species. Ann. Trop. Med. Parasit., *27*:135-156, 327-345, 397-430.

Patton, W. S., and Cragg, F. W.
1913. A textbook of medical entomology. Christian Literature Society for India; London, Madras and Calcutta. 768 pp.

Paul, J. R.
1955. Epidemiology of poliomyelitis. WHO Monograph Ser. No. 26, pp. 9-29.

Paul, J. R., Horstmann D. M., Riordan, J. T., Opton, E. M.,
Niederman, J. C., Isacson, E. P., and Green, R. A.
1962. An oral poliovirus vaccine trial in Costa Rica. Bull. Wld. Health Org., *26*:311-329.

Paul, J. R., Trask, J. D., Bishop, M. B., Melnick, J. L., and Casey, A. E.
1941. The detection of poliomyelitis virus in flies. Science, *94*:395-396.

Paullini, C. F.
1706. *Observationes medico—physicae*. Joh. H. Richteri. Leipzig. See pp. 247-249.

Pavan, M.
1949. Richerche sugli antibiotici di origine animale. Nota riassuntiva. La Richerca Scientifica, *19*:1011-1017.

Pavan, M.
1952. "Iridomyrmecin" as insecticide. Trans. Ninth Intl. Congr. Entom., *1*:321-327.

Pavillard, E. R., and Wright, E. A.
1957.   An antibiotic from maggots. Nature, *180*:916-917.
Pavlovskiĭ, E. N., and Bychkov-Oreshnikov, V. A.
1934.   O mikroflore mukh nekotorykh lagereĭ i vozmozhnoĭ roli ikh v rasseivanii kishechnykh infektisiĭ. *In*: Trudy Voenno-med. akad. RKKA, Sbornik, No. 1: Leningrad.
Pearl, R., Miner, J. R., and Parker, S. L.
1927.   Experimental studies on the duration of life. XI. Density of population and life duration in *Drosophila*. Amer. Nat., *61*:289-318.
Peffly, R. L.
1953.   A summary of recent studies on house flies in Egypt. Jour. Egypt. Publ. Health Assoc., pp. 55-74.
Peffly, R. L., and Labrecque, G. C.
1956.   Marking and trapping studies on dispersal and abundance of Egyptian house flies. Jour. Econ. Entom., *49*:214-217.
Peluffo, C. A.
1964.   Salmonellosis in South America. *In*: World problems of salmonellosis, van Oye (Ed.), pp. 476-506.
Peña Chavarría, A., and Shipley, P. G.
1925.   Contribución al estudio de los carates de América Tropical. Rev. Med. Lat. Amer., *10*:648-721.
Penner, L. R., and Melnick, J. L.
1952.   Methods for following the fate of infectious agents fed to single flies. Jour. Exp. Med., *96*:273-280.
Penning, C. A.
1904.   Les trypanosomoses aux Indes Neérlandaises. Janus, *9*:514-522.
Peppler, H. J.
1944.   Usefulness of microorganisms in studying dispersion of flies. Bull. U. S. Army Med. Dept., *75*:121-122.
Pérez, C.
1910.   Recherches histologiques sur la métamorphose des muscides *Calliphora erythrocephala* Mg. Arch. Zool. Exper. Gen., *4*:1-274.
Périès, J., Printz, P., Canivet, M., and Chuat, J. C.
1966.   Multiplication du virus de la stomatite vésiculaire chez *Drosophila melanogaster*. Compt. Rend. Acad. Sci., *262*:2106-2107.
Perraju, A., and Tirumalaro, V.
1956.   *Stomoxys calcitrans* L. (stable fly). Occurrence as a pest of cattle. Indian Jour. Entom., *18*:8 pp.
Perry, A. S., and Miller, S.
1965.   The essential role of folic acid and the effect of antimetabolites on growth and metamorphosis of housefly larvae, *Musca domestica* (L.). Jour. Insect. Physiol., *11*:1277-1287.
Perry, H. M., and Bensted, H. J.
1928.   *Bacillus dysenteriae* Sonne as an aetiological agent of dysentery in Egypt. Trans. Roy. Soc. Trop. Med. Hyg., *21*:417-420.
Perry, H. M., and Bensted, H. J.
1929.   Investigations in Egypt of some acute bacillary intestinal infections. Trans. Roy. Soc. Trop. Med. Hyg., *22*:511-522.
Petit, P. J.
1923.   Recherches sur le trachome en Tunisie. Arch. Inst. Pasteur Tunis, *12*: 82-89.
Petragnani, G.
1925.   La mosca è sempre batterifera sin dalla nascita. Ig. Moderna, *18*:33-41.

Petrilla, A.
1933. Epidemiologiai adatok a typhus abdominalis földrajzi elterjedtségéhez. [Epidemiological data on the geographical distribution of typhoid.] Népegészségügy, 7:3-12.

Petrov, V. P.
1964. O sezonnosti zabolevaemosti dizenterei v Dnepropetrovskoi oblasti za poslednie gody. [Seasonal incidence of dysentery in Dnepropetrovsk region in recent years.] Zhurn. Mikrob. Epidem. Immun., 41:19-21.

Pfeiler, W., Schlaak, and Thomsen
1927. Die Weidesenche in Schleswig-Holstein (bösartige Euterentzündung der Rinder) und ihre Behandlung durch Impfung nekst einigen Bermerkungen über die Behandlung von Euterentzündungen überhaupt. Tierärztl. Rundschau, 33:318-321.

Philip, C. B.
1948. Observations on experimental Q fever. Jour. Parasit., 34:457-464.

Philip, C. B., Davis, G. E., and Parker, R. R.
1932. Experimental transmission of tularemia by mosquitoes. Publ. Health Rept., 47:2077-2088.

Piana, G. P.
1896. Osservazioni sul Dispharagus nasutus Rud. dei polli e sulle larve nematoelmintiche delle mosche e dei porcellioni. Atti. Soc. Ital. Sci. Nat. Milano, 36:239-262.

Picado, C.
1935. Sur le principe bactéricide des larves des mouches (Myiases des plaies et myiases des fruits). Bull. Biol. France Belgique, 69:409-438.

Pickens, L. G., Morgan, N. O., Hartsock, J. G., and Smith, J. W.
1967. Dispersal patterns and populations of the house fly affected by sanitation and weather in rural Maryland. Jour. Econ. Entom., 60:1250-1255.

Piercy, P. L.
1956. Transmission of anaplasmosis. Ann. New York Acad. Sci., 64:40-48.

Pimentel, D., and Fay, R. W.
1955. Dispersion of radioactively tagged Drosophila from pit privies. Jour. Econ. Entom., 48:19-22.

Pipkin, A. C.
1943. Experimental studies on the role of filth flies in the transmission of certain helminthic and protozoan infections of man. Absts. theses Tulane Univ., 44:9-13.

Pipkin, A. C.
1949. Experimental studies on the role of filth flies in the transmission of Endamoeba histolytica. Amer. Jour. Hyg., 49:255-275.

Pletneva, N. A.
1937. Rol'mukh kak mekhanicheskikx perenoschikov tsist Protozoa i ĭaitz glist Ashkhabade. [The role of flies as the mechanical carriers of Protozoa and of the eggs of the tapeworm in Ashkhabad.] Problemy parazitologii i fauna Turkmenii, pp. 117-120.

Pliny
23-79 A.D. Historia naturalis (10:40).

Plowright, W.
1956. Cutaneous streptothricosis of cattle: I. Introduction and epizootiological features in Nigeria. Vet. Rec., 68:350-355.

Pod'ĭapol'skaĭa, V. P., and Gnedina, M. P.
1934. O roli mukhi v epidemiologii glistnykh zabolevaniĭ. [On the role of flies in the epidemiology of tapeworm infestation.] Med. Parazit., 3:179-185.

402 BIBLIOGRAPHY

Poinar, G. O., Jr.
　1969.　Arthropod immunity to worms. *In*: Immunity to parasitic animals. Edited by G. J. Jackson, R. Herman, and I. Singer. Vol. I. Appleton-Century-Crofts, New York. See pp. 173-210.

Pokrovskiĭ, S. N., and Zima, G. G.
　1938.　Mukhi kak perenoschiki ĩaits glist v estestvennȳkh usloviĩakh. [Flies as carriers of tapeworm eggs under natural conditions.] Med. Parazit., *7*: 262-264. [Rev. Appl. Entom., *26*:244.]

Pollitzer, R.
　1955.　Cholera studies. 6. Pathology. Bull. World Health Org., *13*:1075-1199.

Pollitzer, R., and Meyer, K. F.
　1961.　The ecology of plague. *In*: May, J. M. (Ed.), Studies in disease ecology. Hafner Publ. Co., Inc., New York, pp. 433-501.

Ponghis, G.
　1957.　Quelques observations sur le rôle de la mouche dans la transmission des conjonctivites saisonnières dans le Sud-Marocain. Bull. World Health Org., *16*:1013-1027.

Post, F. J., and Foster, Jr., F. J.
　1965.　Distribution and characterization of fecal streptococci in muscoid flies. Jour. Invert. Path., *7*:22-28.

Power, M. E., and Melnick, J. L.
　1945.　A three-year survey of the fly population in New Haven during epidemic and non-epidemic years for poliomyelitis. Yale Jour. Biol. Med., *18*: 55-69.

Prado, E., and Jimenez, L. A.
　1955.　Estudio bacteriológico de la flora enterica transportada por las moscas de la ciudad de Santiago. Biol. Inst. Bact. Chile, *8*:14-18.

Prakash, O.
　1962.　Epidemiology and techniques for isolation and identification of entero-pathogenic *Escherichia coli*. Indian Jour. Med. Res., *50*:599-606.

Pratt, H. C.
　1908.　Distribution of certain species of biting flies in the Federated Malay States. Fed. Malay States Inst. Med. Res. Studies, pp. 36-38.

Price, D. A., and Hardy, W. T.
　1954.　Isolation of blue-tongue virus from Texas sheep—*Culicoides* shown to be a vector. Jour. Amer. Vet. Med. Assoc., *124*:255-258.

Pringle, Sir J.
　1810.　Observations on the diseases of the army. E. Earle, Philadelphia, 411 pp. See pp. 223, 224.

Přívora, M., Vranova, J., and Kudova, J.
　1969.　On the problem of mechanical transmission of microorganisms on tarsi of the fly *M. domestica* L. [in Czech]. Cesk. Epidem. Mikrobiol. Imunol., *18*:353-359.

Quarterman, K. D., Baker, W. C., and Jensen, J. A.
　1949.　The importance of sanitation in municipal fly control. Amer. Jour. Trop. Med., *29*:973-982.

Quarterman, K. D., Kilpatrick, J. W., and Mathis, W.
　1954a.　Fly dispersal in a rural area near Savannah, Georgia. Jour. Econ. Entom., *47*:413-419.

Quarterman, K. D., Mathis, W., and Kilpatrick, J. W.
　1954b.　Urban fly dispersal in the area of Savannah, Georgia. Jour. Econ. Entom., *47*:405-412.

Quevedo, F., and Carranza, N.
　1966.　Le role des mouches dans la contamination des aliments au Perou. Ann. Inst. Pasteur, Lille, *17*:199-202.

BIBLIOGRAPHY　**403**

Raccuglia, G.
1952. La toxoplasmosi. Ann. Sanità Pubbl., *13*:425-516.

Radeleff, R. D.
1949. Clinical encephalitis occurring during an outbreak of vesicular stomatitis in horses. Vet. Med., *44*:494-496.

Radvan, R.
1956. Přežívání bakterií při metamorfose *Musca domestica* L. Sborník Vědeckých Prací Voj. Lék. Akad. JEvP., *4*:104-111.

Radvan, R.
1960a. Persistence of bacteria during development in flies. I. Basic possibilities of survival. Folia Microb., *5*:50-56.

Radvan, R.
1960b. Persistence of bacteria during development in flies. II. The number of surviving bacteria. Folia Microb., *5*:85-92.

Radvan, R.
1960c. Persistence of bacteria during development in flies. III. Localization of the bacteria and transmission after emergence of the fly. Folia Microb., *5*:149-156.

Raimbert, M. A.
1869. Recherches expérimentales sur la transmission du charbon par les mouches. Compt. Rend. Acad. Sci., *69*:805-812.

Raimbert, A.
1870. Recherches expérimentales sur la transmission du charbon par les mouches. L'Union Med. (Paris), *9*:209-210, 350-352, 507-509, 709-710.

Ranade, D. R.
1956. Some observations on the range of flight of the common Indian housefly, *Musca domestica nebulo*, Fabr. Jour. Anim. Morph. Physiol., *3*: 104-108.

Ransom, B. H.
1913. The life-history of *Habronema muscae* (Carter), a parasite of the horse transmitted by the house fly. U.S. Dept. Agric. Bull., No. 163, 36 pp.

Rao, B.R.R.
1931. A note on the seasonal prevalence of catarrhal ophthalmia in Bangalore. Proc. All India Ophthal. Soc., *2*:62-63.

Ratner
1931. Iak perenosiat mukhi iaitsia glistiv. [How flies carry helminth eggs.] Profil. Med., No. 5-6, p. 143 (not seen).

Reed, W., Vaughn, V. C., and Shakespeare, E. O.
1904. Report on the origin and spread of typhoid fever in the United States military camps during the Spanish War of 1898. U.S. Surgeon-General's Office, *1*: see pp. 427-428, 665-666.

Reid, W. M., and Ackert, J. E.
1937. The cysticercoid of *Choanotaenia infundibulum* and the housefly as its host. Trans. Amer. Microscop. Soc., *56*:99-104.

Reinhards, J.
1970. Personal communication.

Reinhards, J., Weber, A., Nižetič, B., Kupka, G., and Maxwell-Lyons, F.
1968. Studies in the epidemiology and control of seasonal conjunctivitis and trachoma in Southern Morocco. Bull. World Health Org., *39*:497-545.

Reinstorf, A.P.N.
1923. Uebertragung der Ruhr durch Fliegen. Inaug. Dissert. Giessen. Berlin, Morenhoven, 21 pp.

Reiter, H., and Ramme
1916. Beiträge zur Aetiologie der Weilschen Krankheit. Deut. Med. Woch-schr., *42*:1282-1284.

Rendtorff, R. C., and Francis, T.
1943. Survival of the Lansing strain of poliomyelitis virus in the common house fly, *Musca domestica* L. Jour. Infect. Dis., *72*:198-205.

Rendtorff, R. C., and Holt, C. J.
1954. The experimental transmission of human intestinal protozoan parasites. III. Attempts to transmit *Endamoeba coli* and *Giardia lamblia* cysts by flies. Amer. Jour. Hyg., *60*:320-326.

Reyes, H., Hevia, H., Schenone, H., and Sapunar, J.
1967. Myiasis humana por *Phaenicia sericata* (Meigen, 1826) en Chile. (Diptera, Calliphoridae). Bol. Chileno Parasitol., *22*:168-171.

Richard, J. L.
1963. Studies on the possible transmission of *Trichophyton equinum* by the common stable fly. Proc. Iowa Acad. Sci., *70*:114-120.

Richard, J. L., and Pier, A. C.
1966. Transmission of *Dermatophilus congolensis* by *Stomoxys calcitrans* and *Musca domestica*. Amer. Jour. Vet. Res., *27*:419-423.

Richards, C. S., Jackson, W. B., DeCapito, T. M., and Maier, P. P.
1961. Studies on rates of recovery of *Shigella* from domestic flies and from humans in southwestern United States. Amer. Jour. Trop. Med. Hyg., *10*:44-48.

Richardson, U. F.
1924. Report of Veterinary Pathologist for the year 1923. Ann. Rept. Vet. Dept. Uganda Protectorate, 1923, pp. 11-14.

Richardson, U. F.
1928. Notes on trypanosomiasis of cattle in Uganda. Trans. Roy. Soc. Trop. Med. Hyg., *22*:137-146.

Ricque
1865. Des accidents déterminés par les piqueurs des mouches. Rec. Mem. Med. Milit. 3rd Ser. *14*:472-480.

Ringertz, O., and Mentzing, L.-O.
1968. *Salmonella* infection in tourists. I. An epidemiological study. Acta Path. Microb. Scandinav., *74*:397-404.

Rinonapoli, G.
1930. Contributo allo studio epidemiologico della infezione carbonchiosa. Med. Prat., *15*:281-287.

Riordan, J. T., Paul, J. R., Yoshioka, I., and Horstmann, D. M.
1961. The detection of poliovirus and other enteric viruses in flies. Results of tests carried out during an oral poliovirus vaccine trial. Amer. Jour. Hyg., *74*:123-136.

R.M.
1937. Sotsialist. Nauk i Tekhnika, *5*: No. 9. (Seen as: Mouches et infections intestinales. Bull. Inst. Pasteur, *37*:418, 1939.)

Roberg, D. N.
1915. The rôle played by the insects of the dipterous family Phoridae in relation to the spread of bacterial infections. II. Experiments on *Aphiochaeta ferruginea* Brunetti with the cholera vibrio. Philip. Jour. Sci. B. Trop. Med., *10*:309-336.

Roberts, D. S.
1963. The release and survival of *Dermatophilus dermatonomus* zoospores. Austral. Jour. Agric. Res., *14*:386-399 (not seen).

Roberts, E. W.
1947. The part played by the faeces and vomit-drop in the transmission of *Entamoeba histolytica* by *Musca domestica*. Ann. Trop. Med. Parasit., *41*:129-142.

Roberts, F.H.S.
   1934.   The large roundworm of pigs, *Ascaris lumbricoides* L., 1758: Its eco-
            nomic importance in Queensland, life cycle, and control. Animal Health
            Stat., Yeerongpilly Bull. No. 1, pp. 1-81.
Roberts, F.H.S.
   1952.   Insects affecting livestock. Angus and Robertson, Sydney and London,
            267 pp.
Roberts, M. J.
   1971.   The structure of the mouth parts of some calypterate dipteran larvae in
            relation to their feeding habits. Acta Zool., *52*:171-188.
Roberts, R. H., Dicke, R. J., Hanson, R. P., and Ferris, D. H.
   1956.   Potential insect vectors of vesicular stomatitis in Wisconsin. Jour. Infect.
            Dis., *98*:121-126.
Robertson, A.
   1908.   Flies as carriers of contagion in yaws (framboesia tropica). Jour. Trop.
            Med. Hyg., *11*:213.
Robertson, C. R., and Pollitzer, R.
   1938.   Cholera in Central China during 1938—its epidemiology and control.
            Trans. Roy. Soc. Trop. Med. Hyg., *33*:213-223.
Robinson, W., and Norwood, V. H.
   1933.   The role of surgical maggots in the disinfection of osteomyelitis and
            other infected wounds. Jour. Bone Joint Surg., *15*:409-412.
Robinson, W., and Norwood, V. H.
   1934.   Destruction of pyogenic bacteria in the alimentary tract of surgical mag-
            gots implanted in infected wounds. Jour. Lab. Clin. Med., *19*:581-586.
Roch Marra
   1908.   Etudes sur la fiévre aphteuse. Rev. Gén. Méd. Vet., *11*:49-57.
Roger, L.
   1901.   The transmission of the *Trypanosoma evansi* by houseflies and other
            experiments pointing to the probable identity of surra of India and
            Nagana or Tsetse fly disease of Africa. Proc. Roy. Soc., London, *68*:
            163-170.
Rohrer, H., and Möhlmann, H.
   1950.   Experimentelle Beitrag zur Frage der natürlichen Ansteckung bei infek-
            tiösen Anämie der Pferde. Mh. Vet. Med., *5*:179-182.
Romanova, V. P.
   1947.   Role of bloodsucking flies in epidemiology of tularemia. Zhurn. Mikro-
            biol. Epidemiol. Immunobiol., *7*:42-46.
Root, F. M.
   1921.   Experiments on the carriage of intestinal Protozoa of man by flies. Amer.
            Jour. Hyg., *1*:131-153.
Rosenau, M. J.
   1914.   Further experiments in poliomyelitis. (Infantile paralysis in Massachu-
            setts.) Mass. St. Bd. Health, 1907-1912, p. 62.
Rosenau, M. J., and Brues, C. T.
   1912.   Some experimental observations upon monkeys concerning the transmis-
            sion of poliomyelitis through the agency of *Stomoxys calcitrans*. 15th
            Internat. Congr. Hyg. Demog., *1*:616-623.
Rosenow, E. C., South, L. H., and McCormack, A. T.
   1937.   Bacteriologic and serologic studies in the epidemic of poliomyelitis in
            Kentucky. 1935. Kentucky Med. Jour., *35*:437-446.
Ross, C.
   1929.   Observations on the hydatid parasite and the control of hydatid disease
            in Australia. Council Scien. Indus. Res., Melbourne Bull. No. 40, pp.
            1-61.

Ross, T. S.
1916.   Flies in a jail. Indian Med. Gaz., *51*:133-134.
Ross, W. C.
1927-8.   The epidemiology of cholera. Ind. Jour. Med. Res., *15*:951-964.
Roth, A. R., and Hoffman, R. A.
1952.   A new method of tagging insects with P^{32}. Jour. Econ. Entom., *45*:1091.
Roubaud, E.
1911a.   Sur la biologie et la viviparité poecilogonique de la mouche des bestiaux (*Musca corvina* Fab.) en Afrique tropicale. Compt. Rend. Acad. Sci., Paris, *152*:158-160.
Roubaud, E.
1911b.   Compléments biologiques sur quelques stomoxydes de l'Afrique occidentale. Bull. Soc. Path. Exot., *5*:544-549.
Roubaud, E.
1918.   Le rôle des mouches dans la dispersion des amibes dysentériques et autres protozoaires intestinaux. Bull. Soc. Path. Exot., *11*:166-171.
Roubaud, E., and Descazeaux, J.
1921.   Contribution à l'histoire de la mouche domestique comme agent vecteur des habronémoses d'équidés. Cycle evolutif et parasitisme de l'*Habronema megastoma* (Rudolphi 1819) chez la mouche. Bull. Soc. Path. Exot., *14*:471-506.
Roubaud, E., and Descazeaux, J.
1922.   Deuxième contribution à l'étude des mouches, dans leurs rapports avec l'évolution des habronèmes d'équidés. Bull. Soc. Path. Exot., *15*:978-1001.
Round, M. C.
1961.   Observations on the possible role of filth flies in the epizootiology of bovine cysticercosis in Kenya. Jour. Hyg., *59*:505-513.
Rowe, B., Taylor, J., and Bettelheim, K. A.
1970.   An investigation of travellers' diarrhoea. Lancet, *1*(7636):1-5.
Roy, D. N.
1928a.   Report on investigation into aetiology and prevention of Naga sore in Assam. Indian Med. Gazette, *63*:673-687.
Roy, D. N.
1928b.   A note on the breeding and habits of the eye-fly, *Siphonella funicola*, Meij. Indian Med. Gaz., *63*:369-370.
Roy, D. N.
1938.   On the number of eggs of the common house-frequenting flies of Calcutta. Indian Jour. Med. Res., *26*:531-533.
Rozov, A. A., and Prokhorova, E. M.
1968.   Izuchenie vyzhivaemosti virusa yashchura na mukhakh s primeneniem kul'tury tkanei. [A study of the survival rate of foot and mouth disease virus on flies using a tissue culture.] Tr. Vses. Nauch-Issled. Inst. Vet. Sanit., *30*:28-31.
Ruchkovskiǐ, S. N.
1929.   The fly *Lucilia caesar* as a possible carrier of *B. botulinus*. Mikrobiol. Zh., *8*: No. 3 (not seen).
Rueger, M. E., and Olson, T. A.
1969.   Cockroaches (*Blattaria*) as vectors of food poisoning and food infection organisms. Jour. Med. Entom., *6*:185-189.
Ruhland, H. H., and Huddleson, I. F.
1941.   The rôle of one species of cockroach and several species of flies in the dissemination of *Brucella*. Amer. Jour. Vet. Res., *2*:371-372.
Ruiz Sandoval, G.
1881.   Memoria sobre el mal del pinto. Gaceta Med. Mex., *16*, see pp. 36-45, 49-64, 65-80, 81-94, 103-112.

Ruprah, N. S., and Treece, R. E.
  1968. Further studies on the effect of bovine diet on face fly development. Jour. Econ. Entom., *61*:1147-1150.
Russo, C.
  1930. Recherches expérimentales sur l'épidémiogenése de la peste bubonique par les insectes. Bull. Men. Off. Intl. Hyg. Publ., *22*:2108-2120.
Ryan, A. F.
  1954. The sheep blowfly problems in Tasmania. Austral. Vet. Jour., *30*:109-113.
Sabin, A. B., and Ward, R.
  1941. Flies as carriers of poliomyelitis virus in urban epidemics. Science, *94*:590-591.
Sabin, A. B., and Ward, R.
  1942. Insects and epidemiology of poliomyelitis. Science, *95*:300-301.
Sabrosky, C. W.
  1941. The *Hippelates* flies or eye gnats: preliminary notes. Canad. Entom., *73*:23-27.
Sabrosky, C. W.
  1951. Nomenclature of the eye gnats (*Hippelates* spp.). Amer. Jour. Trop. Med., *31*:257-258.
Sabrosky, C. W.
  1961. Our first decade with the face fly, *Musca autumnalis.* Jour. Econ. Entom., *54*:761-763.
Salmon, D. E., and Smith, T.
  1885. Report on swine plague. U. S. Dept. Agric., Bur. Anim. Ind., 2nd Ann. Rept.
Salt, G.
  1932. The natural control of the sheep-blowfly, *Lucilia sericata* Meigen. Bull. Entom. Res., *23*:235-245.
Salt, G.
  1961. The haemocytic reaction of insects to foreign bodies. *In*: The cell and the organism. Cambridge University Press. See pp. 175-192.
Salt, G.
  1963. The defense reactions of insects to metazoan parasites. Parasitology, *53*:527-642.
Salt, G.
  1970. The cellular defense reactions of insects. Cambridge Monographs in Experimental Biology 16. 124 pp. Cambridge University Press.
Sanborn, C. E., and Moe, L. H.
  1934. Anaplasmosis investigations. Rept. Okla. Agr. Exp. Sta., 1932-1934: 275-279.
Sanborn, C. E., Stiles, G. W., Moe, L. H., and Orr, H. W.
  1931. Transmission of anaplasmosis by flies. Okla. Agr. Exp. Sta. Bull. no. 204.
Sanders, D. A.
  1933. Notes on the experimental transmission of bovine anaplasmosis in Florida. Jour. Amer. Veter. Med. Assoc., *88*:799-805.
Sanders, D. A.
  1940a. *Musca domestica* a vector of bovine mastitis (Preliminary Report). Jour. Amer. Veter. Med. Assoc., *97*:120-123.
Sanders, D. A.
  1940b. *Hippelates* flies as vectors of bovine mastitis (Preliminary Report). Jour. Amer. Veter. Med. Assoc., *97*:306-308.
Sanders, E., Brachman, P. S., Friedman, E. A., Goldsby, J., and McCall, C. E.
  1965. Salmonellosis in the United States. Results of nationwide surveillance. Amer. Jour. Epidem., *81*:370-384.

Sandes, T. L.
1911.   The mode of transmission of leprosy. British Med. Jour., 2:469-470.
Sandoval, C. G., and Galvan, B. M.
1965.   Etiología de las diarreas bacterianas en el niño. Rev. Mex. Pediat., 34:
         66-70.
Satchell, G. H., and Harrison, R. A.
1953.   II. Experimental observations on the possibility of transmission of yaws
         by wound-feeding Diptera, in Western Samoa. Trans. Roy. Soc. Trop.
         Med. Hyg., 47:148-153.
Sauer, E.
1908.   Können ohne veterinär-polizeiliche Bedenken die Häute rauschbrand-
         kranker Tiere zu Gerbereizwecken verwendet werden? Zeitsch. Tiermed.,
         12:34-71.
Saunders, E. W., Meisenbach, R., and Wisdom, W. E.
1914.   The causation and prevention of infantile paralysis. Jour. Missouri State
         Med. Assoc., 10:305-317.
Saunders, G. M., Kumm, H. W., and Rerrie, J. I.
1936.   The relationship of certain environmental factors to the distribution of
         yaws in Jamaica. Amer. Jour. Hyg., 23:558-579.
Savage, E. P., and Schoof, H. F.
1955.   The species composition of fly populations at several types of problem
         sites in urban areas. Ann. Entom. Soc. Amer., 48:251-257.
Sawyer, W. A., and Herms, W. B.
1913.   Attempts to transmit poliomyelitis by means of the stable-fly (Stomoxys
         calcitrans). Jour. Amer. Med. Assoc., 61:461-466.
Schiller, E.
1954.   Studies on the helminth fauna of Alaska. XIX. An experimental study
         on blowflow [blowfly] (Phormia regina) transmission of hydatid disease.
         Exp. Parasit., 3:161-166.
Schliessmann, D. J.
1959.   Diarrhoeal disease and the environment. Bull. Wld. Health Org., 21:
         381-386.
Schlossberger, H., Betzel-Langbein, H., and Kreuz, G.
1954.   Über die Epidemiologie der Leptospirosen. Acta Trop., 11:300-302.
Schlossberger, H., and Langbein, H.
1952.   Übertragung von Leptospira icterohaemorrhagiae durch Ornithodorus
         moubata. Zeitschr. Immunitätsforsch., 109:366-370.
Schmit-Jensen, H. O.
1927.   Eksperimentelle undersøgelser over tovingede insekters (Diptera's)
         betydning som smittespredere ved mundog Klovesyge. I. Virus' skaebne
         i stikfluen, Stomoxys calcitrans' organisme, belyst ved forsøg paa marsvin.
         Maanedsskrift Dyrlaeger, 39:1-39.
Schneider, I.
1963.   In vitro culture of Drosophila organs and tissues. Genetics, 48:908.
Schneider, I.
1964a. Differentiation of larval Drosophila eye-antennal discs in vitro. Jour.
         Exp. Zool., 156:91-103.
Schneider, I.
1964b. Culture in vitro of Drosophila ovarian tissue. Genetics, 50:284.
Schofield, F. W.
1949.   Virus enteritis in mink. North Amer. Veter., 30:651-654.
Scholtz, K.
1933.   Magyarország egészségügyi helyzetéröl és felkészültségéröl. [Sanitary
         conditions and precautions in Hungary.] Orvosi Hetilap, 77:957-975.

Schoof, H. F., and Mail, G. A.
 1953. Dispersal habits of *Phormia regina* in Charleston, West Virginia. Jour. Econ. Entom., *46*:258-262.
Schoof, H. F., Mail, G. A., and Savage, E. P.
 1954. Fly production sources in urban communities. Jour. Econ. Entom., *47*: 245-253.
Schoof, H. F., and Siverly, R. E.
 1954. Multiple release studies on the dispersion of *Musca domestica* at Phoenix, Arizona. Jour. Econ. Entom., *47*:830-838.
Schoof, H. F., Siverly, R. E., and Jensen, J. A.
 1952. House fly dispersion studies in metropolitan areas. Jour. Econ. Entom., *45*:675-683.
Schuberg, A., and Böing, W.
 1913. Weitere Untersuchungen über die Übertragung von Krankheitserregern durch einheimische Stechfliegen. Centralbl. Bakt. Parasit. Infekt., I. Abt., *57*:301-303.
Schuberg, A., and Kuhn, P.
 1909. Über die Übertragung von Krankheiten durch einheimische stechende Insekten. I. Teil. Arb. Kaiser. Gesundheitsamte, *31*:377-393.
Schuberg, A., and Kuhn, P.
 1912. Über die Übertragung von Krankheiten durch einheimische stechende Insekten. II. Teil. Arb. Kaiser. Gesundheitsamte, *40*:209-234.
Schumann, H.
 1961. Die Bedeutung symboviner Fliegen als Verbreiter von Mastitis-Erregern. Mh. Vetmed., 16 Pt., pp. 624-626.
Schwarz, E. A., Stiles, Riley and Hubbard
 1895. Exhibition of *Hippelates pusio* and subsequent discussion on the relation of similar flies to the spread of Florida "sore-eye," etc. Proc. Entom. Soc. Washington, *3*:178-180.
Schwartz, P. H.
 1965. Some observations of the reproduction of the eye gnat, *Hippelates pusio*. Jour. Med. Entom., *2*:141-144.
Scott, J. R.
 1917a. Studies upon the common house-fly (*Musca domestica*, Linn.). I. A general study of the bacteriology of the house-fly in the District of Columbia. Jour. Med. Res., *37*:101-119.
Scott, J. R.
 1917b. Studies upon the common house-fly (*Musca domestica*, Linn.). II. The isolation of *B. cuniculicida*, a hitherto unreported isolation. Jour. Med. Res., *37*:121-124.
Scott, J. W.
 1915. Insect transmission of swamp fever. Science, *42*:659.
Scott, J. W.
 1920. Experimental transmission of swamp fever or infectious anemia by means of insects. Jour. Amer. Veter. Med. Assoc. n.s. *9*:448-454.
Scott, J. W.
 1922. Insect transmission of swamp fever or infectious anemia of horses. Univ. of Wyoming Agric. Exp. Sta. Bull. No. 133:57-111.
Scudder, H. I.
 1947. A new technique for sampling the density of housefly populations. Publ. Health Repts., *62*:681-686.
Sedee, P.D.J.W.
 1953. Quantitative vitamin requirements for growth of larvae of *Calliphora erythrocephala* (Meigen). Experientia, *9*:142-143.

Seecof, R. L.
1966. Sigma virus content and hereditary transmission in *Drosophila melanogaster*. Virology, *29*:1-7.

Seecof, R. L., and Unanue, R. L.
1968. Differentiation of embryonic *Drosophila* cells *in vitro*. Exp. Cell. Res., *50*:654-660.

Sehgal, B. S., and Kumar, P.
1966. A study of the seasonal fluctuations in fly populations in two villages near Lucknow. Indian Jour. Med. Res., *54*:1175.

Seiderer, H.
1839. *De anthrace sive pustula maligna*. Berlin.

Semenov, V. D.
1945. Observations on the seasonal variations in room fly (*Musca domestica*) population in relation to the dynamics of certain gastroenteric infections at Gor'kii. Med. Parazit., *14*:50-54.

Sen, S. K.
1925-6. Experiments on the transmission of rinderpest by means of insects. Mem. Dept. Agric. India, Entom. Ser., *9*:59-63, 166-181, 184-185.

Sen, S. K., and Minett, F. C.
1944. Experiments on the transmission of anthrax through flies. Indian Jour. Veter. Sci. Anim. Husb., *14*:149-158.

Sen, S. K., and Salem, A.
1937. Experiments on the transmission of rinderpest through the agency of *Stomoxys calcitrans* Linn. Indian Jour. Veter. Science. Anim. Husb., *5*:219-224.

Sergent, E., and Sergent, E.
1905. Trypanosomiase des dromadaires de l'Afrique du Nord. Ann. Inst. Pasteur, *19*:17-48.

Sergent, E., and Donatien, A.
1922a. Les *Stomoxes*, propagateurs de la trypanosomiase des dromadaires. Compt. Rend. Acad. Sci., Paris, *174*:582-584.

Sergent, E., and Donatien, A.
1922b. Transmission naturelle et expérimentale de la trypanosomiase des dromadaires par les *Stomoxes*. Arch. Inst. Pasteur Afr. Nord., *2*:291-315.

Shapiro, M.
1969. Immunity of insect hosts to insect parasites. *In*: Immunity to parasitic animals. Edited by G. J. Jackson, R. Herman, and I. Singer. Vol. I. Appleton-Century-Crofts, New York. See pp. 211-228.

Sheffield, H., and Melton, M.
1969. *Toxoplasma gondii*: transmission through feces in absence of *Toxocara cati* eggs. Science, *164*:431-432.

Shelton, G. C.
1954. Giardiasis in the chinchilla. II. Incidence of the disease and results of experimental infections. Am. Jour. Veter. Res., *15*:75-78.

Sheremet'ev, N. N.
1964. Dinamika vydeleniya vaktsinnykh shtammov virusa poliomielita ot mukh posle provedeniya vakstinatsii. [Dynamics of the isolation of poliomyelitis vaccinal virus strains from flies after vaccination.] Zh. Mikrobiol. Epidemiol. Immunobiol., *41*:102-106.

Shillito, J. F.
1947. Notes on insects visiting diseased elms. Entom. Monthly Mag., *83*:290-292.

Shimizu, F., Hashimoto, M., Taniguchi, H., Oota, W., Kakizawa, H., Takada, R., Kano, R., Tange, H., Kaneko, K., Shinonaga, S., and Miyamoto, K.
  1965. Epidemiological studies on fly-borne epidemics. Report I. Significant role of flies in relation to intestinal disorders. Japanese Jour. Sanit. Zool., *16*:201-211.

Shinedling, S. T., and Greenberg, B.
  1971. Culture of cells of the flesh fly, *Sarcophaga bullata*. *In*: Arthropod cell cultures and their application to the study of viruses. Ed. E. Weiss. Current topics in Microbiology and Immunology, *55*:12-19.

Shircore, T. D.
  1916. A note on some helminthic diseases with special reference to the house-fly as a natural carrier of the ova. Parasit., *7*:239-243.

Shkapo, L. E.
  1958. The transmission of eggs of Ascarida and of the dwarf Cyclophyllideae by certain synanthropic insects. Autoreferat dissertatsii, Khar'kov [in Russian] (not seen).

Shooter, R. A., and Waterworth, P. M.
  1944. A note on the transmissibility of haemolytic streptococcal infection by flies. Brit. Med. Jour., *1*:247-248.

Shope, R. E.
  1927. Bacteriophage isolated from the common house fly (*Musca domestica*). Jour. Exp. Med., *45*:1037-1044.

Shortt, H. E.
  1937. Rept. Sci. Adv. Bd. I.R.F.A., New Delhi, pp. 22-24.

Shterngol'd, E. Ia.
  1949. Komnatnye mukhi kak perenoschiki kishechnykh infektsii. [Roomflies as carriers of intestinal infections.] Trudy Uzbekskogo Inst. Epidem. Microb., *3*:172-180.

Shura-Bura, B. L.
  1952a. Zagriaznenie fruktov synantropnymi mukhami. [Soiling of fruit by flies.] Entom. Obozrenie (Moscow), *32*:117-125.

Shura-Bura, B. L.
  1952b. Experimental study of migration of houseflies using radioactive indicators. Zool. Zhurn., *31*:410-412.

Shura-Bura, B. L.
  1955. On the occurrence of naturally infected flies in the environment of dysenteric patients. Report given at the fourth lecture in honor of N. A. Kholodkovskii, 3 March, 1951. Moscow/Leningrad, pp. 29-52.

Shura-Bura, B. L.
  1957. Emploi des isotopes radioactifs dans l'étude du rôle épidémiologique des mouches. Jour. Hyg. Epidem. Microb. Immun., *1*:249-255.

Shura-Bura, B. L., Ivanova, E. V., Onutshin, A. N., Glazunova, A. Ia., and Shaikov, A. D.
  1956. Migrations of flies of medical importance in Leningrad district. Entom. Obozren., *35*:334-346 [in Russian].

Shura-Bura, B. L., Shaikov, A. D., Ivanova, E. V., Glazunova, A. Ia., Mitriukova, M. S., and Federova, K. G.
  1956. Migration of synanthropic flies to the cities from open fields. Med. Parasit. Parasitarn. Bolezn., *25*:368-372.

Shura-Bura, B. L., Shaikov, A. D., Ivanova, E. V., Glazunova, A. Ia., Mitriukova, M. S., and Federova, K. G.
  1958. The character of dispersion from the point of release in certain species of flies of medical importance. Entom. Rev., *37*:282-290.

Shura-Bura, B. L., Sukhomlinova, O. I., and Isarova, B. I.
  1962. Radioactive tracers as an aid to studying the ability of synanthropic flies to fly over water barriers. Entom. Rev., *41*:55-60; *also in*: Entom. Obozren., *41*:99-108.
Sieber, H., and Gonder, R.
  1908. Übertragung von *Trypanosoma equiperdum*. Arch. Schiffs. Tropen-Hyg., *12*:646.
Sieyro, L.
  1942. Die Hausfliege (*Musca domestica*) als Übertrager von *Entamoeba histolytica* und anderen Darmprotozoen. Deutsche Tropenmed. Zeitsch., *46*: 361-372.
Sigaud, J.F.X.
  1844. Du climat et des maladies du Bresel. Masson, Paris.
Simitch, T., and Kostitch, D.
  1937. Rôle de la mouche domestique dans la propagation du "*Trichomonas intestinalis*" chez l'homme. Ann. Parasit., *15*:323-325.
Simmonds, M.
  1892. Fliegen und Choleraübertragung. Deutsche. Med. Wochen., *41*:931.
Simmons, P.
  1923. A house fly plague in the American Expeditionary Force. Jour. Econ. Entom., *16*:357-363.
Simmons, S. W.
  1935a. The bactericidal properties of excretions of the maggot of *Lucilia sericata*. Bull. Entom. Res., *26*:559-563.
Simmons, S. W.
  1935b. A bactericidal principle in excretions of surgical maggots which destroys important etiological agents of pyogenic infections. Jour. Bact., *30*:253-267.
Simmons, S. W.
  1944. Observations on the biology of the stablefly in Florida. Jour. Econ. Entom., *37*:680-686.
Simmons, S. W.
  1959. The insecticide dichlorodiphenyltrichloroethane and its significance. Vol. II. Human and veterinary medicine. Lehrb. Monog. Geb. exakt. Wiss., chem. Reihe 10, Birkhausen Verlag, Basle and Stuttgart, 570 pp.
Simmons, S. W., and Dove, W. E.
  1941. Breeding places of the stablefly or "dog fly" *Stomoxys calcitrans* (L.) in northwestern Florida. Jour. Econ. Entom., *34*:457-462.
Simmons, S. W., and Dove, W. E.
  1942. Waste celery as a breeding medium for the stablefly or "dog fly," with suggestions for control. Jour. Econ. Entom., *35*:709-715.
Singh, P., and House, H. L.
  1970. Antimicrobials: "Safe" levels in a synthetic diet of an insect, *Agria affinis*. Jour. Insect. Physiol., *16*:1769-1782.
Singh, S. B., and Judd, W. W.
  1966. A comparative study of the alimentary canal of adult calyptrate Diptera. Proc. Entom. Soc. Ont., *96*:29-80.
Sinha, R., Deb, B. C., De, S. P., Abou-Gareeb, A. H., and Shrivastava, D. L.
  1967. Cholera carrier studies in Calcutta in 1966-67. Bull. Wld. Health Org., *37*:89-100.
Siniscal, A. A.
  1955. The trachoma story. Publ. Health Rept., *70*:497-507.
Sivori, F., and Lecler, E.
  1902. La surra americana ou mal de caderas. Buenos Aires. Pp. 1-79.

Skavinskiĭ, N. A.
1960. O zarazhennosti sinantropnykh mukh ĭaitsami gelmintov. [On the contamination of synanthropic flies with helminth eggs.] Zdravookh. Belorussii, 6:54.

Skidmore, L. V.
1932. The transmission of fowl cholera to turkeys by the common house fly (*Musca domestica* Linn.), with brief notes on the viability of fowl cholera microorganisms. Cornell Veter., 22:281-285.

Skinner, H.
1917. Insects and war. Entom. News, Philadelphia, 28:330-331.

Skovgaard, N.
1968. The incidence of hemolytic bacteria in cattle with a special view to *Corynebacterium pyogenes* as the causative agent of "Summer Mastitis." Yearbook, Roy. Veter. Agric. Coll., Copenhagen, pp. 89-108.

Slocum, M. A., McClellan, R. H., and Messer, F. C.
1933. Investigation into the modes of action of blow fly maggots in the treatment of chronic osteomyelitis. Pennsylvania Med. Jour., 36:570-573.

Smillie, G. W.
1916. Epidemiology of bacillary dysentery [and discussion by H. I. Bowditch]. New York Med. Jour., 104:1073-1074.

Smirnov, E. S.
1940. Le probléme des mouches à Tadjikistane. Med. Parasit., 9:515-517.

Smit, B.
1928. Observations on the life history of the sheep maggot fly, *Lucilia sericata*. Sci. Bull. Dept. Agric. So. Africa, No. 68, 12 pp.

Smith, K. M.
1967. Insect virology. Academic Press. 256 pp.

Soares de Sousa, G.
1587. Noticia do Brasil. Introd. coment. e notas pelo Prof. Piraza da Silva. (Biblioteca Hist. Brasil XVI.) São Paulo: Livr. Martins Ed. [1940?] ii, see pp. 124-125.

Soberón y Parra, G., and León y Blanco, F.
1944. Las moscas del genero *Hippelates* como posibles vectores del mal del pinto. Ciencia, 4:299-300.

Soltys, M. A.
1954. Transmission of *T. congolense* by other vectors than tsetse flies. Intl. Scient. Comm. for Trypanosomiasis Research. Meeting. Reunion, 5th, pp. 137-140.

Somov, P., Romanova, V., and Romanova, V.
1937. Bull. Azov-Black Sea District Inst. Microb. Epid. at Rostov-on-Don., Issue 16, pp. 91-100 (original not seen).

Soparker, M. B.
1938. Rept. Sci. Adv. Bd. I.R.F.A., New Delhi, pp. 25-29.

Sorin, M. V., and Rundkvist, V. A.
1956. The problem of flies in connection with dysentery. Sanitarno-gigieneicheskogo Med. Inst. Trudy, 27:129-138.

Soulsby, E.J.L.
1968. Helminths, arthropods and protozoa of domesticated animals (Mönnig). 6th ed., Williams and Wilkins Co., Baltimore. See pp. 594, 824.

de Souza-Araujo, H. C.
1944. Verificação da infecção de moscas da familia Tachinidae pela *Empusa* Cohn 1855. Essas moscas, sugando ulceras lipróticas, se infestaram com o bacilo de Hansen. Mem. Inst. Oswaldo Cruz, 41:201-203.

Spaar, E. C.
1925. Some observations on the common endemic fever of Ceylon. Jour. Trop. Med. Hyg., 28:349-352.

414   BIBLIOGRAPHY

Spanedda, A., and Marchinu, M.
1948. Azione del DDT sull'andamento della morbilità per tifo, paratifi, dissenteria bacillare, gastroenteriti infantili. Rassegna Med. Sarda, *50*:45-49.
Spiller, D.
1966. House flies. *In*: Insect colonization and mass production. Ed. C. N. Smith, Academic Press. See pp. 203-225.
Spillman and Haushalter
1887. Dissémination du bacille de la tuberculose par les mouches. Compt. Rend. Acad. Sci., *105*:352-353.
Spotarenko, S. S., Tikhomirov, E. D., and Kikodze, S. L.
1969. Kotsenke roli mukh v epidemiologii dizenterii i infektsionnogo hepatita. [Evaluation of the role of flies in the epidemiology of dysentery and infectious hepatitis.] Zh. Mikrobiol. Epidem. Immunobiol., *46*:43-48.
Srivastava, H. D., and Dutt, S. C.
1963. Studies on the life history of *Stephanofilaria assamensis*, the causative parasite of "humpsore" of Indian cattle. Indian Jour. Vet. Sci., *33*:173-177.
Stedman, J. G.
1796. Narrative of a five years' expedition, against the Revolted Negroes of Surinam, in Guiana, on the Wild Coast of South America. J. Johnson, St. Paul's Church Yard, & J. Edwards, Pall Mall, London. See p. 274.
Stein, C. D., Lotze, J. C., and Mott, L. O.
1942. Transmission of equine infectious anemia or swamp fever by the stable fly, *Stomoxys calcitrans*, the horsefly, *Tabanus sulcifrons* (Macquart) and by injections of minute amounts of virus. Amer. Jour. Veter. Res., *3*:183-193.
Stein, C. D., Lotze, J. C., and Mott, L. O.
1943. Evidence of transmission of inapparent (subclinical) form of equine infectious anemia by mosquitoes (*Psorophora columbiae*) and by injection of the virus in extremely high dilution. Jour. Amer. Veter. Med. Assoc., *102*:163-169.
Steiner, L. F.
1965. A rapid method for identifying dye-marked fruit flies. Jour. Econ. Entom., *58*:374-375.
Steinhaus, E. A.
1946. Insect microbiology. Hafner Publ. Co., Inc. N. Y. Re-issued, 1967.
Steinhaus, E. A., and Dineen, J.
1960. Observations on the role of stress in a granulosis of the variegated cutworm. Jour. Insect Path., *2*:55-65.
Steiniger, F.
1957. Untersuchungen von Fliegen aus Tierfuttermittelrohstoff-Lagern auf Typhus-Paratyphusbakterien. Desinfekt. Schädigungsbekämpfung, *49*:91.
Stephens, J. M.
1963. Bactericidal activity of hemolymph of some normal insects. Jour. Insect Path., *5*:61-65.
Steve, P. C.
1960. Biology and control of the little house fly, *Fannia canicularis*, in Massachusetts. Jour. Econ. Entom., *53*:999-1004.
Steve, P. C., and Lilly, J. H.
1965. Investigations on transmissibility of *Moraxella bovis* by the face fly. Jour. Econ. Entom., *58*:444-446.
Stewart, W.
1944. On the viability and transmission of dysentery bacilli by flies in North Africa. Jour. Roy. Army Med. Corps, *83*:42-46.

Stewart, W. H., McCabe, Jr., L. J., Hemphill, E. C., and DeCapito, T.
  1955.   IV. Diarrheal disease control studies. The relationship of certain environ-
          mental factors to the prevalence of *Shigella* infection. Amer. Jour. Trop.
          Med. Hyg., *4*:718-724.
St. Germaine, J.
  1955.   Flies of public health interest in Santa Clara County. Calif. Vector
          Views, *2*:39, 54.
Stiles, C. W., and Keister, W. S.
  1913.   Flies as carriers of *Lamblia* spores. The contamination of food with
          human excreta. Public Health Rept., *28*:2530-2534.
Stirrat, J. H., McLintock, J., Schwindt, G. W., and Depner, K. R.
  1955.   Bacteria associated with wild and laboratory-reared horn flies, *Siphona
          irritans* (L.) (Diptera: Muscidae). Jour. Parasit., *41*:398-406.
St. John, J. H., Simmons, J. S., and Reynolds, F.H.K.
  1930.   The survival of various microorganisms within the gastro-intestinal tract
          of *Aëdes aegypti*. Amer. Jour. Trop. Med., *10*:237-241.
Stoddard, H. L.
  1931.   The bobwhite quail. Its habits, preservation and increase. Charles Scrib-
          ner's Sons, New York, see p. 311.
Stoffolano, Jr., J. G.
  1970.   Nematodes associated with the genus *Musca* (Diptera: Muscidae). Bull.
          Entom. Soc. Amer., *16*:194-203.
Stoffolano, Jr., J. G., and Matthysse, J. G.
  1967.   Influence of photoperiod and temperature on diapause in the face fly,
          Musca autumnalis (Diptera: Muscidae). Ann. Entom. Soc. Amer., *60*:
          1242-1246.
Stohler, H. R.
  1961.   The peritrophic membrane of blood-sucking Diptera in relation to their
          role as vectors of blood parasites. Acta. Trop., Basel, *18*:263-266.
Streams, F. A.
  1968.   Defense reactions of *Drosophila* species (Diptera: Drosophilidae) to the
          parasite *Pseudeucoila bochei* (Hymenoptera: Cynipidae). Ann. Entom.
          Soc. Amer., *61*:158-164.
Strong, R. P.
  1944.   Stitt's diagnosis, prevention and treatment of tropical diseases. The
          Blakiston Co., Philadelphia, 7th Edition, Vol. 1, pp. 164, 174.
Suenaga, O., and Fukuda, M.
  1963.   Ecological studies of flies. 7. On the species and seasonal prevalance of
          flies breeding out from a privy and a urinary pit in a farm village.
          Endemic Dis. Bull. Nagasaki Univ., *5*:72-80.
Sukhacheva, E. I.
  1963.   Rol'nekotorykh vidov lichinok nasekomykh v èpidemiologii askaridoza.
          [The role of certain insect larvae in the epidemiology of ascariasis.] Med.
          Parazit. Parazitarn. Bolezni, *32*:600-604.
Sukhova, M. N.
  1950.   Novye dannye po ekologii i èpidemiologicheskomu znacheniiū sinikh
          miāsnykh mukh *Calliphora uralensis* Vill. and *Calliphora erythrocephala*
          Meig. (Diptera, Calliphoridae). [New data on the ecology and epidemio-
          logical significance of the blue meat flies, *Calliphora uralensis* Vill. and
          Calliphora erythrocephala Meig.] Entom. Obozrenie, *31*:90-94.
Sukhova, M. N.
  1951.   K voprosu ob epidemiologicheskom znachenii koprobiontnykh èksofil'nyk
          vidov sinantropnykh mukh. [On the problem of the epidemiological sig-
          nificance of coprobiontic exophilous species of synanthropic flies.] Zool.
          Zhurnal., *30*:188-190.

Sukhova, M. N.
1953. O znachenie bazarnoĭ mukhi (*Musca sorbens* Wied.) v èpidemiologii ostrago èpidemicheskogo kon'iūnktivita v zapadnoi Turkmanii. [On the significance of the bazaar fly (*M. sorbens* Wied.) in the epidemiology of acute epidemic conjunctivitis in western Turkmenia.] Gigiena Sanitaria, 7:40-42.

Sukhova, M. N.
1954. Voprosy kommunal' noĭ gigeny v usloviĭakh zharkogo klomata sredneĭ azii. Medgiz Akad. Med. Nauk., Moscow, pp. 126-141. [Flies in population centers in west Turkmenia, their sanitary and epidemiological significance and their control. Anthology . . . *in* Problems of communal hygiene in the hot-climate conditions of Central Asia.]

Sukhova, M. N.
1957. Synanthrope Fliegen einzelner Landschaftszonen der USSR. [Synanthropic flies of the different landscape-zones of the USSR.] Autoreferat Disser., Moskau, pp. 1-26.

Surcouf, J.M.R.
1921. 4° et 5°. Notes biologiques sur certains diptères. Bull. Mus. Hist. Nat., Paris, 27:67-74.

Surcouf, J.M.R.
1923. Deuxième note sur les conditions biologiques du *Stomoxys calcitrans* L. Bull. Mus. Hist. Nat., Paris, 29:168-172.

Sychevskaĭa, V. I.
1954. Materialy k biologii i ekologii sinantropnych mukh roda *Fannia* R. D. v Samarkande. [Material on the biology and ecology of synanthropic flies of the genus *Fannia* R. D. in Samarkand.] Med. Parazit. Parazitarn. Bolezni, 23:45-54.

Sychevskaĭa, V. I.
1960. Methods and practice of phenological observations of synanthropic flies (experiences gathered in Uzbekistan). Med. Parazit., Moscow, 29:712-720.

Sychevskaĭa, V. I.
1962. On changes in the daily dynamics of the specific composition of flies associated with man in the course of the season. Entom. Obozren., 41:545-553.

Sychevskaĭa, V. I.
1965. Biology and ecology of *Calliphora vicina* R.-D. in central Asia. Zool. Zhurn., 44:552-560.

Sychevskaĭa, V. I., Grudtsina, M. V., and Vyrvikhvost, L. A.
1955. Spontaneous infection of flies with dysenteric microflora. Soveshch. Parazit. Prob., Moskva, pp. 148-149.

Sychevskaĭa, V. I., Grudtsina, M. V., and Vyrvikhvost, L. A.
1959. The epidemiological significance of synanthropic flies (Diptera) in Bukhara. Entom. Obozren., 38:568-578.

Sychevskaĭa, V. I., and Petrova, T. A.
1958. O roli mukh rasprostranenii yaits gel'mintov v Uzbekistane. [The role of flies in spreading the eggs of helminths in Uzbekistan.] Zool. Zhurn., 37:563-569.

Sychevskaĭa, V. I., Skopina, N. P., and Petrova, Z. F.
1959. Contamination of synanthropic flies with dysentery bacilli and helminth eggs in Fergana. Trudy Uzbekistanskogo Instituta Malĭarii i Meditsinskoĭ Parazitologii, 4:225-235. *In*: Works of Uzbek Inst. Malaria and Parasit. Uzbek Dept. Publ. Health, Samarkand.

Syddiq, M. M.
1938. *Siphunculina funicola* (eye-fly). Indian Med. Gaz., 73:17-19.

Sydenham, T.
1666. Works of Thomas Sydenham, from Latin Ed. of Dr. Greenhill. Thomas Sydenham Society, 1848, Vol. I, see pp. 166, 269, 271.

Sykes, G. F.
1910. Twenty-eighth annual report of the Superintendent of Health of the city of Providence for the year 1910. 27 pp.

Tacal, Jr., J. V., and Meñez, C. F.
1967. *Salmonella* studies in the Philippines. VII. The isolation of *Salmonella derby* from abattoir flies and chicken ascarids. Philippine Jour. Veter. Med., *6*:106-111.

Takaki (Baron)
1906. The preservation of health amongst the personnel of the Japanese army. Jour. Roy. Army Med. Corps, *7*:54-62.

Takeuchi, T., Mori, I., and Maeda, M.
1966. Studies on the persistency of bacteria in flies and the quantity of bacteria transmitted. Jap. Jour. Sanit. Zool., *17*:226-231.

Talice, R. V.
1943. El por qué de la lucha contra la "mosca doméstica." Arch. Uruguayos Med., Cirugia Especialidades, *22*:658-601.

Tanada, Y., Holdaway, F. G., and Quisenberry, J. H.
1950. DDT to control flies breeding in poultry manure. Jour. Econ. Entom., *43*:30-36.

Tao, C. S.
1936. Transmission of helminths ova by flies. Jour. Shanghai Science Inst., Sect. 4, *2*:109-116.

Taplin, D., Zaias, N., and Rebell, G.
1967. Skin infections in a military population. Develop. Ind. Microbiol., *8*:3-12.

Tarasov, V. A., and Chaïkin, V. N.
1941. On the frequency of occurrence of enteric bacilli in the room fly. Zhurn. Mikrob. Epidem. Immun., *9*:23-27.

Tashiro, H., and Schwardt, H. H.
1953. Biological studies of horse flies in New York. Jour. Econ. Entom., *46*:813-822.

Tatchell, R. J.
1958. The physiology of digestion in larvae of the horse bot-fly, *Gasterophilus intestinalis* (DeGeer). Parasit., *48*:448-458.

Tauber, M. J.
1968. Biology, behavior, and emergence rhythm of two species of *Fannia* (Diptera: Muscidae). Univ. of Calif. Publ. in Entom., *50*:1-77.

Taylor, A. W.
1930. Experiments on the mechanical transmission of West African strains of *Trypanosoma brucei* and *T. gambiense* by *Glossina* and other biting flies. Trans. Roy. Soc. Trop. Med. Hyg., *23*:289-303.

Taylor, B. C., and Nakamura, M.
1964. Survival of shigellae in food. Jour. Hyg., *62*:303-311.

Taylor, E. L.
1935. An attempt to transmit anaplasmosis by British biting flies. Veter. Jour., *91*:4-11.

Taylor, J.
1961. Host specificity and enteropathogenicity of *Escherichia coli*. Jour. Appl. Bact., *24*:316-325.

Taylor, J. F.
1919. The role of the fly as a carrier of bacillary dysentery in the Salonica command. London Med. Res. Coun. Spec. Rept. Series No. 40, Med. Res. Commit., pp. 68-83.

418 BIBLIOGRAPHY

Tebbutt, H.
1913. On the influence of the metamorphosis of *Musca domestica* upon bacteria administered in the larval stage. Jour. Hyg., *12*:516-526.

Teodoro, G.
1916. Persistenza e resistenza del bacilli dell'ileo tifo nell'apparato digerente delle mosche. Atti Reale Ist. Veneto, *75*:1559-1568.

Terry, C. E.
1913. Fly-borne typhoid fever and its control in Jacksonville. U.S. Publ. Health Repts., *28*:68-73.

Tesch, J. W.
1937. Investigations into the epidemiology of various communicable diseases. *In*: The hygiene study ward center at Batavia. Leyden, University Press. See pp. 140-148.

Teskey, H. J.
1960. A review of the life history and habits of *Musca autumnalis* DeGeer (Diptera: Muscidae). Canadian Entom., *92*:360-367.

Teskey, H. J.
1969. On the behavior and ecology of the face fly, *Musca autumnalis* (Diptera: Muscidae). Canadian Entom., *101*:561-576.

Testi, F.
1909. Ricerche batteriologiche sull'intestino degli insetti. Riv. Ig. Sanità. Publ., *20*:491-498.

Thomas, H. T.
1951. Some species of the blow-fly genera *Chrysomyia* R-D., *Lucilia* R-D., *Hemipyrellia* Tnsd., and *Calliphora* R-D. from south-eastern Szechwan, China. Proc. Zool. Soc. London, *121*:147-200.

Thomsen, M.
1938. The housefly (*Musca domestica*) and the stable fly (*Stomoxys calcitrans*). Investigations on their biology and control, with a review of other species of flies associated with domestic animals or dwellings [in Danish]. Veretn. Vet. -og Landbohojsk, København, No. 176, 352 pp.

Thomsen, M., and Hammer, O.
1936. The breeding media of some common flies. Bull. Entom. Res., *27*:559-587.

Thomson, D., and Thomson, J. G.
1916. Protozoological researches, including investigations on the sand in Egypt, undertaken to elucidate the mode of spread of amoebic dysentery and the flagellate diarrhoeas: with conclusions regarding the sanitary measures necessary to prevent these diseases. Jour. Roy. Army Med. Corps., *27*:1-30.

Thomson, F. W.
1912. The house-fly as a carrier of typhoid infection. Jour. Trop. Med. Hyg., *15*:273-277.

Thomson, J. G., and Lamborn, W. A.
1934. Mechanical transmission of trypanosomiasis, leishmaniasis, and yaws through the agency of non-biting haematophagous flies. British Med. Jour., *2*:506-509.

Thomson, R.C.M.
1937. Observations on the biology and larvae of the Anthomyidae. Parasitology, *29*:273-358.

Thorwald, J.
1963. Science and secrets of early medicine, Egypt, Mesopotamia, India, China, Mexico, Peru. Transl. by R. and C. Winston. Harcourt, Brace & World, Inc., N. Y. 1st. American ed. 331 pp. See pp. 88-89, 140.

Tiensuu, L.
  1936.   Insect life on plants attacked by aphids. Ann. Entom. Fenn., *2*:161-169.
Tiensuu, L.
  1938.   Beiträge zur Kenntnis den Musciden (Dipt.) Finlands. Ann. Entom.
          Fenn., *4*:21-33.
Tilden, J. W.
  1957.   Flies from major sources. Calif. Vector Views, *4*:24, 27.
Tishchenko, T. A.
  1955.   Comparative study of the viability of the bacteria of dysenteric grippe
          in the organism of some synanthropic insects. Diss., Univ. Karhov, 14
          pp. (not seen).
Titze, C.
  1921.   Die Probleme der Maul-und Klauenseuchenforschung unter Berücksichti-
          gung des letzten Seuchenzuges. Arch. Wiss. Tierheilk., *47*:273-291.
Tizzoni, G., and Cattani, J.
  1886.   Untersuchungen über Cholera. Centralbl. Med. Wissen., *24*:769-771.
Todd, D. H.
  1964.   The biting fly *Stomoxys calcitrans* (L.) in dairy herds in New Zealand.
          New Zealand Jour. Agric. Res., *7*:60-79.
Tokugawa, H.
  1941.   Untersuchungen über die Beziehungen zwischen Fliegen and Bakterien.
          Zentralbl. Gesamte Hyg. Bakt. Immun., *47*:670-671.
Tolstīak, I. E.
  1956.   Transmission of swine erysipelas through the bite of the stable fly [in
          Russian]. Vet., *33*:73-75.
Toomey, J. A., Pirone, P. P., Takacs, W. S., and Schaeffer, M.
  1947.   Can *Drosophila* flies carry poliomyelitis virus? Jour. Infect. Dis., *81*:135-
          138.
Toomey, J. A., Takacs, W. S., and Tischer, L. A.
  1941.   Poliomyelitis virus from flies. Proc. Soc. Exp. Biol. Med., *48*:637-639.
Tooth, H. H., and Calverley, J.E.G.
  1901.   A civilian war hospital: being an account of the work of the Portland
          hospital and of experience of wounds and sickness in South Africa, 1900.
          John Murray, London.
Torrey, J. C.
  1912.   Numbers and types of bacteria carried by city flies. Jour. Infect. Dis.,
          10:166-177.
Trager, W.
  1959.   Tsetse-fly tissue culture and the development of trypanosomes to the
          infective stage. Ann. Trop. Med. Parasit., *53*:473-491.
Trask, J. D., and Paul, J. R.
  1943.   The detection of poliomyelitis virus in flies collected during epidemics
          of poliomyelitis. II. Clinical circumstances under which flies were col-
          lected. Jour. Exp. Med., *77*:545-556.
Trask, J. D., Paul, J. R., and Melnick, J. L.
  1943.   The detection of poliomyelitis virus in flies collected during epidemics
          of poliomyelitis. I. Methods, results, and types of flies involved. Jour.
          Exp. Med., *77*:531-544.
Trawiński, A., and Trawińska, J.
  1958.   Studies on the transmission of salmonellae by the developmental stages
          of flies [in Polish]. Ann. Univ. Mariae Curie-Sklodowska, Lublin, *13*:
          31-40.
Treece, R. E.
  1960.   Distribution, life-history and control of the face fly in Ohio. Proc. North
          Central Branch Entom. Soc. Amer., *15*:107.

420   BIBLIOGRAPHY

Treece, R. E.
1964. Research on *Musca autumnalis* Deg. in the United States. Proc. XII Int. Congr. Ent. London.

Trofimov, G. K.
1963. The bazaar fly *Musca sorbens* Wd. (Diptera, Muscidae) in Azerbaijan. Entom. Obozren., *42*:757-764.

Trofimov, G. K.
1965. A brief review of synanthropic flies (Muscidae, Calliphoridae and Sarcophagidae) of the Talysh region of the Caucasus. Entom. Obozren., *44*: 357-361.

Trofimov, G. K., and Engel'hardt, L. S.
1948. O roli sinantropnykh mukh v èpidemiologii glistnykh zubolevaniǐ v Baku. [On the role of synanthropic flies in the epidemiology of tapeworm illness in Baku.] Med. Parazit., *17*:247-252 (not seen).

Tsuzuki, J.
1904. Bericht über meine epidemiologischen Beobachtungen und Forschungen während der Choleraepidemie in Nordchina in Jahre 1902 und über die im Verlaufe derselben von mir durchgeführten prophylaktischen Massregeln mit besonderer Berücksichtigung der Choleraschutzimpfung. Arch. Schiffs. Tropenhyg., *8*:71-81.

Tucker, E.F.G.
1903. A contribution to the discussion on the aetiology of Lepra. Indian Lancet, *21*:830-831.

Tugwell, P., Burns, E. C., and Witherspoon, B.
1966. Notes on the flight behavior of the horn fly, *Haematobia irritans* (L.) (Diptera: Muscidae). Jour. Kansas Entom. Soc., *39*:561-565.

Tukhmanyants, A. A., Shakhurina, E. A., and Eskina, G. V.
1963. K ekologii *Musca larvipara* (Portsch, 1910), promezhutochnogo khozyaina *Thelazia rhodesi* (Desmarest, 1827) krupnogo rogatogo skota. [Contribution to the ecology of *Musca larvipara* (Portsch, 1910) an intermediate host of *Thelazia rhodesi* (Desmarest, 1827) from horned cattle.] Uzbeksk. Biol. Zh., *7*:57-62 (original not seen).

Tulinius, S.
1943. Shiga-Kruse-dysenteri-epidemi i Aarhus og odder 1943. Ugeskrift Laeger, *105*:1325-1331.

Tulloch, F.M.G.
1906. The internal anatomy of *Stomoxys*. Jour. Roy. Army. Med. Corps, 7: 154-162.

Udall, D. H.
1954. The practice of veterinary medicine. Published by the author. 6th Ed., pp. 504-514. Ithaca, N. Y.

Uemoto, K.
1960. Studies on the dispersion of flies at winter season. Japanese Jour. Sanit. Zool., *11*:95-101.

Uffelmann, J.
1892. Beiträge zur Biologie des Cholerabacillus. Berliner Klinische Wochen., *29*:1209-1215.

Uhlenhuth, P., and Fromme, W.
1916. Untersuchungen über die Aetiologie Immunität und spezifische Behandlung der Weilschen Krankheit (*Icterus infectiosus*). Zeitschr. Immunität., *25*:317-434.

Uhlenhuth, P., and Kuhn, P.
1917. Experimentelle Übertragung der Weilschen Krankheit durch die Stallfliege (*Stomoxys calcitrans*). Zeitschr. Hyg. Infekt. Krankh., *84*:517-540.

Uvarova, A. I.
1958. On the role of flies in the seasonality of dysentery [in Russian]. Zhurn. Mikrob. Epidem. Immun., No. 3, 124-129.

Vainshtein, B. A., and Rodova, R. A.
1940. Les lieux de développement des mouches de fumier dans les conditions du Tadjikistan montagneux [in Russian]. Med. Parasit., 9:364-368. (Rev. Appl. Entom. B., *31*:126, 1943.)

Van Buren, E. D.
1936-7. Mesopotamian fauna in the light of the monuments. Archeological remarks upon Landsberger's "Fauna des alten Mesopotamien." Arch. Orientforschung, *11*:1-37.

Van Buren, E. D.
1939. The fauna of ancient Mesopotamia as represented in art. *Pontificium Institutum Biblicum*, Rome. 113 pp. See pp. 108, 109.

Van Es, L.
1910. Report of progress on swamp fever. 10th Rept. North Dakota Agric. Exp. Sta., p. 193.

Van Ness, G. B.
1971. Ecology of anthrax. Science, *172*:1303-1306.

Vanni, V.
1946. Sul meccanismo infettante della mosca domestica. Ann. Ig., *56*:151-155.

van Oye, E. (editor)
1964. The world problem of salmonellosis. Dr. W. Junk, Pub., The Hague, 606 pp.

Vanskaĩa, R. A.
1943. Sezonnye izmeneniĩa obiliĩa komnatnoi mukhi sviaz' ikh s dvizheniem kishechnykh infektsiĩ za trigoda. [Seasonal changes in the abundance of the room fly in city conditions and their connection with the course of intestinal infections for the past three years.] Zhurn. Mikrob. Epidem. Immun., *4*:59.

Vanskaĩa, R. A.
1947. Number of *Musca domestica* L. and incidence of enteric infections in urban conditions, based on observations over 5 years [in Russian]. Zhurn. Mikrob. Epidem. Immun., *10*:46-49.

Vanskaĩa, R. A.
1957. Sezonnyĩ khod chislennosti mukh *Musca domestica* L. i zabolevaemost' dizenteriĩ i briŭshnym tiphom. [Seasonal course of the quantity of *Musca domestica* L. and the incidence of dysentery and intestinal typhus.] Med. Parazit. Parazitar. Bolezni., Moscow, *26*:75.

van Saceghem, R.
1918. Cause étiologique et traitement de la dermite granuleuse. Bull. Soc. Path. Exot., *11*:575-578.

van Saceghem, R.
1922. Mécanisme de la propagation des trypanosomiases par les stomoxes. Ann. Soc. Belge. Med. Trop., *2*:161-164.

van Thiel, P. H.
1949. The transmission of toxoplasmosis and the role of *Calliphora erythrocephala* Meig. Doc. Neerl. Indo. Morb. Trop., *1*:264-269.

van Voorst Vader, P.J.A.
1941. Immunity against bacillary dysentery and "infectious enteritis." Connection with flies and amoebic dysentery. Doc. Neerl. Indo. Morb. Trop., *3*:83-90.

van Zijl, W. J., Wolff, H. L., Timochine, D., Han, G. K., and Roy, M.
1966. Studies on diarrhoeal diseases by the World Health Organization diarrhoeal diseases advisory team in co-operation with the Ministry of Health Venezuela. WHO/ENT/66.7, pp. 1-164.

422 BIBLIOGRAPHY

Varela, G., and Zavala, J.
1961. Ensayos de transmission de la toxoplasmosis por insectos. Rev. Inst. Salubr. Enferm. Trop. (Mex.), 21:141-148.

Vargas Cuéllar, P. I.
1941. El pian en el Departamento del Valle del Cauca, Colombia. Bol. San. Panamer., 20:897-913.

Varwich
1577. Tractatlin von der pestilenz (not seen).

Vashchinskaĩa, N. V.
1956. Migration, distance and speed of the housefly. Akad. Nauk. Armianskoi SSR, Erian Izvestiia [in Russian]. Biol. Sel' Skokhoz Iastvennye, 9:73-77.

Vasil'ev, A.
1935. Mikroflora "dikikh mukh" m. Alchevs'ka. [Microflora of "wild flies" in the town of Alchevska.] Mikrob. Zhurn., 2:85-107.

Vaughan, V. C.
1900. Conclusions reached after a study of typhoid fever among the American soldiers in 1898. Jour. Amer. Med. Assoc., 34:1451-1459.

Vedder, E. B.
1928. A discussion of the etiology of leprosy, with especial reference to the possibility of the transference of leprosy by insects, and the experimental inoculation of three men. Philippine Jour. Sci., 37:215-243.

Verhoestraete, L. J., and Puffer, R. R.
1958. Diarrhoeal disease with special reference to the Americas. Bull. Wld. Health Org., 19:23-51.

Világiová, I.
1962. Význam múch pre vývin očných parazitov—pôvodcov teláziózy hovädzieho dobytka. Biológia, 17:297-299.

Világiová, I.
1967. Results of experimental studies on the development of preinvasive stages of worms of the genus Thelazia Bosc. 1819 (Spirurata:Nematoda), parasitic in the eye of cattle. Folia Parasit., 14:275-280.

Vine, M. J.
1947. The anti-malaria campaign in Greece, 1946. Bull. Wld. Health Org., 1: 197-204.

Viswanathan, D. K., and Rao, T. R.
1948. Control of rural malaria with D.D.T. indoor residual spraying in Kanara and Dharwar districts, Bombay Province: Second year's results, 1947-48. Indian Jour. Malar., 2:157-210.

Vogelsang, E. G., Jallo, P., and Mayandon, T. H.
1955. Elementos bacteriostaticos en las larvas de gasterophilus. Rev. Med. Vet. Parasit., 14:41-44.

Von Linstow, O. F.
1875. Beobachtungen an neuen und bekannten Helminthen. Arch. Naturg. Berlin, 1:183-207.

Waddell, A. H.
1969. A survey of Habronema spp. and the identification of third-stage larvae of Habronema megastoma and Habronema muscae in section. Aust. Vet. Jour., 45:20-21.

Waldmann, O., and Hirschfelder, H.
1938. Die epizootische Bedeutung der Ratten, des Wildes, der Vögel und der Insekten für die Verbreitung der maul-und Klauenseuchen. Berl. Tierarztl. Wschr., 46:229-234.

Walker, A. R., and Davies, F. G.
1971. A preliminary survey of the epidemiology of bluetongue in Kenya. Jour. Hyg., 69:47-60.

Walker, J.
    1930.   East African swine fever. Thesis, University Zurich, Bailliere, Tindall and Cox, London.

Wallace, G. D.
    1971.   Experimental transmission of *Toxoplasma gondii* by filth-flies. Amer. Jour. Trop. Med. Hyg., *20*:411-413.

Wallis, R. C., and Lite, S. W.
    1970.   Primary monolayer cell culture from the house fly, *Musca domestica.* Ann. Entom. Soc. Amer., *63*:1788-1790.

Walsh, R.
    1830.   Notices of Brazil in 1828 and 1829. London, Vol. 1, see pp. 403-404.

Walz, G. H.
    1803.   Ueber die Natur und Behandlungsweise der Rinderpest. Stuttgart (not seen).

Wang, C. H.
    1964.   Laboratory observations on the life history and habits of the face fly, *Musca autumnalis* (Diptera: Muscidae). Ann. Entom. Soc. Amer., *57*: 563-569.

Ward, R., Melnick, J. L., and Horstmann, D. M.
    1945.   Poliomyelitis virus in fly-contaminated food collected at an epidemic. Science, *101*:491-493.

Wardle, R. A.
    1930.   Significant variables in the blowfly environment. Ann. Appl. Biol., *17*: 554-574.

Washburn, F. L.
    1911.   The typhoid fly on the Minnesota Iron Range. Popular Sci. Monthly, *79*:137-150.

Watt, J.
    1949.   Fly control and the acute diarrheal diseases. Bol. Oficina Sanit. Panamer., March 1949, pp. 249-258.

Watt, J., and De Capito, T.
    1950.   The frequency and distribution of *Salmonella* types isolated from man and animals in Hidalgo County, Texas. Amer. Jour. Hyg., *51*:343-352.

Watt, J., and Lindsay, D. R.
    1948.   Diarrheal disease control studies. I. Effect of fly control in a high morbidity area. Publ. Health Rept., *63*:1319-1334.

Wave, H. E., Henneberry, T. J., and Mason, H. C.
    1963.   Fluorescent biological stains as markers for *Drosophila.* Jour. Econ. Entom., *56*:890-891.

Wawrinsky, R.
    1888.   Bemerkungen über eine kleine Pockenepidemie in Stockholm während der Jahres 1884. Arch. Hyg., *8*:367-368.

Wayson, N. E.
    1914.   Plague and plague-like disease. A report on their transmission by *Stomoxys calcitrans* and *Musca domestica.* U. S. Publ. Health Rept., *29*: 3390-3393.

Webb, J. E., Jr., and Graham, H. M.
    1956.   Observations on some filth flies in the vicinity of Fort Churchill, Manitoba, Canada, 1953-54. Jour. Econ. Entom., *49*:595-600.

Weil, G. C., Simon, R. J., and Sweadner, W. R.
    1933.   A biological, bacteriological and clinical study of larval or maggot therapy in the treatment of acute and chronic pyogenic infections. Amer. Jour. Surg., *19*:36-48.

Weir, J. M., Wasif, I. M., Hassan, F. R., Attia, S.D.M., and Kader, M. A.
    1952.   An evaluation of health and sanitation in Egyptian villages. Jour. Egyptian Publ. Health Assoc. *27*:55-114.

424   BIBLIOGRAPHY

Weiser, J.
1969.   Immunity of insects to Protozoa. *In*: Immunity to parasitic animals. Edited by G. J. Jackson, R. Herman, and I. Singer. Vol. I. Appleton-Century-Crofts, New York. See pp. 129-147.

Weismann, R.
1964.   Untersuchungen über die Bedeutung der Sinnesorgane an Russels der Stubenfliege. Mitt. Schweiz. Entom. Ges., *36*:249-274.

Weiss, A.
1869.   Beobachtungen über den Milzbrand bei Menschen. Bayer. arztliches Intell.-Bl. No. 25 [reviewed in: Virchow-Hirsch Jahresber. I, p. 490].

Weiss, E.
1971.   Arthropod cell cultures and their application to the study of viruses. *In*: Current topics in microbiology and immunology, 55. Springer-Verlag, New York. 288 pp.

Welander, E.
1896.   Beiträge zur Frage der Uebertragung der Gonococcen bei Augenblennorrhoe. Wien. Klin. Rundschau., *10*:883-885.

Welch, E. V.
1939.   Insects found on aircraft at Miami, Fla. in 1938. Publ. Health Repts., *54*:561-566.

Wellmann, G.
1948.   Übertragung des Rotlaufs auf Schweine durch den gemeinen Wadenstecher (*Stomoxys calcitrans*). Zentralbl. Bakt. Parasit., *153*:200-203.

Wellmann, G.
1949a. Die gemeine Stechfliege (*Stomoxys calcitrans*) als Überträger des Rotlaufs. Zentralbl. Bakt. Parasit., *153*:185-199.

Wellmann, G.
1949b. Die Übertragung des Schweinerotlaufs durch den Saugakt der gemeinen Stechfliege (*Stomoxys calcitrans*) und ihre epidemiologische Bedeutung. Berliner Münchener Tierärztl. Wschr., No. 4, 39-46.

Wellmann, G.
1950a. Rotlaufübertragung durch verschiedene blutsaugende Insektenarten auf Tauben. Zentralbl. Bakt. Parasit., *155*:109-115.

Wellmann, G.
1950b. Blutsaugende Insekten als Mechanische Brucellenüberträger. Zentralbl. Bakt. Parasit. Infekt. Hyg., *156*:414-426.

Wellmann, G.
1954.   Die Übertragung der Schweinrotlaufinfektion durch die Stubenfliege (*Musca domestica*). Zentralbl. Bakt. Parasit., Infekt. Hyg., *162*:261-264.

Wellmann, G.
1959.   Stechfliegen als Überträger von Zoonosen. Intl. Symposium über schädliche Fliegen. Zeitsch. Angew. Zool., *46*:328-331.

Wenyon, C. M.
1911a. Oriental sore in Bagdad, together with observations on a gregarine in *Stegomyia fasciata*, the haemogregarine of dogs and the flagellates of house flies. Parasit., *4*:273-340.

Wenyon, C. M.
1911b. Experimental work on oriental sore. Kala Azar Bulletin No. 1: 36-58.

Wenyon, C. M., and O'Connor, F. W.
1917a. The carriage of cysts of *Entamoeba histolytica* and other intestinal Protozoa and eggs of parasitic worms by house-flies with some notes on the resistance of cysts to disinfectants and other agents. Jour. Roy. Army Med. Corps, *28*:522-527.

Wenyon, C. M., and O'Connor, F. W.
  1917b. An inquiry into some problems affecting the spread and incidence of intestinal protozoal infections of British troops and natives in Egypt, with special reference to the carrier question, diagnosis and treatment of amoebic dysentery, and an account of three new human intestinal Protozoa. Part IV. Experimental work with the human intestinal Protozoa, their carriage by house-flies and the resistance of their cysts to disinfectant and other agents. Jour. Roy. Army Med. Corps, *28*:686-698.

Werner, G. H., Latte, B., and Contini, A.
  1964. Trachoma. Sci. Amer., *210*:79-86.

Werner, H.
  1909. Studies regarding pathogenic amoebae. Indian Med. Gaz., July, pp. 241-245.

Wesselinoff, W., and Toneff, M.
  1968. Salmonellen in eingefuhrtem Tierkorpermehl. [Salmonella in imported animal meal.] Berlin Munch. Tierar. Woch., *81*:426-428.

West, L. S.
  1951. The house fly. Its natural history, medical importance, and control. Comstock Publ. Co., Inc. Ithaca, New York. 584 pp.

West, L. S.
  1953. Fly control in the Eastern Mediterranean and elsewhere. Report of a survey and study. WHO/Insecticides/19, pp. 1-139, May 12.

Wetzel, R.
  1936. Neuere Ergebnisse über die Entwicklung von Hühnerbandwürmern. Verhandl. Deutschen Zool. Ges., pp. 195-200.

Wharton, R. H., Seow, C. L., Ganapathipillai, A., and Jabaratnam, G.
  1962. House fly populations and their dispersion in Malaya with particular reference to the fly problem in the Cameron Highlands. Med. Jour. Malaya, *17*:115-131.

Wherry, W. B.
  1908. Notes on rat leprosy and on the fate of human and rat lepra bacilli in flies. Publ. Health Repts., *23*:1481-1487.

Whitfield, F.G.S.
  1939. Air transport, insects and disease. Bull. Entom. Res., *30*:365-442.

Wijesundra, D. P.
  1957a. The life-history and bionomics of *Chrysomyia megacephala* (Fab.). Ceylon Jour. Sci., *25*:169-185.

Wijesundra, D. P.
  1957b. On the longevity of the adults of *Chrysomyia megacephala* (Fab.) under controlled humidity. Ceylon Jour. Sci., *25*:187-192.

Wilhelmi, J.
  1917. Die gemeine Stechfliege (Wadenstecher); Untersuchungen über die Biologie der *Stomoxys calcitrans* (L.). Zeitschr. Angew. Entom., Monograph 2, 110 pp.

Wilhelmi, J.
  1918. Zur Frage der Übertragung der Maul-und Klauenseuche durch stechende Insekten, unter besonderer Berücksichtigung von *Stomoxys calcitrans*. Verh. Dtsch. Ges. Angew. Entom., 2 Mitgl-vers., 156-167.

Wilhelmi, J.
  1922. Die Ueberträgerfrage bei der infektiösen Anaemie der Pferde. Berl. Tierärztztl. Wochens. No. 24, pp. 227-280.

Wilhelmi, J.
  1927. Untersuchungen über die Übertragung der Maul-und Klauenseuche durch die Stechfliege *Stomoxys calcitrans* L. Zeitsch. Desinfekt. Gsndhtsw., *19*:104-111.

Wilkins, S. D., and Dutcher, R. A.
  1920. Limberneck in poultry. Jour. Amer. Veter. Med. Assoc., *57*:653-685.
Wilkoff, L. J., Westbrook, L., and Dixon, G. J.
  1969. Persistence of *Salmonella typhimurium* on fabrics. Appl. Microb., *18*: 256-261.
Willett, K. C.
  1962. Recent advances in the study of tsetse-borne diseases. *In* Maramorosch, K., Biological transmission of disease agents. Academic Press, New York. Pp. 109-121.
Willett, K. C.
  1963. Trypanosomiasis and the tsetse fly problem in Africa. Ann. Rev. Entom., *8*:197-214.
Williams, A. J.
  1913. Notes on an outbreak of horsesickness connected with the presence of a *Lyperosia* as a possible transmitter. Veter. Jour., *69*:382-386.
Williams, C. B., Common, I.F.B., French, R. A., Muspratt, V., and Williams, M. C.
  1956. Observations on the migration of insects in the Pyrenees in the autumn of 1953. Trans. Roy. Entom. Soc. London, *108*:385-407.
Williams R. W.
  1954. A study of the filth flies in New York City—1953. Jour. Econ. Entom., *47*:556-563.
Williamson, D. L.
  1966. Atypical transovarial transmission of sex ratio spirochetes by *Drosophila robusta* Sturtevant. Jour. Exp. Zool., *161*:425-430.
Williamson, J.
  1817. Medical and miscellaneous observations relative to the West Indian Islands. Edinburgh, Vol. 2, see p. 142.
Wilson, L. F.
  1968. Dead alewives and black blowflies discourage bathers at Lake Michigan beaches. Mich. Entom., *1*:282-283.
Wilson, P. W., and Mathis, M. W.
  1930. Epidemiology and pathology of yaws. A report based on a study of 1,423 consecutive cases in Haiti. Jour. Amer. Med. Assoc., *94*:1289-1292.
Wilson, R. P.
  1936. 11th Annual Report, Memorial Ophthalmic Laboratory, Giza. Cairo, Egypt, 1937; pp. 95-112.
Wilton, D. P.
  1961. Refuse containers as a source of flies in Honolulu and nearby communities. Proc. Hawaiian Entom. Soc., *17*:477-481.
Winterbottom, T.
  1803. An account of the native Africans in the neighborhood of Sierra Leone. London, Vol. 2, see pp. 141-142.
Woke, P. A., Jacobs, L., Jones, F. E., and Melton, M. L.
  1953. Experimental results on possible arthropod transmission of toxoplasmosis. Jour. Parasit., *39*:523-532.
Wolfinsohn, M.
  1953. The rearing, marking and trapping of houseflies (*Musca domestica vicina*) for dispersal studies. Bull. Res. Counc. Israel, *3*:263-264.
Wollman, E.
  1921. Le rôle des mouches dans le transport des germes pathogènes étudié par la méthode des élevages aseptiques. Ann. Inst. Pasteur, *35*:431-449. [Also in Compt. Rend. Acad. Sci., *172*:298-301.]
Wollman, E.
  1927. Le rôle des mouches dans le transport de quelques germes importants pour la pathologie tunisienne. Arch. Inst. Pasteur Tunis, *16*:347-364.

Womeldorf, D. J., and Mortenson, E. W.
  1962. Occurrence of eye gnats (*Hippelates* spp.) in the Central San Joaquin Valley, California. Jour. Econ. Entom., *55*:457-459.
Womeldorf, D. J., and Mortenson, E. W.
  1963. *Hippelates* gnats in Central California. Calif. Vector Views, *10*:57-62.
Woo, P., Soltys, M. A., and Gillick, A. C.
  1970. Trypanosomes in cattle in southern Ontario. Canad. Jour. Comp. Med., *34*:142-147.
Woodcock, H. M.
  1919. Note on the epidemiology of amoebic dysentery. Jour. Roy. Army Med. Corps, *32*:231-235.
Wright, M.
  1945. Dragonflies predaceous on the stablefly, *Stomoxys calcitrans* (L.). Florida Entom., *28*:31-32.
Wright, W.
  1828. *Dissertatio medica inauguralis de framboesia. In* Memoir of the late William Wright, M.D., Edinburgh and London, see p. 408.
Wuttge
  1828. Ein tödtlicher Fliegenstich. Mag. Ges. Heilk., Berlin: *25*, see p. 111.
Yakunin, B. M.
  1966. Ecology of the blood-sucking flies (Diptera, Muscidae) of south-eastern Kazakhistan. Tr. Inst. Zool. Akad. Nauk. Kaz. SSR, *25*:121 and on.
Yao, H. Y., Yuan, I. C., and Huie, D.
  1929. The relation of flies, beverages and well water to gastro-intestinal diseases in Peiping. Natl. Med. Jour. China, *15*:410-418.
Yasuyama, K.
  1928. Viability of *Treponema pertenue* outside of the body and its significance in the transmission of yaws. Philippine Jour. Sci., *35*:333-349.
Yates, W. W., and Lindquist, A. W.
  1952. Further studies of dispersion of flies tagged with radioactive phosphoric acid. Jour. Econ. Entom., *45*:547-548.
Yavrumov, V. A.
  1947. Epidemiological importance of flies in the transmission of enteric illness [in Russian]. Zhurn. Mikrob. Epidem. Immun., *10*:49-53.
Yerington, A. P., and Warner, R. M.
  1961. Flight distances of *Drosophila* determined with radioactive phosphorus. Jour. Econ. Entom., *54*:425-428.
Yersin
  1894. La peste bubonique à Hong-Kong. Ann. Inst. Pasteur, *8*:667.
Ylppö, A., Hallman, N., Donner, M., Louhivuori, K., and Yliruokanen, A.
  1950. The rôle of insects in the spreading of infantile diarrhea in Finland. Ann. Med. Int. Fenn. *39*:149-150.
Yurkiewicz, W. J.
  1968. Flight range and energetics of the sheep blowfly during flight at different temperatures. Jour. Insect. Physiol., *14*:335-339.
Zaïdenov, A. M.
  1960. On the study of the dispersal of house-flies (Diptera, Muscidae) by means of the luminescent method of marking in the city of Chita [in Russian]. Rev. Entom. U.R.S.S., *39*:574-584.
Zaïdenov, A. M.
  1961. Opyt izucheniĭa epidemiólogicheskogo znacheniĭa cinantropnykh mukh v usloviĭakh goroda. [The experience of the study of the epidemiological role of synanthropic flies under the conditions of a town.] Entom. Obozren., *40*:554-567.

428   BIBLIOGRAPHY

Zardi, O.
1964.   Importanza di *Musca domestica* nella trasmissione dell'agente del tracoma. Nuovi Ann. Ig. Microb., *15*:587-590.
Zetek, J.
1914.   Dispersal of *Musca domestica* Linne. Ann. Entom. Soc. Amer., *7*:70-72.
Zhovtyi, I. F.
1954.   A connection between the seasonal rises in dysentery morbidity and the phenology of the house-fly, *Musca domestica* L. [in Russian]. Med. Parazit., *1*:43-45.
Zhuzhikov, D. P.
1963.   The possibility of bacteria surviving house fly metamorphosis [in Russian]. Med. Parazit. Parazitar. Bolezni, *32*:558-562.
Zhuzhikov, D. P.
1964.   Function of the peritrophic membrane in *Musca domestica* L. and *Calliphora erythrocephala* Meig. Jour. Insect Physiol., *10*:273-278.
Zimin, L. S.
1944.   The bazaar fly (*Musca sorbens* Wd.) of Tadzhikistan and its possible role in the spread of eye infections. Izvestiya Tadzhikskogo filiala Akad. Nauk SSSR, No. 5, pp. 144-151.
Zinchenko, V. S., and Nestervodskaya, E. M.
1956.   Problem of the role of fly factor in seasonal outbreaks of dysentery. Zhurn. Mikrob. Epidem. Immun., *27*:33 and on.
Zlotnick, I.
1955.   Cutaneous streptothricosis in cattle. Veter. Rec., *67*:613-614.
Zmeev, G. Ia.
1936.   O znachenie nekotorykh sinantropnykh nesekomykh kak perenoschikov i promezhutochnykh khoziaev parasiticheskikh chervei v Tadzhikistame. [On the importance of certain synanthropic insects as carriers and intermediate hosts of parasitic worms in Tadzhikistan.] *In* Malaria and other parasitological problems of Southern Tadzhikistan, Trudy (Procs.) Tadzhikskoi bazy Akad. Nauk. SSSR, *6*:241-248.
Zmeev, G. Ia.
1943.   Opyt bakteriologicheskogo obsledovaniĩa razlichnykh vidov mukh s tsel'iũ otsenki ikh epidemiológicheskogo znacheniĩa. [Experimental observation of various species of flies to evaluate their epidemiologic importance.] *In* Problemy kishechnykh infektsiĭ, Izd. AN SSSR, Stalinabad (Dushanbe), pp. 118-122.
Zumpt, F.
1965.   Myiasis in man and animals in the Old World. Butterworths, London, 267 pp.
Zumpt, F., and Patterson, P. M.
1952.   Flies visiting human faeces and carcasses in Johannesburg, Transvaal. South African Jour. Clin. Science, *3*:92-106.
Zwick, K. G.
1914.   Massnahmen gegen die Uebertragung von Infektionskrankheiten durch die Hausfliegen. Schweizerisch Runds. Med., *14*:491-508.

Index

bacillary dysentery, 173, 177-214
Bacillus, 129; *aegyptius* (=*Bacterium conjunctivitides*), 220; *anthracis*, 121, 138, 300, 304; *cereus*, 132; *cuniculicida*, 283; *thuringiensis*, 144
bactericides, 127, 172; in flies, 129, 132
bacteriocins, 131
bacteriophage, 130
bacteriostatic activity in fly gut, 131, 132
Bagdad, 253
Bahama, 81
bakery personnel and *Salmonella*, 195
Baltimore, 187
Bancroftian filariasis, 262
Bangalore, 217, 223
Bangkok, 67
Bangladesh, 167
Barca, 12
Bari, Italy, 251
bat, neotropical, *Salmonella* from, 174
bat, neotropical insectivorous, *Shigella* from, 177
Batavia, 170, 201
Bay of Biscay, 183
Bdellolarynx latifrons, 323
bedbugs, 156, 230; leprosy, 228; plague, 285; poliomyelitis, 156
beef, *Salmonella* from, 175
beef tapeworm, *see Taenia saginatus*
beetles and plague, 285
Belgian Congo, 272
Beloruss SSR, 266
Benares, 182
Bengal, 167
Berlin, 56
Bermuda, 182
Beta-hemolytic streptococci, 225
Bible, 3, 12
bifid ring, 100
bifurcation, foregut, 105
Bihar, 170
Bilharzia mansoni, 265
biltong, *Salmonella* from, 174
birds, *Brucella* in, 290; limberneck in, 305
Birmingham, 195
Bishop Knud, 14
biting stable fly, *see Stomoxys calcitrans*
black-headed gull, *Salmonella* from, 174
blackleg, 307-308
blindness, 214. *See also* eye diseases
blood meal, *Salmonella* from, 175

blood osmotic level, fly, 105
blood parasites, peritrophic membrane, penetration of, 101
blow fly, 74, 81, 101, 105, 112, 129. *See also* specific species
bluetongue, 273-74
bobcats, vesicular stomatitis in, 275
Boer War, 181
Bolivia, 325
Bologna, 170
Bombay, 169
bone meal, *Salmonella* from, 175
Book of the Ten Treatises on the Eye, 216
Borborus equinus, 80; *punctipennis*, 264
Borellia recurrentis, 235
Boston, 187
Bothriocephalus marginatus, 257
botulism in birds, *see* limberneck
boubas, *see* framboesia tropica
bovids, 320; *Salmonella* from, 174, 192-94
bovine mastitis, 293-97
bovine venereal trichomoniasis, 330
Brazil, 231, 237, 239, 241
Brazzaville, 44
breeding and oviposition, *see* specific species
Brewer's blackbird, *Salmonella* from, 174
British Guiana, 201
Brucella abortus, 289, 290; *bovis*, 290; *melitensis*, 289, 290; *suis*, 289, 290
brucellosis, 288-90
Budapest, 195
budgerigars, pathogenic *Escherichia coli* from, 179
buffalo epidemic, *see* pasteurellosis
buffalo, pasteurellosis in, 284; rinderpest in, 276; *Thelazia rhodesii* in, 339
Bufolucilia silvarum, 86, 157
Bukhara, 195, 265
Bulgaria, 175, 336
bullfinch, *Salmonella* from, 174
Burma, 18, 339
bursati, 335
bush fly, 38, 40, 42. *See also Musca sorbens*

Cairo, 66, 198
cake mixes, *Salmonella* from, 175
Calcutta, 173
calf milk replacer, *Salmonella* from, 175

Maccabees, 12
Madagascar, 50, 228
Madras, 26, 46
maggot; gut, 109-114; host for
   microbes, 127-34; microbes, *see*
   microbes, and microbial interactions;
   therapy, 16, 18. *See also* digestive
   tract
Makhach Kala, 265
mal de caderas, *see Trypanosoma
   equinum*
mal de coit, *see* dourine
mal del pinto, *see* pinto
malaria control, in suppression of
   enteric diseases, 207-12
maligant pustule, *see* anthrax
malnutrition and diarrheal disease, 214
malpighian tubules, 105
Malta, 169, 289
Malta fever, 288
Manchuria, 50, 191
mango gnats, 217. *See also
   Siphunculina funicola*
Manila, 40, 195
Manitoba, 76
Margaropus annulatus, 271
Marje, 207
markets, outdoor, fly-borne enteric
   infections, 193-95, 265, 266
marking techniques, dispersal, 67-71
marmots, pathogenic *Escherichia coli*
   from, 179
Maryland, 35, 76
Massachusetts, 27, 33, 52, 56, 185
mating, 35. *See also* oviposition and
   breeding under specific species
Mauritius, 213, 292, 321
mbori, 319
Mbow, 204
meadow mouse, tularemia in, 286
meat scraps, *Salmonella* from, 175
mechanical transport, 134-37, 259
meconium, 119
Mediterranean, 39
Melophagus ovinus, 271, 274
meningococci, 216
mentum, 102
mesenteron, 105
Mesopotamia, 6, 7, 180
metamorphosis, 9, 118, 119, 127
metazoan parasite, fly response to, 132
meteorological factors, dispersal, 75-79
Mexico, 21, 68, 89, 232, 235, 236, 275
Mexico City, 52, 61

mice, *Entamoeba histolytica* in, 250;
   Giardia in, 245, 250; vesicular
   stomatitis virus in, 275
Michigan, 52, 54, 56, 71, 163, 197
Mic-Mac Indians, 11
microbes, 127; tags, *see* dispersal
   marking techniques; external vs.
   internal transport in flies, 134, 136;
   in maggot nutrition, 117; mechanical
   transport in adult flies, 134-37;
   multiplication in flies, 139-42;
   persistence in adults, 137; survival in
   preadults, 118-32; transformation in
   flies, 172
microbial competition in adults, 140,
   141; interactions in flies, 121-27, 171
Microcalliphora varipes, 87
micrococci, hemolytic, 219
Micrococcus lysodeikticus, 132
Microfilaria sanguinus equi africano,
   261
microfilariae, 261
Microsporum canis, 313; *gypseum*, 314
Middle East, 6, 214
midgut, 105, 111, 127, 129, 241, 253;
   bactericides in, 127, 129; microbes
   in, 127
migration, 65, 66
milk, *Salmonella* from, 175; *Shigella*
   from, 179
milzbrand, *see* anthrax
mink, enteritis virus infection, 280, 281;
   farm and salmonellosis, 192;
   Salmonella from, 174
Minnesota Iron Range, 188
mirabilis-typhimurium interaction, 126
Missouri, 37
mites, plague, 285
Miyagawanella pecoris, 311
Mohammed, 13
Mohand, India, 320
Molotovsk region, U.S.S.R., 192, 299
Mombasa, 265
Mongolia, 262, 285
monkey, *Brucella* in, 289; *Entamoeba
   histolytica* in, 244; pathogenic
   Escherichia coli from, 179; *Shigella*
   from, 177; *Trichomonas hominis* in,
   246; trypanosomes in, 326-28
Montevideo, 198
moose, *Salmonella* from, 174
Mopti, 315
Morax-Axenfeld bacillus, 215, 220. *See
   also Moraxella lacunata*

440   INDEX

Oregon, 68, 76
Orestes, 5, 6
oriental sore, 252, 254. *See also*
leishmaniasis
Orissa, 170
Orlando, Florida, 23
Ornithodoros, 236, 242, 310; *crossi*,
317; *turicata*, 310
Ostfriesland, 294
Ottawa, 75
ovarian follicle, 341
overwintering, 224, 266. *See also* species
involved and bibliographic
addendum, end of Chapter 2
oviposition, 63
ox, *Giardia* in, 245

Pacific fence lizard, *Salmonella* from,
174
painted turtle, *Salmonella* from, 174
paints, *see* dispersal marking
techniques
Pakistan, 168
Palestine, 216
Panama, 224, 225, 275, 319
papillae, hindgut, 106
Papyrus Gizeh, 9
Parabronema skrjabini, 262
Paraguay, 325
parakeet, *Salmonella* from, 174
Parascaris equorum, 257
paratyphoid, 194, 201, 208
Paré, Amboise, 3, 16
Pasteurella, 284; *multocida*, 283;
pestis, 121, 122, 285
pasteurellosis, 283, 284
pathogenic staphylococci, 296
peacock, swine erysipelas organism in,
298
peafowl, *Salmonella* from, 174
Pediculus humanus, 334
Peiping, 195, 196, 250
pepsin-like enzyme, 112
Periplaneta americana, 334
peritrophic membrane, 105, 106, 111
Pernambuco, 193
Persia, 9, 13
Peru, 175, 194
pH, 105, 131, 168
Phaenicia, 6, 55, 111, 159; *cuprina*, 86,
145, 261; *pallescens*, 86; *sericata*,
55-60, 70, 75-77, 78, 86, 129, 132,
139, 140, 145, 149, 157-60, 164,

165, 192-94, 196, 206, 248, 259,
262, 271, 310, 333
Phaonia scutellaris, 81
pharyngeal pump, 101; receptors, 105;
sclerites, 111
pharynx, 102
pheasant, 274, 307; limberneck in, 306;
Salmonella from, 174. *See also*
Phormia regina
phenology of flies and enteric disease,
199-204
Philadelphia, 186, 187, 335
Philippines, 38, 50, 56, 167, 172, 237,
320, 323
Philistine, 9
Philornis, 274
Phlebotomus, 155, 252, 276, 324, 334
Phoenicia, 9
Phoenix dactylifera, 43
Phormia, 130, 159; *regina*, 52, 53, 75,
85, 100, 107, 121, 129, 139, 140, 142,
145, 157-60, 164, 165, 193, 260, 274
photoreceptors in maggots, 111
physical sterilizing agents, 143, 144
pian, *see* framboesia tropica
Pieris rapae, 116
pig, *Entamoeba histolytica* and *Giardia*
in, 250; foot and mouth disease, 269;
infectious keratoconjunctivitis,
290; *Leptospira* in, 309; pathogenic
Escherichia coli from, 179;
trypanosomes in, 326-28
pigeon meat, *Salmonella* from, 175
piggeries, 192
pink-eye, 215, 223, 291, 292. *See also*
infectious keratoconjunctivitis
pinto, 235
Piophila casei, 306
pizza dough, *Salmonella* from, 175
plague, 284-86
plague flies (*Musca ochrapesus* =
Chyromya flava), 169
planta, 134
Plasmodium, 106, 316
Plautus, 9
Pliny, 9
Plutarch, 13
pneumococci, 215, 217, 222
Polietes lardaria, 82
poliomyelitis, 155-62, 193
poliovirus, 121
pollen, 222
Pollenia rudis, 266
polyarthritis, 298
Pondicherry, 228

442   INDEX

444 INDEX